CISM Courses and Lectures

The series presents lecture notes, monographs, edited works and proceedings in the field of Mechanics, Engineering, Computer Science and Applied Mathematics.
Purpose of the series is to make known in the international scientific and technical community results obtained in some of the activities organized by CISM, the International Centre for Mechanical Sciences.

International Centre for Mechanical Sciences

Courses and Lectures Vol. 562

For further volumes:
www.springer.com/series/76

Davide Bigoni
Editor

Extremely Deformable Structures

Springer

Editor

Davide Bigoni
University of Trento

ISSN 0254-1971
ISBN 978-3-7091-1876-4 ISBN 978-3-7091-1877-1 (eBook)
DOI 10.1007/ 978-3-7091-1877-1
Springer Wien Heidelberg New York Dordrecht London

All contributions have been typeset by the authors
Printed in Italy

Printed on acid-free paper

Springer is part of Springer Science+Business Media (www.springer.com)

PREFACE

Structures have traditionally been designed to work below their critical load, because any instability was normally identified as connected to failure or loss of functionality. Instability and bifurcation were viewed simply as potentially dangerous phenomena and hence structural deformations under load were required to be small. Recently, a variety of soft structures have been considered in mechanics. These are structures that work in a large deformation regime, where elastic elements are subject to extreme deformations and loads well beyond the critical values for buckling. Examples of structures that exhibit excellent mechanical performance even under severe deformation conditions are found in biological systems, deployable space structures, and a variety of devices in everyday use.

The lesson from nature is that the possibility of exploiting highly deformable structures, made for instance of rubber or gel, may open new and unexpected technological possibilities.

The exploration of these possibilities is the focus of this volume and of the so-called Extreme Mechanics, an emerging branch of the instability of solids and structures. This branch is aimed at the investigation of instabilities as related to pattern formation and the subsequent nonlinear behaviour of large deformations.

Here the challenge is the design of deformable and bi-stable mechanisms, which can give superior mechanical performance and which will have an impact on many high tech applications such as stretchable electronics, nanotube serpentines, deployable structures for aerospace engineering, cable deployment in the ocean, as well as on sensors and flexible actuators and vibration absorbers.

This monograph is the collection of the Lecture Notes for the Advanced School 'Extremely Deformable Structures' held at the International Centre for Mechanical Sciences (CISM) in Udine, Italy, June 2-6 2014. The course was given by six lecturers and attended by nearly fifty participants from eight European and four extra European countries. The chapters are devoted to an introduction to the methods used in the study of the stability of elastic structures in the finite dimensional case (A. Lazarus, C. Maurini and S. Neukirch), in the infinite dimensional case for the Euler elastica (D. Bigoni, F. Bosi, D. Misseroni, F. Dal Corso, and G. Noselli), and to the advanced problem of the dynamics of a naturally curved elastica (B. Audoly, A. Callan-Jones, and P.-T. Brun). Two-dimensional problems are introduced with the purpose of analyzing the mechanics of two-dimensional periodic and highly deformable cellular structures (K. Bertoldi) and of the folding and deployment of thin-shells (S. Pellegrino). Structural concepts are linked to the modelling of growth processes occurring in biology (A. Erlich, Th. Lessinnes, D. E. Moulton, and A. Goriely). It is believed that the volume can represent a valid introduction to the field of extreme mechanics.

I wish to thank the Rectors of the CISM Professors E. Guazzelli, F. Pfeiffer, and F.G. Rammerstorfer, the Secretary General Professor B.A. Schrefler and all the staff for the warm hospitality and kind assistance during the course. Finally, I would like to gratefully acknowledge support from the FP7-PEOPLE-IDEAS-ERC-2013-ADG-340561-INSTABILITIES.

Davide Bigoni

CONTENTS

Stability of discretized nonlinear elastic systems

Arnaud Lazarus[*†], Corrado Maurini[*†], and Sébastien Neukirch[*†]

[*] Sorbonne Universités, UPMC Univ Paris 06, UMR 7190,
Institut Jean Le Rond d'Alembert, F-75005, Paris, France

[†] Centre National de la Recherche Scientifique, UMR 7190,
Institut Jean Le Rond d'Alembert, F-75005 Paris, France

Abstract. These notes give a short introduction to the methods for the study of stability of elastic structures. We consider only the finite-dimensional case, where the state of the system is represented by a discrete set of variables. The core of the exposition focuses on the illustration of energetic methods where equilibrium and stability are found by studying the point of stationarity and minima of a scalar function of the state variables. After three introductory sections presenting the links between stability and energy minimization (Section 2), potential energy (Section 3) and discretization methods (Section 4), we detail the mathematical methods required to minimize a function of n variables (Section 5-8). We include the theory and recipes to deal with equality and inequality constraints, providing several examples of applications to simple structures. We then show how to classify regular and singular points (bifurcations) in force-displacement diagrams (Section 9) and give a fully worked example with several degrees of freedom (Section 10). Section 11 presents, through an example, the dynamical theory of stability including Floquet theory for systems with periodic solutions. Finally, Section 12 shows how energetic methods can be applied to the study of material instabilities, by considering the case of springs with irreversible damage.

D. Bigoni (Ed.), *Extremely Deformable Structures*,
CISM International Centre for Mechanical Sciences
DOI 10.1007/ 978-3-7091-1877-1_1 © CISM Udine 2015

Contents

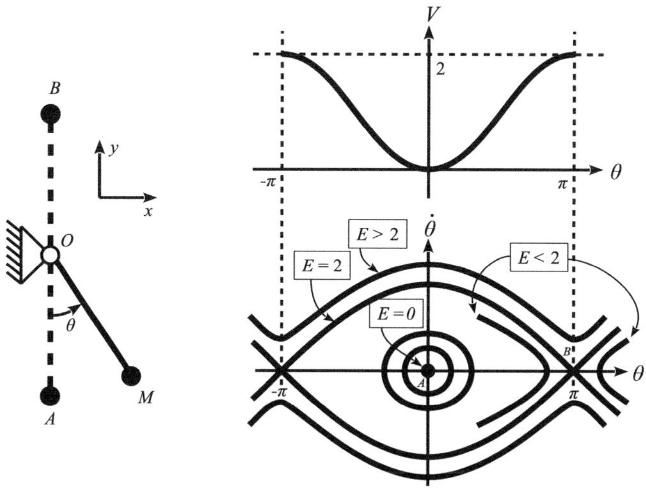

Figure 1. Dynamics of a simple pendulum. Associated potential energy $V(\theta)$ and phase plane $(\theta, \dot{\theta})$.

1 Introduction

We study the equilibrium and stability of nonlinear systems. Nonlinearity has two main consequences: (*i*) closed-form solutions are the exception and numerical methods have to be used; and (*ii*) unicity is lost, that is several solutions may exist for the same load. While the ubiquity of computers reduces the importance of the first consequence, the second raises questions such as: Which solution is to be considered? Which ones are stable?

Apart from Section 11, where dynamical stability is introduced and damping is present, we restrict to the conservative case, where for example there is no friction. Hence internal and external forces derivate from potential energies, for example $1/2 \int_0^L \kappa^2 ds$ (curvature strain energy in a beam), mgz (gravitational energy), $-\boldsymbol{F} \cdot \boldsymbol{OA}$ (work of external load). We therefore do not consider follower or gyroscopic loads (see for example Ziegler (1977) or Bigoni (2012)).

2 Energy minimization and dynamical systems

We follow the principle stating that in conservative systems a stable equilibrium corresponds to a local minimum of the potential energy. In this section we relate this energy-based stability criterion to the study of stability through the dynamics of the system.

We consider a simple pendulum, a punctual mass M fixed at the end of a massless rigid bar of length L, see Figure 1. The dynamics of the system obeys

$$\frac{\mathrm{d}^2\theta}{\mathrm{d}t^2} = -\sqrt{\frac{g}{L}}\sin\theta \tag{1}$$

where g is the acceleration of gravity. Using the dimensionless time $\tau := t\sqrt{L/g}$, Eq. (1) writes $\ddot{\theta} + \sin\theta = 0$, where $\dot{(\ )} := \mathrm{d}/\mathrm{d}\tau$. This equation is integrated once to yield

$$\frac{1}{2}\dot{\theta}^2 + 1 - \cos\theta = E \tag{2}$$

where the integration constant E is the total mechanical energy of the system, the sum of the kinetic energy $T = \frac{1}{2}\dot{\theta}^2$ and the potential energy $V = 1 - \cos\theta$, that is $E = T + V$. In figure 1 the stable (respectively unstable) equilibrium position A (resp. B) clearly appears as the local minimum (resp. maximum) of the potential energy V. Trajectories corresponding to the dynamics of the system for all possible initial conditions are drawn in the phase space $(\theta, \dot{\theta})$. There, equilibrium are points and the stable equilibrium A has closed trajectories in its neighborhood: perturbation of the equilibrium A leads to small vibrations around the position $\theta_A = 0$. The unstable equilibrium B has escaping trajectories in its neighborhood: some perturbations lead to evolution of the system far from B.

3 Potential energy for an elastic beam

In order to illustrate the theory, we focus on a simple elastic system : an elastic beam bent in the (x, y) plane, see Figure 2. The potential energy is then given by the functional (Audoly and Pomeau, 2010)

$$V[\theta] = \frac{1}{2}EI\int_0^L \theta'(s)^2\mathrm{d}s + P\int_0^L \cos\theta(s)\,\mathrm{d}s \tag{3}$$

the first term being the internal strain energy and the second term the work done by the external applied load P. For the two cases in Figure 2, in the minimization process we have to take boundary (4) and/or integral conditions (5) into account:

$$\theta(0) = 0 \tag{4}$$

$$y(L) = 0 \text{ that is } \int_0^L \sin\theta(s)\,\mathrm{d}s = 0 \tag{5}$$

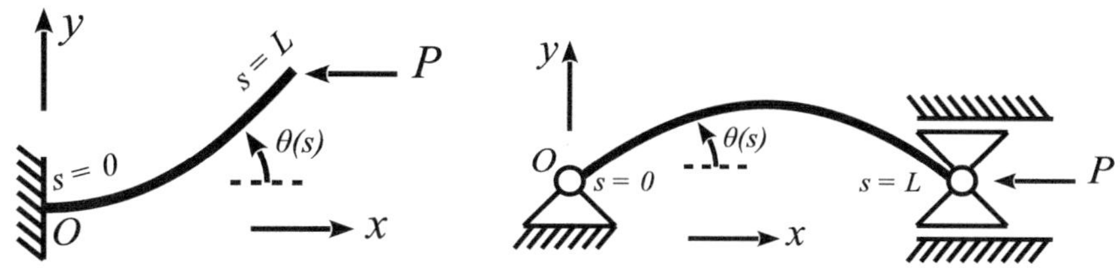

Figure 2. Elastic beams buckling in the plane. (Left) Clamped-free boundary conditions. (Right) Pinned-pinned boundary conditions.

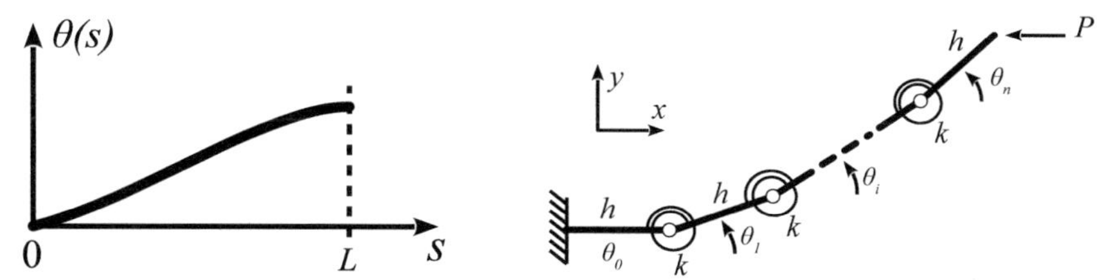

Figure 3. (Left) Solution $\theta(s)$ for the configuration in Figure 2-Left. (Right) A system of rigid bars linked by spiral springs.

4 Three ways to discretize a continuous system

4.1 Finite differences

Using $s_i = ih$, with $i \in (0, n+1)$, and $s_{n+1} = L$, we apply a finite differences scheme for the derivative of $\theta(s)$, $\theta'(s) \simeq (\theta_{i+1} - \theta_i)/h$. We then apply the rectangle rule for integration and the energy functional (3) becomes:

$$V \simeq V_{\text{fd}} = \frac{1}{2} EI \sum_{i=0}^{n-1} \left(\frac{\theta_{i+1} - \theta_i}{h} \right)^2 h + P \sum_{i=0}^{n} \cos\theta_i h \tag{6}$$

We introduce dimensionless variables:

$$v = \frac{1}{n+1} \frac{V_{\text{fd}} L}{EI} = \frac{1}{2} \sum_{i=0}^{n-1} (\theta_{i+1} - \theta_i)^2 + \frac{p}{(n+1)^2} \sum_{i=0}^{n} \cos\theta_i \tag{7}$$

where $p = \frac{PL^2}{EI}$.

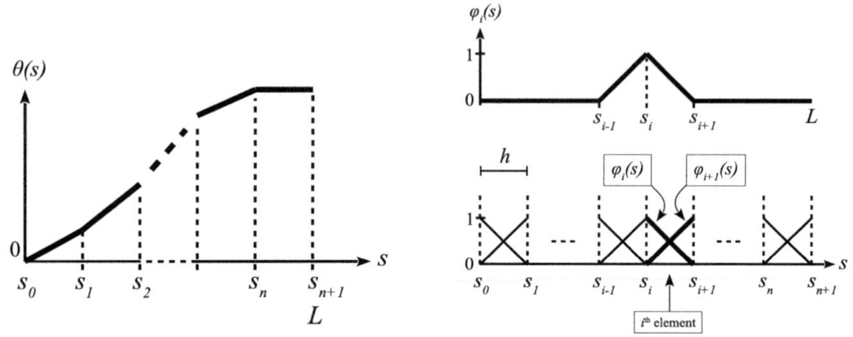

Figure 4. (Right) Piecewise linear approximation of the solution $\theta(s)$ shown in Figure 3-Left. (Left) A triangle base function, and the partitioning of the interval $(0, L)$ into $n+1$ subintervals (or elements).

This is equivalent to the energy of a system of $n+1$ rigid bars linked by spiral springs (see Figure 3):

$$V_{\text{bars}} = \frac{1}{2} k \sum_{i=0}^{n-1} (\theta_{i+1} - \theta_i)^2 + Ph \sum_{i=0}^{n} \cos\theta_i \tag{8}$$

where the spiral spring stiffness is k and the rigid bars have length h. The equivalence $V_{\text{bars}} = V_{\text{fd}}$ is obtained using $EI = k\,h$ and $L = (n+1)h$.

4.2 Finite elements

In the former section, the solution $\theta(s)$ was approximated by a discontinuous piecewise constant function. Here we look for a continuous piecewise linear approximation. We therefore introduce triangle functions that take the value 1 at their center and have a compact support, see Figure 4. The interval $(0, L)$ is split into $n+1$ elements of size h. The ith element corresponds to $s \in (s_i, s_{i+1})$, with $s_i = i\,h$. The triangle functions $\varphi_i(s)$ span over elements $i-1$ and i, with $\varphi_i(s_i) = 1$. Consequently, in the ith element the two basis functions that are non zero are:

$$\varphi_i(s) = \frac{s_{i+1} - s}{h} \tag{9a}$$

$$\varphi_{i+1}(s) = \frac{s - s_i}{h} \tag{9b}$$

And in this element the approximation to the function $\theta(s)$ is given by

$$\theta_{\rm fe}(s) = c_i\,\varphi_i(s) + c_{i+1}\,\varphi_{i+1}(s) \tag{10}$$

where the boundary condition $\theta'(L) = 0$ imposes $c_{n+1} = c_n$. Putting this ansatz in the functional (3) we obtain

$$\begin{aligned} v &= \frac{1}{n+1}\frac{VL}{EI} \\ &= \frac{1}{2}\sum_{i=0}^{n-1}(c_{i+1}-c_i)^2 + \frac{p}{(n+1)^2}\sum_{i=0}^{n-1}\frac{\sin c_{i+1} - \sin c_i}{c_{i+1}-c_i} + \frac{p}{(n+1)^2}\cos c_n \end{aligned} \tag{11}$$

At the discretization gets finer, the second term tends toward $\cos c_i$ and

$$v \xrightarrow{h\to 0} \frac{1}{2}\sum_{i=0}^{n-1}(c_{i+1}-c_i)^2 + \frac{p}{(n+1)^2}\sum_{i=0}^{n}\cos c_i \tag{12}$$

which is the same result as (7) with $\theta_i = c_i$.

4.3 Spectral decomposition

A third way to discretize the system in Figure 2-Left is to use the first n buckling modes as a functional basis and decompose the solution in this basis, see Figure 5. Here we use $n = 2$, the buckling modes are

$$\theta_1(s) = \sin\frac{\pi}{2}s \tag{13a}$$

$$\theta_2(s) = \sin\frac{3\pi}{2}s \tag{13b}$$

The approximate solution is then written as

$$\theta(s) = \alpha\,\sin\frac{\pi}{2}s + \beta\,\sin\frac{3\pi}{2}s \tag{14}$$

We inject this ansatz into the functional (3) and obtain:

$$\begin{aligned} v(\alpha,\beta) &= \frac{1}{2}\int_0^1\left(\alpha\,\frac{\pi}{2}\cos\frac{\pi}{2}s + \beta\,\frac{3\pi}{2}\cos\frac{3\pi}{2}s\right)^2 \mathrm{d}s \\ &\quad + P\int_0^1 \cos\left[\alpha\,\sin\frac{\pi}{2}s + \beta\,\sin\frac{3\pi}{2}s\right]\mathrm{d}s \end{aligned} \tag{15}$$

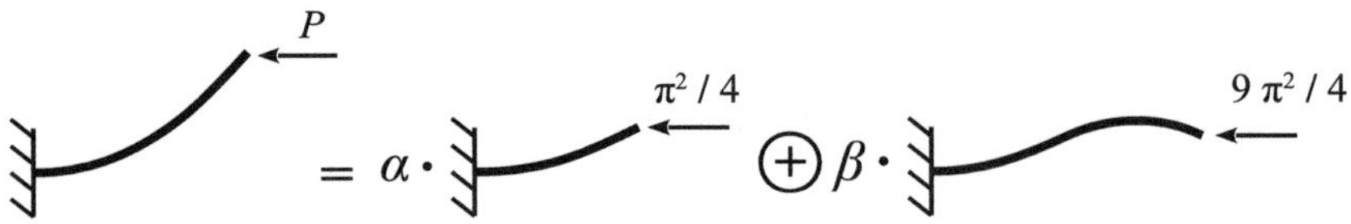

Figure 5. Spectral discretization based on the first two buckling modes.

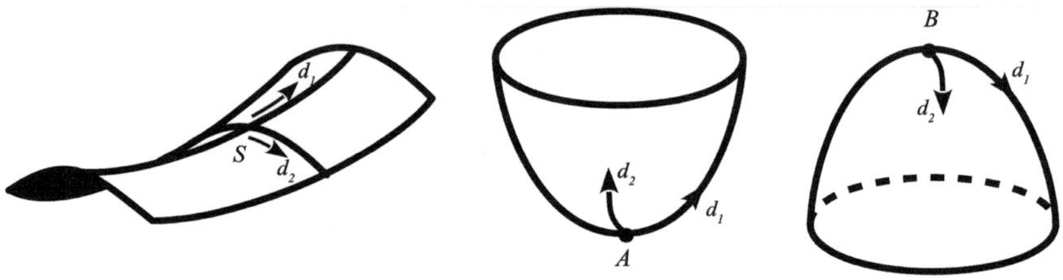

Figure 6. A saddle point (S), a minimum (A), and a maximum (B)

The first integral is easily evaluated but not the second one. Gauss-Legendre integration is then used, with 2 points. We obtain

$$v(\alpha,\beta) = \frac{\pi^2}{16}(\alpha^2 + 9\beta^2) + \frac{P}{2}\left[\cos(0.95\alpha - 0.54\beta) + \cos(0.33\alpha - 0.84\beta)\right] \tag{16}$$

5 Minimum, maximum, extremum

Once the discretization is achieved, we are left with a function $V(\boldsymbol{x})$, where $V \in \mathbb{R}$ and $\boldsymbol{x} \in \mathbb{R}^n$, and we look for the point(s) for which V is locally minimum. Examples are:

- $V(x_1, x_2) = x_1^2 + x_2^2$ has a *global* minimum at $(x_1, x_2) = (0, 0)$.
- $V(x_1, x_2) = x_1^2 - x_2^2$ has no *global* or *local* minimum. The point $(x_1, x_2) = (0, 0)$ is a saddle point.
- $V(x) = x^2 - x^4$ has a *local* minimum at $x = 0$.

Stated in this way the problem of the equilibrium and stability of an elastic structure becomes a problem of nonlinear optimization. We therefore use methods of nonlinear programming, see for example Luenberger and Ye (2008).

5.1 Values of a function in one direction

We are given a function $V(\boldsymbol{x})$, where $V \in \mathbb{R}$ and $\boldsymbol{x} \in \mathbb{R}^n$. We are interested to know if the point $\boldsymbol{x_0}$ is a local minimum. Instead of looking at the values of V all around $\boldsymbol{x_0}$, we *restrict* to the values taken by V in the direction $\boldsymbol{d}$. That is we study the one-variable function

$$V(\boldsymbol{x}_0)(\boldsymbol{d}) := V(\boldsymbol{x}_0 + \epsilon\boldsymbol{d}) = V(\epsilon) \quad \text{for } \epsilon \in \mathbb{R} \tag{17}$$

Example: For the function $V(x_1, x_2) = x_1^2 - x_2^2$, the point $(x_1, x_2) = (0, 0)$ is a saddle point:

- If we study $V(\boldsymbol{x}_0)(\boldsymbol{d})$ with $x_0 = (0, 0)$ and $\boldsymbol{d} = (1, 0)$. We have $V(\boldsymbol{x}_0)(\boldsymbol{d}) = V(\epsilon) = \epsilon^2$. In this direction we have a local minimum.
- If we study $V(\boldsymbol{x}_0)(\boldsymbol{d})$ with $x_0 = (0, 0)$ and $\boldsymbol{d} = (0, 1)$. We have $V(\boldsymbol{x}_0)(\boldsymbol{d}) = V(\epsilon) = -\epsilon^2$. In this direction we have a local maximum.

5.2 Derivative of a function in one direction

We now want to study the derivative with regard to ϵ of $V(\epsilon)$, that is $V'(\epsilon)$. This is the directional derivative of $V(\boldsymbol{x})$ in the direction $\boldsymbol{d}$, at the point $\boldsymbol{x_0}$:

$$V'(\boldsymbol{x}_0)(\boldsymbol{d}) := \left.\frac{d}{d\epsilon} V(\boldsymbol{x}_0 + \epsilon\boldsymbol{d})\right|_{\epsilon=0} \tag{18}$$

Example : for the function $V(x_1, x_2) = x_1^2 - x_2^2$,

- we compute the directional derivative at the point $(x_1, x_2) = (0, 0)$ in the direction $d = (1, 0)$. We have $V(\epsilon) = V(\boldsymbol{x}_0 + \epsilon\boldsymbol{d}) = \epsilon^2$, $V'(\epsilon) = 2\epsilon$, and finally $V'(\boldsymbol{x}_0)(\boldsymbol{d}) = V'(\epsilon)|_{\epsilon=0} = 0$.
- we compute the directional derivative at the point $(x_1, x_2) = (1, 0)$ in the direction $d = (1, 2)$. We have $V(\epsilon) = V(\boldsymbol{x}_0 + \epsilon\boldsymbol{d}) = -3\epsilon^2 + 2\epsilon + 1$, and $V'(\boldsymbol{x}_0)(\boldsymbol{d}) = V'(\epsilon)|_{\epsilon=0} = 2$.

5.3 Directional derivative and gradient

We consider $V(\boldsymbol{x}_0 + \epsilon\boldsymbol{d})$ with $\boldsymbol{x}_0 = (x_1^0, x_2^0)$ and $\boldsymbol{d} = (d_1, d_2)$. The Taylor expansion around $\boldsymbol{x}_0$ for small ϵ writes:

$$\begin{aligned} V(x_1^0 + \epsilon d_1, x_2^0 + \epsilon d_2) &= V(x_1^0, x_2^0) + \epsilon \left.\frac{\partial V}{\partial x_1}\right|_{\boldsymbol{x}=\boldsymbol{x}_0} d_1 + \epsilon \left.\frac{\partial V}{\partial x_2}\right|_{\boldsymbol{x}=\boldsymbol{x}_0} d_2 + O(\epsilon^2) \\ &= V(x_1^0, x_2^0) + \epsilon\, \boldsymbol{\nabla} V|_{\boldsymbol{x}_0} \cdot \boldsymbol{d} + O(\epsilon^2) \end{aligned} \tag{19}$$

where $\boldsymbol{\nabla} := \frac{\partial}{\partial \boldsymbol{x}}$.

5.4 Taylor expansion up to order two

The Taylor expansion up to second order is:

$$V(x_1^0+\epsilon d_1, x_2^0+\epsilon d_2) = V(x_1^0, x_2^0) + \epsilon \boldsymbol{\nabla} V|_{\boldsymbol{x}_0} \cdot \boldsymbol{d} + \frac{1}{2}\epsilon^2 d_1^2 \left.\frac{\partial^2 V}{\partial x_1^2}\right|_{\boldsymbol{x}=\boldsymbol{x}_0}$$
$$+\epsilon^2 d_1\, d_2 \left.\frac{\partial^2 V}{\partial x_1 \partial x_2}\right|_{\boldsymbol{x}=\boldsymbol{x}_0} + \frac{1}{2}\epsilon^2 d_2^2 \left.\frac{\partial^2 V}{\partial x_2^2}\right|_{\boldsymbol{x}=\boldsymbol{x}_0} \qquad (20)$$

The second order term can be written as:

$$\frac{1}{2}\epsilon^2\, \boldsymbol{d} \cdot \boldsymbol{H}_0 \cdot \boldsymbol{d} \qquad (21)$$

where $\boldsymbol{H} = \boldsymbol{H}(\boldsymbol{x})$ is the Hessian matrix, $H_{ij} = \frac{\partial^2 V}{\partial x_i \partial x_j}$, and $\boldsymbol{H}_0$ means $\boldsymbol{H}$ when evaluated on $\boldsymbol{x} = \boldsymbol{x}_0$.

6 Minimization without constraints

6.1 First derivative

We are given a function $V(\boldsymbol{x})$, where $V \in \mathbb{R}$ and $\boldsymbol{x} \in \mathbb{R}^n$ and we require the point $\boldsymbol{x}_0$ to be a local minimum. That is we want

$$V(\boldsymbol{x}) \geq V(\boldsymbol{x}_0) \quad \forall \boldsymbol{x} \text{ near } \boldsymbol{x}_0 \qquad (22)$$

Hence we write $\boldsymbol{x} = \boldsymbol{x}_0 + \epsilon \boldsymbol{d}$ and consider small ϵ and all possible directions $\boldsymbol{d}$ (but not $\boldsymbol{d} = \boldsymbol{0}$). Using (19), the condition (22) yields

$$\epsilon \left.\frac{\partial V}{\partial \boldsymbol{x}}\right|_{\boldsymbol{x}_0} \cdot \boldsymbol{d} \geq 0 \quad \forall \epsilon \text{ and } \forall \boldsymbol{d} \qquad (23)$$

As positive and negative ϵ can be considered this implies $\left.\frac{\partial V}{\partial \boldsymbol{x}}\right|_{\boldsymbol{x}_0} \cdot \boldsymbol{d} = 0\ \forall \boldsymbol{d}$, and finally

$$\left.\frac{\partial V}{\partial \boldsymbol{x}}\right|_{\boldsymbol{x}_0} = \boldsymbol{0} \qquad (24)$$

This is a necessary condition in order to have a minimum at $\boldsymbol{x}_0$. If V is a potential energy of a system, then these n equations for the n unknowns $(x_1, x_2, \ldots, x_n)$ are the equilibrium equations.

6.2 Second derivative

Having solved the (nonlinear) equilibrium equations (24) and found one or more solutions $\boldsymbol{x}_A$, $\boldsymbol{x}_B$, ..., we want to test whether these solutions are stable, that is whether they correspond to local minimum of the energy. As for these solutions the first derivative vanishes, we need to look at the second derivative term in (20). The requirement (22) for a minimum now yields:

$$\boldsymbol{d} \cdot \boldsymbol{H}_0 \cdot \boldsymbol{d} \geq 0 \quad \forall \boldsymbol{d} \tag{25}$$

A sufficient condition is

$$\boldsymbol{d} \cdot \boldsymbol{H}_0 \cdot \boldsymbol{d} > 0 \quad \forall \boldsymbol{d} \tag{26}$$

In the case where there is a $\boldsymbol{d}$ for which $\boldsymbol{d} \cdot \boldsymbol{H}_0 \cdot \boldsymbol{d} = 0$, we cannot conclude and would have to compute higher orders. In the case there is one (or more) $\boldsymbol{d}$ for which $\boldsymbol{d} \cdot \boldsymbol{H}_0 \cdot \boldsymbol{d} < 0$, then the point $\boldsymbol{x}_0$ is not a minimum.

Example: Minimize the function $V(x_1, x_2) = x_1^2 - x_2^2 + x_2^4$ in $\mathbb{R}^2$.

6.3 Recipe

As the matrix $\boldsymbol{H}_0$ is symmetric, we can write it in its eigenspace. It is then diagonal with real entries, its eigenvalues λ_i with $i \in (1, n)$. In this representation, the condition (26) writes:

$$d_1^2\,\lambda_1 + d_2^2\,\lambda_2 + \ldots + d_n^2\,\lambda_n > 0 \quad \forall d_i \tag{27}$$

which means that we must have $\lambda_i > 0\ \forall i$, that is all eigenvalues must be strictly positive.

Here is the procedure to follow when we have the potential energy $V(x)$ of a system to minimize:

- compute the equilibria, that is the solution(s) of $\frac{\partial V}{\partial \boldsymbol{x}} = \mathbf{0}$

then, for each equilibrium solution:

- compute the matrix $H_{ij} = \frac{\partial^2 V}{\partial x_i \partial x_j}$, and evaluate it on the equilibrium solution, then compute its eigenvalues λ_i
- if $\lambda_i > 0\ \forall i$ then the equilibrium is stable
- if there is one (or more) i such that $\lambda_i < 0$, then the equilibrium is unstable
- if the λ_i are all positive, but there is one (or more) i such that $\lambda_i = 0$, we cannot at this stage conclude on the stability.

6.4 Discrete vs. Continuous

If on the one hand we consider the continuous energy (3) and write Euler-Lagrange equations for it, we obtain

$$EI\,\theta''(s) + P\sin\theta(s) = 0 \tag{28}$$

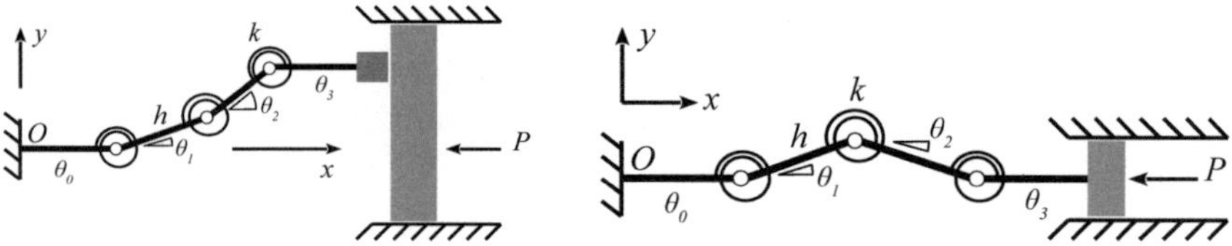

Figure 7. System of 4 bars linked by spiral springs. (Left) Ends are horizontal. (Right) Ends are horizontal and aligned

If now on the other hand we consider the discretized energy (6) and write the first derivative condition (24) we obtain

$$-EI\,\frac{\theta_{i+1}-2\theta_i+\theta_{i-1}}{h} - Ph\sin\theta_i = 0 \tag{29}$$

and we see that (29) is the finite differences version of (28), showing that the discretization and minimization procedures are commutative operations:

$$\begin{array}{ccc}
V[\theta(s)] & \xrightarrow{\text{Euler-Lagragne}} & EI\theta'' + P\sin\theta = 0 \\
\Big\downarrow \text{finite diff.} & & \Big\downarrow \text{finite diff.} \\
V(\theta_i) & \xrightarrow{\text{First derivative}} & EI\,\frac{\theta_{i+1}-2\theta_i+\theta_{i-1}}{h^2} + P\sin\theta_i = 0
\end{array}$$

6.5 Example: The clamped beam

We here compute the buckling load of the two degrees of freedom system of Figure 7-Left. The energy of the system is

$$V = \frac{1}{2}k\,(\theta_1-\theta_0)^2 + \frac{1}{2}k\,(\theta_2-\theta_1)^2 + \frac{1}{2}k\,(\theta_3-\theta_2)^2 \\ + Ph\,(\cos\theta_0+\cos\theta_1+\cos\theta_2+\cos\theta_3) \tag{30}$$

and the boundary conditions reads: $\theta_0 = 0 = \theta_3$. The variables are then (θ_1, θ_2). We introduce dimensionless variables $v = V/k$ and $p = 16Ph/k$ in order to follow notations of equation (7). Equilibrium equations are

$$\frac{\partial v}{\partial\theta_1} = 2\theta_1 - \theta_2 - \frac{p}{16}\sin\theta_1 = 0 \tag{31a}$$

$$\frac{\partial v}{\partial\theta_2} = -\theta_1 + 2\theta_2 - \frac{p}{16}\sin\theta_2 = 0 \tag{31b}$$

One trivial solution is $\theta_1 = 0 = \theta_2$. We study its stability. The Hessian matrix is

$$\boldsymbol{H}_0 = \begin{pmatrix} 2 - p/16 & -1 \\ -1 & 2 - p/16 \end{pmatrix} \tag{32}$$

and has $\lambda_1 = 1 - p/16$ and $\lambda_2 = 3 - p/16$ as eigenvalues. We conclude the trivial solution is stable for $p < 16$ and unstable for $p > 16$. Buckling therefore happens at $p = 16$.

7 Minimization with equality constraints

Given a function $V(\boldsymbol{x})$, where $V \in \mathbb{R}$ and $\boldsymbol{x} \in \mathbb{R}^n$, we look for the point(s) for which V is locally minimum, but only for the point(s) that satisfy one (or more) constraints $g(\boldsymbol{x}) = 0$.

Example: Minimize $V(x_1, x_2) = x_1^2 + x_2^2$ in the subset where $g(x_1, x_2) = x_1 - 1 = 0$.

7.1 First derivative

We are given a function $V(\boldsymbol{x})$ that is to be minimized in the subset where $g(\boldsymbol{x}) = 0$. We require the point $\boldsymbol{x}_0$ to be a local minimum by stating that for small ϵ

$$V(\boldsymbol{x} = \boldsymbol{x}_0 + \epsilon\boldsymbol{d}) \geq V(\boldsymbol{x}_0) \quad \forall\boldsymbol{x} \text{ such that } g(\boldsymbol{x}_0 + \epsilon\boldsymbol{d}) = 0 \tag{33}$$

This means that not all $\boldsymbol{d}$ directions will be tested. Only the directions $\boldsymbol{d}$ fulfilling

$$g(\boldsymbol{x}_0 + \epsilon\boldsymbol{d}) = g(\boldsymbol{x}_0) + \epsilon\boldsymbol{\nabla}g|_{\boldsymbol{x}_0} \cdot \boldsymbol{d} + O(\epsilon^2) = 0 \tag{34}$$

will be tested. As in the unconstrained case, positive and negative ϵ can be considered. Therefore we need to have

$$\boldsymbol{\nabla}V|_{\boldsymbol{x}_0} \cdot \boldsymbol{d} = 0 \quad \forall\boldsymbol{d} \text{ such that } \boldsymbol{\nabla}g|_{\boldsymbol{x}_0} \cdot \boldsymbol{d} = 0 \tag{35}$$

This implies that $\boldsymbol{\nabla}V$ and $\boldsymbol{\nabla}g$ are colinear, that is there is a real number λ such that

$$\boldsymbol{\nabla}V|_{\boldsymbol{x}_0} = \lambda\,\boldsymbol{\nabla}g|_{\boldsymbol{x}_0} \tag{36}$$

The function g is given and $\boldsymbol{\nabla}g$ defines a direction in space. The vectors $\boldsymbol{d}$ perpendicular to the gradient of g are in the tangent (hyper-)plane of the surface $g = 0$ at point $\boldsymbol{x}_0$, see Figure 8. The equilibrium equations are the n equations (36) plus $g(\boldsymbol{x}) = 0$, and the unknowns are the n components of $\boldsymbol{x}$ plus the real number λ.

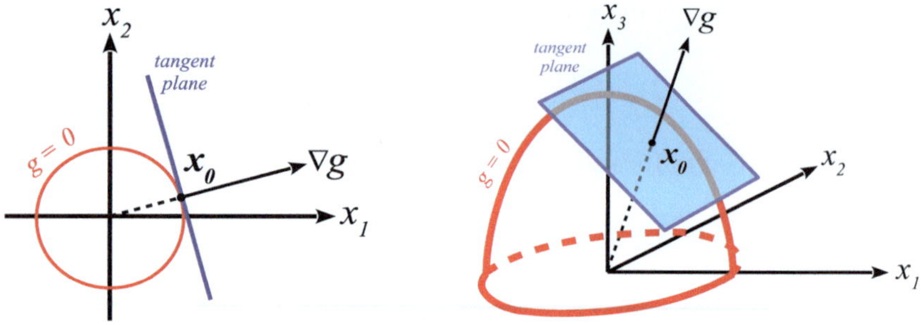

Figure 8. Tangent planes in 2D and 3D

A way to memorize Eq. (36) is through the introduction of the augmented energy $W = V - \lambda g$ and solve for $\boldsymbol{\nabla} W = \boldsymbol{0}$. In such a view λ is called a Lagrange multiplier.

Example: We minimize $V(x_1, x_2) = x_1^2 + x_2^2$ under the constraint $g(x_1, x_2) = (x_1 - 1)^2 + (x_2 - 2)^2 - 1 = 0$. The first order condition $\boldsymbol{\nabla} V = \lambda \, \boldsymbol{\nabla} g$ has two solutions. The first one is point A such that $x_1^A = 1 - 1/\sqrt{5}$, $x_2^A = 2 - 2/\sqrt{5}$, $\lambda_A = 1 - \sqrt{5}$, and the second one is point B such that $x_1^B = 1 + 1/\sqrt{5}$, $x_2^B = 2 + 2/\sqrt{5}$, $\lambda_B = 1 + \sqrt{5}$.

7.2 Second derivative

For the second derivative, we need to consider a perturbation to the equilibrium solution $\boldsymbol{x}_0$ such that $\boldsymbol{x}$ stays on the surface $g = 0$. Contrary to what was done in the unconstrained case we need to develop the perturbation to second order:

$$\boldsymbol{x} = \boldsymbol{x}_0 + \epsilon \, \boldsymbol{d} + \frac{1}{2} \epsilon^2 \, \boldsymbol{e} + O(\epsilon^3) \tag{37}$$

where $|\epsilon| \ll 1$ and where $g(\boldsymbol{x}) = O(\epsilon^3)$. Now $\boldsymbol{d}$ and $\boldsymbol{e}$ cannot be freely chosen, as they are related by the condition

$$g(\boldsymbol{x}) - g(\boldsymbol{x}_0) = 0 = \epsilon \boldsymbol{\nabla} g|_{\boldsymbol{x}_0} \cdot \boldsymbol{d} + \frac{1}{2} \epsilon^2 \boldsymbol{\nabla} g|_{\boldsymbol{x}_0} \cdot \boldsymbol{e} + \frac{1}{2} \epsilon^2 \boldsymbol{d} \cdot \boldsymbol{G}_0 \cdot \boldsymbol{d} + O(\epsilon^3) \tag{38}$$

where $\boldsymbol{G} = \boldsymbol{G}(\boldsymbol{x})$ is the matrix with second derivative of g: $G_{ij} = \frac{\partial^2 g}{\partial x_i \partial x_j}$, and $\boldsymbol{G}_0$ means $\boldsymbol{G}$ when evaluated on $\boldsymbol{x} = \boldsymbol{x}_0$. The Taylor expansion of V

to second order now writes:

$$V(\boldsymbol{x}) - V(\boldsymbol{x}_0) = \epsilon \boldsymbol{\nabla} V|_{\boldsymbol{x}_0} \cdot \boldsymbol{d} + \frac{1}{2}\epsilon^2 \boldsymbol{\nabla} V|_{\boldsymbol{x}_0} \cdot \boldsymbol{e} + \frac{1}{2}\epsilon^2 \boldsymbol{d} \cdot \boldsymbol{H}_0 \cdot \boldsymbol{d} + O(\epsilon^3) \quad (39)$$

Note that in the unconstrained case, the term involving $\boldsymbol{e}$ vanishes because we had $\boldsymbol{\nabla} V|_{\boldsymbol{x}_0} = 0$, but this is no longer the case here. Using (36), (38), and (39), the condition (33) writes:

$$V(\boldsymbol{x}) - V(\boldsymbol{x}_0) = \frac{1}{2}\epsilon^2 \boldsymbol{d} \cdot (\boldsymbol{H}_0 - \lambda \boldsymbol{G}_0) \cdot \boldsymbol{d} \geq 0 \quad \forall \boldsymbol{d} \text{ such that } \boldsymbol{\nabla} g \cdot \boldsymbol{d} = 0 \quad (40)$$

We note $\hat{\boldsymbol{H}} := \boldsymbol{H} - \lambda \boldsymbol{G}$ the *augmented* Hessian matrix. As in the unconstrained case, a sufficient condition is obtained by requiring the matrix to be positive definite in the tangent plane of the surface $g = 0$ at point $\boldsymbol{x}_0$ (that is the $\geq$ sign is replaced by a $>$ sign in (40)).

Example: We study the potential energy $V(x_1, x_2) = x_1^2 + x_2^2$ under the constraint $g(x_1, x_2) = (x_1 - 1)^2 + (x_2 - 2)^2 - 1 = 0$. Equilibrium are point A, such that $x_1^A = 1 - 1/\sqrt{5}$, $x_2^A = 2 - 2/\sqrt{5}$, $\lambda_A = 1 - \sqrt{5}$, and B, such that $x_1^B = 1 + 1/\sqrt{5}$, $x_2^B = 2 + 2/\sqrt{5}$, $\lambda_B = 1 + \sqrt{5}$. The augmented Hessian matrix is

$$\hat{\boldsymbol{H}}_A = \boldsymbol{H}_A - \lambda_A \boldsymbol{G}_A = \begin{pmatrix} +2\sqrt{5} & 0 \\ 0 & +2\sqrt{5} \end{pmatrix} \quad (41a)$$

$$\hat{\boldsymbol{H}}_B = \boldsymbol{H}_B - \lambda_B \boldsymbol{G}_B = \begin{pmatrix} -2\sqrt{5} & 0 \\ 0 & -2\sqrt{5} \end{pmatrix} \quad (41b)$$

We first note that this matrix is positive definite for point A and negative definite for point B. We now study them in the tangent plane of the surface $g = 0$ at points A and B. At point A, $\nabla V = -2/\sqrt{5}\,(1,2)$. The perpendicular directions $\boldsymbol{d}$ write $\alpha\,(-2,1)$ with $\alpha \in \mathbb{R}$. Consequently $\boldsymbol{d} \cdot \hat{\boldsymbol{H}}_A \cdot \boldsymbol{d} = \alpha^2\, 10\sqrt{5} > 0$. Point A is a local minimum, a stable equilibrium. At point B, the perpendicular directions $\boldsymbol{d}$ also write $\alpha\,(-2,1)$ and $\boldsymbol{d} \cdot \hat{\boldsymbol{H}}_B \cdot \boldsymbol{d} = -\alpha^2\, 10\sqrt{5} > 0$. Point B is not a local minimum, it corresponds to an unstable equilibrium.

We note that the restriction $\boldsymbol{\nabla} g \cdot \boldsymbol{d} = 0$ did not change anything to the result: in this example stability could have been correctly inferred from (41).

Example: We study the potential energy $V(x_1, x_2) = x_1^2 - x_2^2$ under the constraint $g(x_1, x_2) = x_2 + 1 = 0$. The equilibrium point is A with $x_1^A = 0$, $x_2^A = -1$, $\lambda_A = 2$. The augmented hessian matrix is:

$$\hat{\boldsymbol{H}}_A = H_A - \lambda_A G_A = \begin{pmatrix} 2 & 0 \\ 0 & -2 \end{pmatrix} \quad (42)$$

This matrix is not positive definite, as there is one negative eigenvalue. Nevertheless positive definitiveness has to be tested under the restriction $\boldsymbol{\nabla} g \cdot \boldsymbol{d} = 0$. The perpendicular directions $\boldsymbol{d}$ write $\alpha\,(1,0)$ with $\alpha \in \mathbb{R}$. Consequently $\boldsymbol{d} \cdot \hat{\boldsymbol{H}}_A \cdot \boldsymbol{d} = 2\,\alpha^2 > 0$. Point A is then a local minimum, a stable equilibrium.

We note that the restriction $\boldsymbol{\nabla} g \cdot \boldsymbol{d} = 0$ was important to consider, as without it we would have (incorrectly) concluded that A was an unstable equilibrium.

7.3 The projected Hessian matrix

In the case there are m equality constraints $g_j(\boldsymbol{x}) = 0$, with $j \in (1, m)$, we show how to verify that the augmented Hessian matrix $\hat{\boldsymbol{H}} = \boldsymbol{H} - \sum_j \lambda_j \boldsymbol{G}_j$ is positive definite in the tangent plane.

• We compute the gradients ∇g_j and place them as the lines of a matrix $\boldsymbol{T}$:

$$\boldsymbol{T} = \begin{pmatrix} - & \boldsymbol{\nabla} g_1 & - \\ - & \boldsymbol{\nabla} g_2 & - \\ - & \vdots & - \\ - & \boldsymbol{\nabla} g_m & - \end{pmatrix} \tag{43}$$

• We compute the kernel of $\boldsymbol{T}$. Generically this yields a set of $(n-m)$ orthonormal vectors $\{k_1, k_2, \ldots, k_{n-m}\}$ which span the tangent plane. We place these vectors in the lines of the matrix $\boldsymbol{K}$:

$$\boldsymbol{K} = \begin{pmatrix} - & \boldsymbol{k}_1 & - \\ - & \boldsymbol{k}_2 & - \\ - & \vdots & - \\ - & \boldsymbol{k}_{n-m} & - \end{pmatrix} \tag{44}$$

All possible vectors $\boldsymbol{d}$ such that $\boldsymbol{\nabla} g_j \cdot \boldsymbol{d} = 0\ \forall j$ are generated with the help of the basis $\{\boldsymbol{k}_1, \boldsymbol{k}_2, \ldots, \boldsymbol{k}_{n-m}\}$, that is we consider $\boldsymbol{d} = \alpha_1\,\boldsymbol{k}_1 + \alpha_2\,\boldsymbol{k}_2 + \ldots + \alpha_{n-m}\,\boldsymbol{k}_{n-m}$, $\forall \alpha_j$. Writing $\boldsymbol{d}^\star = (\alpha_1, \alpha_2, \ldots, \alpha_{n-m})$, vectors $\boldsymbol{d}$ are given by $\boldsymbol{d} = \boldsymbol{K}^T \cdot \boldsymbol{d}^\star$.

• Consequently we want

$$\boldsymbol{d}^\star \cdot \boldsymbol{K}\,\hat{\boldsymbol{H}}\,\boldsymbol{K}^T \cdot \boldsymbol{d}^\star > 0 \quad \forall \boldsymbol{d}^\star \tag{45}$$

We introduce $\hat{\boldsymbol{H}}^\star := \boldsymbol{K}\,\hat{\boldsymbol{H}}\,\boldsymbol{K}^T$, the *projected* augmented Hessian matrix. This a square $(n-m) \times (n-m)$ symmetric matrix. Having only strictly positive eigenvalues implies stability. One (or more) negative eigenvalue yields instability. And in the case where all eigenvalues are positive but one (or more) is zero, stability is undecided (computations at higher orders are needed).

Example: We study the potential energy $V(x_1, x_2, x_3) = x_1 + x_2^2 + x_2 x_3 + 2x_3^2$ under the constraint $g(x_1, x_2, x_3) = x_1^2 + x_2^2 + x_3^2 - 1 = 0$. One equilibrium point is A with $x_1^A = 1$, $x_2^A = 0$, $x_3^A = 0$, $\lambda_A = 1/2$. The augmented Hessian matrix is

$$\hat{\boldsymbol{H}}_A = \boldsymbol{H}_A - \lambda_A \boldsymbol{G}_A = \begin{pmatrix} -1 & 0 & 0 \\ 0 & 1 & 1 \\ 0 & 1 & 3 \end{pmatrix} \tag{46}$$

Eigenvalues are -1, $2 - \sqrt{2}$, and $2 + \sqrt{2}$. The matrix $\boldsymbol{T}_A = (1, 0, 0)$ and finally

$$\boldsymbol{K}_A = \begin{pmatrix} 0 & 1 & 0 \\ 0 & 0 & 1 \end{pmatrix} \quad \text{and} \quad \hat{\boldsymbol{H}}_A^\star = \begin{pmatrix} 1 & 1 \\ 1 & 3 \end{pmatrix} \tag{47}$$

The projected augmented Hessian matrix $\hat{\boldsymbol{H}}_A^\star$ has eigenvalues $2 - \sqrt{2}$ and $2 + \sqrt{2}$, and consequently point A is a stable equilibrium.

Example: We minimize $V(x_1, x_2, x_3) = x_1^2 - x_2 + x_3^2$ under the two constraints $g_1(x_1, x_2, x_3) = x_1 + 2x_2 + x_3 - 1 = 0$ and $g_2(x_1, x_2, x_3) = 2x_1 - x_2 - 3x_3 - 4 = 0$. The equilibrium point is A with $x_1^A = 2/5$, $x_2^A = 1$, $x_3^A = -7/5$, $\lambda_1^A = -16/25$, and $\lambda_2^A = 18/25$. The Hessian matrix is:

$$\hat{\boldsymbol{H}}_A = \boldsymbol{H}_A - \lambda_A \boldsymbol{G}_A = \begin{pmatrix} 2 & 0 & 0 \\ 0 & -2 & 0 \\ 0 & 0 & 2 \end{pmatrix} \tag{48}$$

The matrix $\boldsymbol{T}_A = (1, 0, 0)$ and finally

$$\boldsymbol{T}_A = \begin{pmatrix} 1 & 2 & 1 \\ 2 & -1 & -3 \end{pmatrix} \text{ and } \boldsymbol{K}_A = \begin{pmatrix} 1 & -1 & 1 \end{pmatrix} \text{ and } \hat{\boldsymbol{H}}_A^\star = 2 \tag{49}$$

Consequently point A is a local minimum.

7.4 Example: The aligned clamped beam

We here compute the buckling load of the two degrees of freedom system of Figure 7-Right. The potential energy of the system is

$$V = \frac{1}{2} k \left(\theta_1 - \theta_0\right)^2 + \frac{1}{2} k \left(\theta_2 - \theta_1\right)^2 + \frac{1}{2} k \left(\theta_3 - \theta_2\right)^2 + Ph \left(\cos\theta_0 + \cos\theta_1 + \cos\theta_2 + \cos\theta_3\right) \tag{50}$$

We introduce dimensionless variables $v = V/k$ and $p = 16Ph/k$ in order to follow notations of Equation (7). A first boundary condition implies that $\theta_0 = 0 = \theta_3$. The variables are then (θ_1, θ_2). The second boundary condition implies a constraint $g(\theta_1, \theta_2) = \sin\theta_1 + \sin\theta_2 = 0$. Equilibrium equations are

$$2\theta_1 - \theta_2 - \frac{p}{16}\sin\theta_1 = \lambda\cos\theta_1 \tag{51a}$$

$$-\theta_1 + 2\theta_2 - \frac{p}{16}\sin\theta_2 = \lambda\cos\theta_2 \tag{51b}$$

$$\sin\theta_1 + \sin\theta_2 = 0 \tag{51c}$$

One trivial solution is $\theta_1 = 0 = \theta_2 = \lambda$. We study its stability. The augmented Hessian matrix is

$$\hat{\boldsymbol{H}}_0 = \begin{pmatrix} 2 - p/16 & -1 \\ -1 & 2 - p/16 \end{pmatrix} \tag{52}$$

The matrices $\boldsymbol{T} = (1, 1)$ and $\boldsymbol{K} = (-1, 1)$ yield the projected Hessian matrix $\hat{\boldsymbol{H}}_0^\star = 6 - p/8$. The trivial solution is then stable for $p < 48$. This threshold $p = 48$ has to be compared with the threshold $p = 16$ found in Section 6.5: different boundary conditions yield different buckling loads.

8 Minimization with inequality constraints

Given a function $V(\boldsymbol{x})$, where $V \in \mathbb{R}$ and $\boldsymbol{x} \in \mathbb{R}^n$, we look for the point(s) for which V is locally minimum, but only for the point(s) that satisfy one (or more) constraints $f(\boldsymbol{x}) \geq 0$, see two examples in Figure 9.

Example: Minimize $V(x) = \frac{1}{4}x^4 - \frac{1}{3}x^3 - 2x^2 + 4x$ in the subset where $f(x) = x \geq 0$.

8.1 First derivative

We are given a function $V(\boldsymbol{x})$ that is to be minimized in the subset where $f(\boldsymbol{x}) \geq 0$. We require the point $\boldsymbol{x}_0$ to be a local minimum by stating that for small ϵ

$$V(\boldsymbol{x} = \boldsymbol{x}_0 + \epsilon\boldsymbol{d}) \geq V(\boldsymbol{x}_0) \quad \forall\boldsymbol{x} \text{ such that } f(\boldsymbol{x}_0 + \epsilon\boldsymbol{d}) \geq 0 \tag{53}$$

This means that not all $\boldsymbol{d}$ directions will be tested. Only the directions $\boldsymbol{d}$ fulfilling

$$f(\boldsymbol{x}_0 + \epsilon\boldsymbol{d}) = f(\boldsymbol{x}_0) + \epsilon\boldsymbol{\nabla} f|_{\boldsymbol{x}_0} \cdot \boldsymbol{d} + O(\epsilon^2) \geq 0 \tag{54}$$

will be tested. There are two cases: (*i*) the point $\boldsymbol{x}_0$ is such that $f(\boldsymbol{x}_0) > 0$ (inactive constraint), or (*ii*) the point $\boldsymbol{x}_0$ is such that $f(\boldsymbol{x}_0) = 0$ (active constraint).

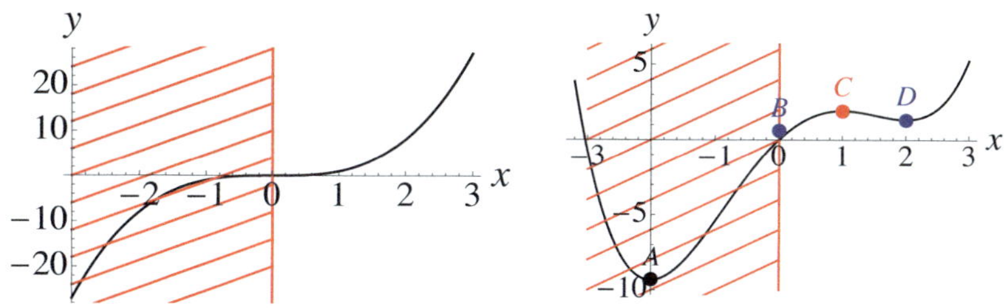

Figure 9. (Left) The function $V(x) = x^3$ with the constraint $x \geq 0$. (Right) The function $V(x) = \frac{1}{4}x^4 - \frac{1}{3}x^3 - 2x^2 + 4x$ with the constraint $x \geq 0$. Points B and D are local minima, point C is a local maximum, and point A is a forbidden minimum.

Inactive constraint ($f(\boldsymbol{x}_0) > 0$): In this case (54) is fulfiled for every direction $\boldsymbol{d}$. Consequently the first order necessary condition is the same as in the unconstrained case: $\boldsymbol{\nabla} V|_{\boldsymbol{x}_0} = \boldsymbol{0}$, this is the case for points C and D in Figure 9-Right.

Active constraint ($f(\boldsymbol{x}_0) = 0$): Contrary to what was done in the case with equality constraints, the sign of ϵ and therefore the direction of $\epsilon\boldsymbol{d}$ cannot be arbitrary. Indeed, here (54) implies that $\boldsymbol{\nabla} f|_{\boldsymbol{x}_0} \cdot \epsilon\boldsymbol{d}$ has to be positive. The first order condition is then

$$\boldsymbol{\nabla} V|_{\boldsymbol{x}_0} \cdot \epsilon\boldsymbol{d} \geq 0 \ \forall \boldsymbol{d} \text{ such that } \boldsymbol{\nabla} f|_{\boldsymbol{x}_0} \cdot \epsilon\boldsymbol{d} \geq 0 \tag{55}$$

which implies that $\boldsymbol{\nabla} V$ and $\boldsymbol{\nabla} f$ are collinear and in the same direction, that is there is a real number μ such that

$$\boldsymbol{\nabla} V|_{\boldsymbol{x}_0} = \mu \boldsymbol{\nabla} f|_{\boldsymbol{x}_0} \text{ with } \mu \geq 0 \tag{56}$$

A way to write the first order condition in both the active and inactive cases is through the so-called Kuhn-Tucker conditions

$$\boldsymbol{\nabla} V|_{\boldsymbol{x}_0} = \mu \boldsymbol{\nabla} f|_{\boldsymbol{x}_0} \tag{57a}$$

$$\mu \geq 0 \tag{57b}$$

$$\mu\, f(\boldsymbol{x}_0) = 0 \tag{57c}$$

8.2 Second derivative

We consider the active and inactive constraint cases separately.

Inactive constraint ($f(\boldsymbol{x}_0) > 0$): as in the first order, every direction $\boldsymbol{d}$ is to be considered and hence we obtain the same sufficient condition as in the unconstrained case: $\boldsymbol{d} \cdot \boldsymbol{H}_0 \cdot \boldsymbol{d} > 0 \ \forall \boldsymbol{d}$. This condition is fulfilled by point D in Figure 9-Right.

For the active constraint case, we consider two sub-cases, (i) when the Lagrange multiplier $\mu > 0$ (strongly active constraint), and (ii) when the Lagrange multiplier $\mu = 0$ (weakly active constraint).

Strongly active constraint ($f(\boldsymbol{x}_0) = 0$ and $\mu > 0$): in this case the perturbations along the gradient of f (i.e. such that $\boldsymbol{\nabla} f \cdot \boldsymbol{d} \neq 0$) are such that, to first order, $V(\boldsymbol{x}) - V(\boldsymbol{x}_0) = \boldsymbol{\nabla} V \cdot \epsilon \boldsymbol{d} > 0$. Consequently we need only worry about the perturbations that let $\boldsymbol{x}$ stay on the surface $f = 0$. As in the case with equality constraints, we develop the perturbation $\boldsymbol{x}$, $f(\boldsymbol{x})$, and $V(\boldsymbol{x})$ to second order, see (37)-(39). We obtain the necessary condition

$$V(\boldsymbol{x}) - V(\boldsymbol{x}_0) = \frac{1}{2}\epsilon^2 \boldsymbol{d} \cdot (\boldsymbol{H}_0 - \mu \boldsymbol{F}_0) \cdot \boldsymbol{d} \geq 0 \quad \forall \boldsymbol{d} \text{ such that } \boldsymbol{\nabla} f \cdot \boldsymbol{d} = 0 \tag{58}$$

where $\boldsymbol{F} = \boldsymbol{F}(\boldsymbol{x})$ is the matrix with second derivative of f: $F_{ij} = \frac{\partial^2 f}{\partial x_i \partial x_j}$, and $\boldsymbol{F}_0$ means $\boldsymbol{F}$ when evaluated on $\boldsymbol{x} = \boldsymbol{x}_0$. A sufficient condition is obtained by requiring the matrix $\boldsymbol{H}_0 - \mu \boldsymbol{F}_0$ to be positive definite in the tangent plane of the surface $f = 0$ at point $\boldsymbol{x}_0$ (that is the $\geq$ sign is replaced by a $>$ sign in (58)).

Weakly active constraint ($f(\boldsymbol{x}_0) = 0$ and $\mu = 0$): in this case the first order of $V(\boldsymbol{x}) - V(\boldsymbol{x}_0)$ vanishes. As a sufficient condition we then ask that the second order $\boldsymbol{d} \cdot \boldsymbol{H}_0 \cdot \boldsymbol{d} > 0$ for all directions $\boldsymbol{d}$, including the direction along $\boldsymbol{\nabla} f$. This sufficient condition is not at all optimal, as can be seen in the example $V(x) = x^3$ with constraint $x \geq 0$ where the point $x = 0$ is clearly a local minimum even if the above sufficient condition is not fulfilled, see Figure 9-Left.

8.3 Summary

When studying the potential energy $V(\boldsymbol{x})$ subject to m inequality constraints $f_j(\boldsymbol{x}) \geq 0$ with $j \in (1, m)$, we first solve the set of nonlinear equations:

$$\mu_1 f_1(\boldsymbol{x}) = 0 \tag{59a}$$
$$\mu_2 f_2(\boldsymbol{x}) = 0 \tag{59b}$$
$$\vdots \quad = 0$$
$$\mu_m f_m(\boldsymbol{x}) = 0 \tag{59c}$$
$$\boldsymbol{\nabla} V(\boldsymbol{x}) - \sum_j \mu_j \boldsymbol{\nabla} f_j(\boldsymbol{x}) = 0 \tag{59d}$$

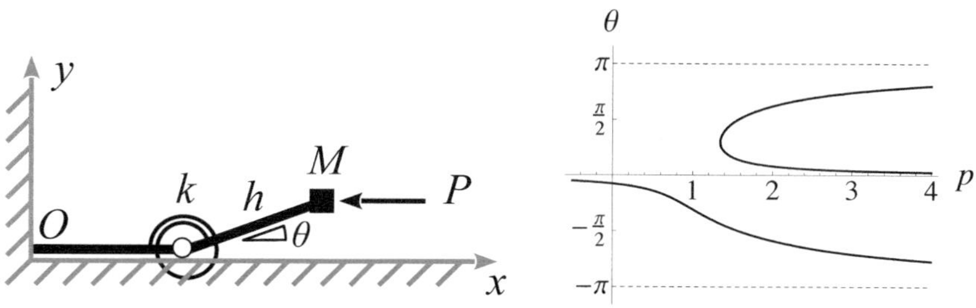

Figure 10. (Left) Two rigid bars linked by a spiral spring, with a punctual mass M at the extremity of the second bar. The system lies on a rigid support and is subject to an external compressive load P. (Right) Solution curves $\theta = \theta(p)$ of equation (62).

This yields a set of candidate local minima $\boldsymbol{x}_0 = \boldsymbol{x}_A, \boldsymbol{x}_B, \ldots$ We review each candidate and eliminate those for which there is one (or more) j such that $f_j(\boldsymbol{x}_0) < 0$. Among the remaining candidates we eliminate those for which there is one (or more) j such that $\mu_j(\boldsymbol{x}_0) < 0$. For each remaining candidate, we build the projected Hessian matrix $\hat{\boldsymbol{H}}^\star := \boldsymbol{K}\,\hat{\boldsymbol{H}}\,\boldsymbol{K}^T$ with $\hat{\boldsymbol{H}} = \boldsymbol{H} - \sum_j \mu_j \boldsymbol{F}_j$. The matrix $\boldsymbol{K}$ is computed, as in Section 7.3, from the kernel of the matrix $\boldsymbol{T}$ whose lines are the gradients $\boldsymbol{\nabla} f_j$ for all j such that $\mu_j > 0$. We then compute the eigenvalues of $\hat{\boldsymbol{H}}^\star$. If there is only strictly positive eigenvalues, the point $\boldsymbol{x}_0$ is a local minimum. But in the case where there is one (or more) zero or strictly negative eigenvalue, we cannot decide whether $\boldsymbol{x}_0$ is a local minimum or not. This shortcoming is due to the fact that when weakly active constraints are present, our sufficient condition is too demanding. As a matter of fact, if no weakly active constraint is present (that is $\mu_j > 0$ for all j such that $f_j(\boldsymbol{x}_0) = 0$) then stability is only undecided when zero eigenvalues are encountered, that is strictly negative eigenvalues yield instability.

8.4 Example: Buckling of heavy beam on rigid support

We study the equilibrium and stability of the system of Figure 10-Left, comprising two rigid bars of length h, linked by a spiral spring of stiffness k. The first bar is strongly anchored at the origin. At the extremity of the second bar lies a punctual mass M, and all the other components of the system are considered massless. An external horizontal force P is applied on the mass. The system is resting on a rigid, impenetrable support. To the internal strain energy $\frac{1}{2}k\theta^2$ of the spiral spring we add the gravitational energy of the mass $M\,g\,h\,\sin\theta$ and the work of the external load $Ph(1+\cos\theta)$ to obtain the total potential energy V of the system. After division by k and subtraction of the constant $2Ph/k$, we obtain

$$v(\theta) = \frac{1}{2}\theta^2 + p\,(\cos\theta - 1) + m\,\sin\theta \tag{60}$$

where $v = V/k - 2Ph/k$, $p = Ph/k$, and $m = M\,g\,h/k$. We study $v(\theta)$ under the constraint $y(M) \geq 0$, that is $f(\theta) = \sin\theta \geq 0$. We fix $m = 1/4$ and look for stable and unstable equilibrium solutions and their evolution as p is varied. We restrict to $p \in (0, 4)$ and $\theta \in (-\pi, \pi)$.

First order (equilibrium) equations are

$$v'(\theta) - \mu\,f'(\theta) = 0 \tag{61a}$$

$$\mu\,f(\theta) = 0 \tag{61b}$$

In the case where $\mu = 0$, (61a) is solved as

$$p = \frac{\theta + m\,\cos\theta}{\sin\theta} \tag{62}$$

and yields two distinct curves, see Figure 10-Right. The lower curve entirely lies in the region $f(\theta) < 0$ and is therefore discarded.

In the case where $f(\theta) = \sin\theta = 0$, we find that either (i) $\theta = 0$ with $m = \mu$, or (ii) $\theta = \pi$ with $\mu = m - \pi = 1/4 - \pi < 0$. This second solution with $\mu < 0$ is discarded.

We now turn to the stability test for the equilibrium we have just found.

In the case where $\mu = 0$, the constraint is passive ($f(\theta) > 0$) and consequently we need only test the second derivative $v''(\theta) = 1 - p\,\cos\theta - m\sin\theta$ for each point on the upper curve of Figure 10-Right. We find that stability changes at the fold point, see Section 9, and that the upper part (respectively the lower part) of the curve is stable (resp. unstable).

In the case where $f(\theta) = 0$, the equilibrium solution $\theta = 0$ has a strictly positive Lagrange multiplier $\mu = m > 0$ which is enough for stability in this one degree-of-freedom system.

The complete bifurcation curve, with stability information, is draw in Figure 11-Left. We remark that the trivial path $\theta = 0$ is stable for any load p and that it is not connected to the path of buckled states. We also remark that if the mass m is increased to $m > \pi$ a stable path $\theta = \pi$ appears for all p. Finally we see in Figure 11-Right that in this one degree-of-freedom problem equilibrium and their stability can readily be assessed by looking at the graph of $v(\theta)$ for different values of p.

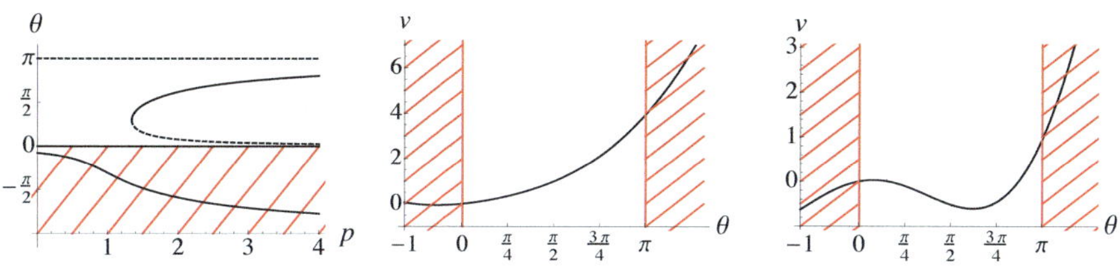

Figure 11. (Left) Complete bifurcation diagram for system of Figure 10-Left with $m = 1/4$, $p \in (0, 4)$, and $\theta \in (-\pi, \pi)$. Plain (respectively dashed) curves correspond to stable (resp. unstable) equilibria. (Center) The graph of $v(\theta)$ for $p = 0.5$ where the unique stable equilibrium is seen at $\theta = 0$. (Right) Same graph, yet for $p = 2$ where one additional unstable and one additional stable equilibria are present.

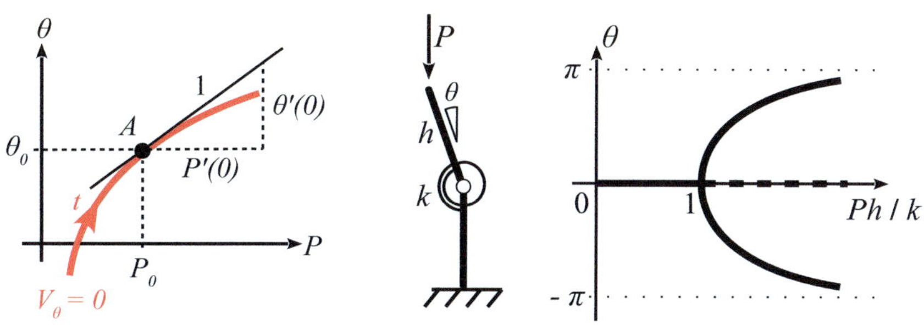

Figure 12. (Left) At point $A = (P_0, \theta_0)$, we look for the tangent approximation to the bifurcation curve $V_\theta(\theta, P) = 0$. (Right) One degree of freedom system and its bifurcation diagram.

9 Bifurcation diagrams

For simplicity reasons, in this section we treat the case of a one degree of freedom system, $V = V(\theta, P)$ where θ is the variable and P the applied load. Equilibrium are given by $V_\theta := \partial V/\partial\theta = 0$, and we want to plot the set of curve(s) implicitly defined by $V_\theta(\theta, P) = 0$ in the plane (P, θ), see Figure 12. This set of curves is called the bifurcation diagram of the system. Treatments of the case with multiple degrees of freedom can be found for example in Iooss and Joseph (1989) or Nguyen (2000).

Example: The system of Figure 12-Right has energy $V(\theta, P) = \frac{1}{2}k\theta^2 + Ph\cos\theta$, its bifurcation diagram exhibits a pitchfork bifurcation point at $(Ph/k, \theta) = (1, 0)$.

We start at a known point on the bifurcation curve, $A = (P_0, \theta_0)$, and we use a parameter t to move about A. We approximate the parametric curve $(P(t), \theta(t))$ locally with the Taylor expansion

$$P(t) = P_0 + t\,P'(0) + O(t^2) \tag{63a}$$

$$\theta(t) = \theta_0 + t\,\theta'(0) + O(t^2) \tag{63b}$$

The goal is to find $P'(0)$ and $\theta'(0)$ in order to construct a tangent approximation to the curve at point A. By definition, point A satisfies $V_\theta(\theta_0, P_0) = 0$. The points in its neighborhood (63) along the bifurcation curve are also equilibrium points, hence they satisfy $V_\theta(\theta(t), P(t)) = 0$. We then have

$$V_\theta(\theta(t), P(t)) = 0 = V_\theta(\theta_0, P_0) + t\,P'(0)\left.\frac{\partial V_\theta}{\partial P}\right|_{t=0} + t\,\theta'(0)\left.\frac{\partial V_\theta}{\partial \theta}\right|_{t=0} + O(t^2) \tag{64}$$

We simply write $V_{\theta P} := \left.\frac{\partial V_\theta}{\partial P}\right|_{t=0}$ and $V_{\theta\theta} := \left.\frac{\partial V_\theta}{\partial \theta}\right|_{t=0}$ and we note that the later is the Hessian at point A. At first order we have

$$P'(0)\,V_{\theta P} + \theta'(0)\,V_{\theta\theta} = 0 \tag{65}$$

As we are given both the potential energy V and point A, we know the second derivatives $V_{\theta P}$ and $V_{\theta\theta}$. Relation (65) is therefore an equation for the two unknowns $P'(0)$ and $\theta'(0)$. There are four cases, as illustrated in Figure 13:

- if $V_{\theta P} \neq 0$ and $V_{\theta\theta} \neq 0$ point A is a regular point of the bifurcation curve, which has local slope $\frac{d\theta}{dP} = -V_{\theta P}/V_{\theta\theta}$.
- if $V_{\theta P} = 0$ and $V_{\theta\theta} \neq 0$ point A is also a regular point. The bifurcation curve has an horizontal tangent at point A: $\frac{d\theta}{dP} = 0$.
- if $V_{\theta P} \neq 0$ and $V_{\theta\theta} = 0$ point A is a singular point. From (65), we see that $P'(0) = 0$ which mean that the curve $\theta = \theta(P)$ has a vertical tangent at point A. Such a point is called a limit point. Considering the bifurcation curve in the neighborhood of A, we see that the Hessian $V_{\theta\theta}$ generically changes sign at a limit point: instabilities arise at limit points.
- if $V_{\theta P} = 0$ and $V_{\theta\theta} = 0$ point A is a singular point. The first order expansion (64) entirely vanishes at such a point. Consequently we push (63) and (64) to second order and find

$$V_{\theta\theta\theta}\,\theta'(0)^2 + 2V_{\theta\theta P}\,P'(0)\,\theta'(0) + V_{\theta PP}\,P'(0)^2 = 0 \tag{66}$$

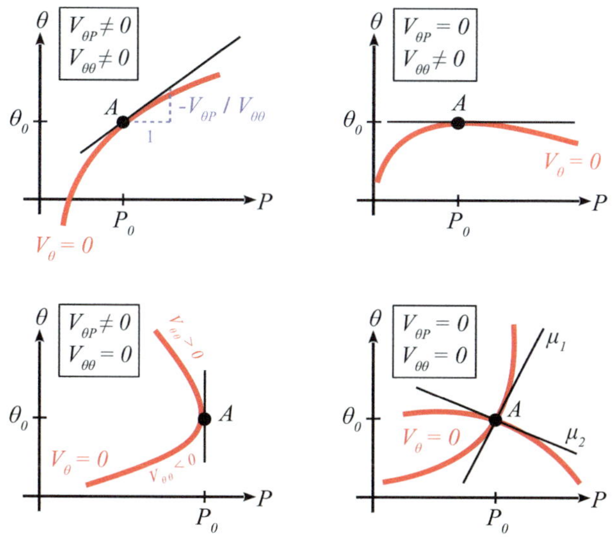

Figure 13. The four possible types of point along a bifurcation curve.

Generically this leads to two solutions for the tangent to the bifurcation curve, meaning that point A is a bifurcation point where two curves cross. From here there are two sub-cases:

+ if $V_{\theta\theta\theta} = 0$, then a first tangent is $P'(0) = 0$ (vertical tangent), and the second one is given by, provided $V_{\theta\theta P} \neq 0$, $\frac{d\theta}{dP} = -(1/2)\, V_{\theta PP}/V_{\theta\theta P}$.

+ if $V_{\theta\theta\theta} \neq 0$ then $P'(0) \neq 0$ and we can set $\mu := \theta'(0) = 0/P'(0) = 0$ and solve $V_{\theta\theta\theta}\, \mu^2 + 2V_{\theta\theta P}\, \mu + V_{\theta PP} = 0$. The two roots μ_1 and μ_2 have to be real, and provided they are distinct, we have the two tangents.

Here also the Hessian is going to change sign at the bifurcation point, instabilities arise at bifurcation points as well.

In conclusion we see that instabilities generically arise at singular points, limit or bifurcation points.

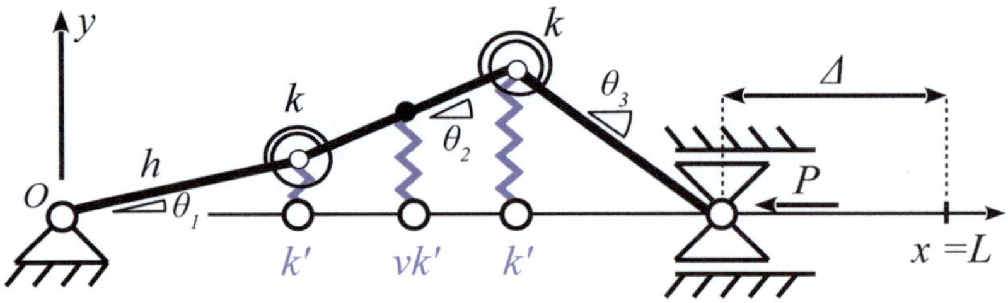

Figure 14. Stein's system: three rigid bars resting of a linear elastic foundation.

10 Buckling and post-bucking of a discrete beam on an elastic foundation

To illustrate the content of the previous sections, we treat the following example of a system of three rigid bars, see Figure 14. The bars have length h and therefore the system has a total contour length $L = 3h$. The three bars are linked together with two spiral springs of stiffness k, the first bar is held by a pivot fixed at the origin, and the third bar is held by a pivot that can slide along the horizontal axis. Moreover the second bar is tied to the ground with three linear springs of stiffness k', $\nu k'$, and k'. These linear springs have zero rest length, and their foot are freely sliding along the horizontal axis. This model has been used by Stein (1959) to illustrate the rich behavior of stiffened elastic plates. Such a system has

$$V_{\text{int}} = \frac{1}{2}k(\theta_2 - \theta_1)^2 + \frac{1}{2}k(\theta_3 - \theta_2)^2 + \frac{1}{2}k'h^2 \sin^2\theta_1 + \frac{1}{2}\nu k' h^2 (\sin\theta_1 + \frac{1}{2}\sin\theta_2)^2 + \frac{1}{2}k'h^2\sin^2\theta_3 \qquad (67)$$

as internal energy. We choose $\nu = 2$. The work of the external load P is

$$V_{\text{ext}} = Ph\,(\cos\theta_1 + \cos\theta_2 + \cos\theta_3) \qquad (68)$$

Boundary conditions impose a first constraint

$$g_1(\theta_1, \theta_2, \theta_3) = \sin\theta_1 + \sin\theta_2 + \sin\theta_3 = 0 \qquad (69)$$

In the case of force-controlled loading the total potential energy is $V = V_{\text{int}} + V_{\text{ext}}$, while in the case of displacement-controlled loading it is simply $V = V_{\text{int}}$. In the latter though, a second constraint is present

$$g_2(\theta_1, \theta_2, \theta_3) = h\,[\cos\theta_1 + \cos\theta_2 + \cos\theta_3] - (L - \Delta) = 0 \qquad (70)$$

where Δ is the end-shortening, see Figure 14. We introduce dimension-

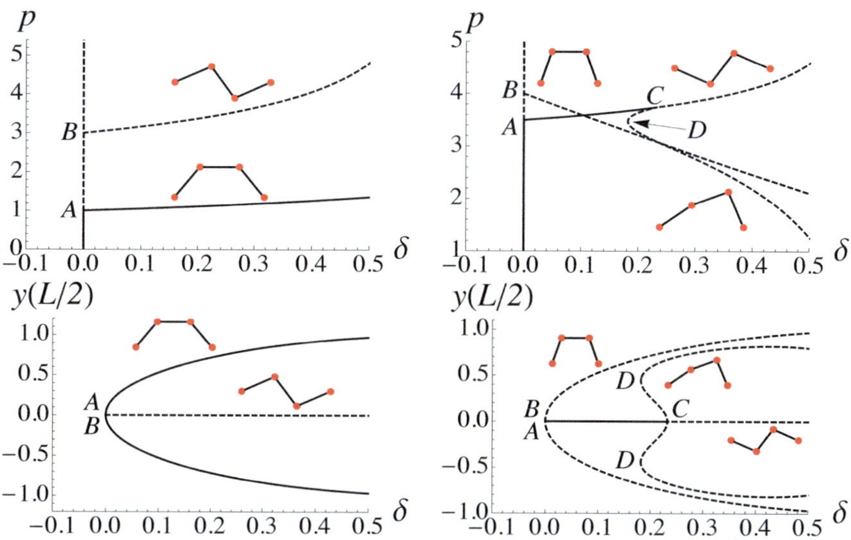

Figure 15. Bifurcation diagram for the system of Figure 14 in the force-controlled case. For both columns, we show two projections of the parameter space, namely (δ, p) for the upper diagram and $(\delta, y(L/2))$ for the lower diagram. Left column: without foundation, $\kappa = 0$. Right column: with foundation, $\kappa = 3/2$. Points A, B, C are pitchfork bifurcation points, and D is a fold point.

less quantities $v = V/k$, $\kappa = k' h^2/k$, $\delta = \Delta/(3h)$, and $p = Ph/k$. The augmented energy is then

$$
\begin{aligned}
v =& \frac{1}{2}(\theta_2 - \theta_1)^2 + \frac{1}{2}(\theta_3 - \theta_2)^2 + \frac{1}{2}\kappa \sin^2\theta_1 + \frac{1}{2}\kappa(\sin\theta_1 + \frac{1}{2}\sin\theta_2)^2 \\
&+ \frac{1}{2}\kappa(\sin\theta_3 + \frac{1}{2}\sin\theta_2)^2 + \frac{1}{2}\kappa\sin^2\theta_3 \\
&+ p\left[\cos\theta_1 + \cos\theta_2 + \cos\theta_3 - 3(1-\delta)\right] - \lambda(\sin\theta_1 + \sin\theta_2 + \sin\theta_3)
\end{aligned}
\tag{71}
$$

where, using (69), we have split the term corresponding to the extension of the center linear spring in two terms, rendering the energy symmetric with regards to the change $\theta_1 \leftrightarrow \theta_3$. We note λ the Lagrange multiplier associated to the constraint g_1. In the case of force-controlled loading, p is the external force, while in the case of displacement-controlled loading it is the Lagrange multiplier associated to the constraint g_2. We then see that the equilibrium equations $\partial v/\partial\theta_i = 0$, and consequently the equilibrium solutions, are the same for the two types of loading. Yet, computing the Hessian matrix, we see that stability depends on the loading type as the projected Hessian matrix is smaller in the case of the displacement-controlled loading: there are 3 variables $(\theta_1, \theta_2, \theta_3)$ and 1 constraint in the force-controlled case, generating a 2×2 projected Hessian matrix, while there are 2 constraints in the displacement-controlled case, generating a 1×1 projected Hessian matrix.

Writing the equilibrium equations $\partial v/\partial\theta_i = 0$, one sees that the straight configuration $(\theta_1, \theta_2, \theta_3) = (0, 0, 0)$ is solution $\forall p$. To test the stability of this trivial solution, we compute the Hessian matrix

$$\hat{\boldsymbol{H}}_0 = \begin{pmatrix} 1+2\kappa-p & -1+\kappa/2 & 0 \\ -1+\kappa/2 & 2-p+\kappa/2 & -1+\kappa/2 \\ 0 & -1+\kappa/2 & 1+2\kappa-p \end{pmatrix} \tag{72}$$

We now restrict to the force-controlled case, and first focus on the case with no foundation, $\kappa = 0$. The augmented Hessian matrix has the three eigenvalues $\lambda_1 = -p$, $\lambda_2 = 1-p$, and $\lambda_3 = 3-p$, but to conclude on the stability the projected Hessian matrix has to be computed. The gradient of the constraint g_1 is evaluated at $(\theta_1, \theta_2, \theta_3) = (0, 0, 0)$ and written in the matrix $\boldsymbol{T} = (1, 1, 1)$. The null-space of $\boldsymbol{T}$ is spanned by $\boldsymbol{k}_1 = (-1, 0, 1)/\sqrt{2}$ and $\boldsymbol{k}_2 = (-1, 2, -1)/\sqrt{6}$, leading to the matrix

$$\boldsymbol{K} = \begin{pmatrix} -1/\sqrt{2} & 0 & 1/\sqrt{2} \\ -1/\sqrt{6} & 2/\sqrt{6} & -1/\sqrt{6} \end{pmatrix} \text{ and } \hat{\boldsymbol{H}}_0^\star(\kappa=0) = \begin{pmatrix} 1-p & 0 \\ 0 & 3-p \end{pmatrix} \tag{73}$$

with $\hat{\boldsymbol{H}}^\star := \boldsymbol{K}\,\hat{\boldsymbol{H}}\,\boldsymbol{K}^T$. The straight solution becomes unstable as p reaches $p = 1$ where the system buckles into a symmetrical $\theta_1 = \theta_3$ shape, while at $p = 3$ a second (anti-symmetrical, $\theta_1 = -\theta_3$) buckling mode appears.

In the case $\kappa > 0$, the matrices $\boldsymbol{T}$ and $\boldsymbol{K}$ are the same, and the projected Hessian matrix is

$$\hat{\boldsymbol{H}}_0^\star(\kappa) = \begin{pmatrix} 1-p+2\kappa & 0 \\ 0 & 3-p+\kappa/3 \end{pmatrix} \tag{74}$$

We have here also two buckling modes $p_1 = 1+2\kappa$ and $p_2 = 3+\kappa/3$, one symmetrical and the other anti-symmetrical. Depending on the κ value (that is on the strength of the foundation) either the symmetrical or the anti-symmetrical one buckles first. Full bifurcation diagrams are shown in Figure 15 where we see that a non-symmetrical solution also exists. The branch containing these non-symmetrical solutions emanates from the secondary bifurcation point, labelled C. In the present force-controlled loading type, the branch comprises only unstable solutions, but in the case of displacement-controlled loading, the branch is stable after the fold point D.

11 Dynamical stability

Here, we generalize the previous concepts to discretized nonlinear elastic systems where dynamical effects are taken into account, that is we deal with dynamical systems, see for example Guckenheimer and Holmes (1983) or Strogatz (1994). We broaden our numerical tools and address not only equilibrium states, that are fixed in time, but periodic states, that repeat in equal interval of time. Due to the complexity added by the presence of inertial terms, we will limit ourselves to mechanical systems without constraints.

We consider a one-degree-of-freedom system comprising two rigid bars of length L, as illustrated in Figure 16. The first bar is fixed on an infinitely rigid and massive base that eventually moves in a harmonic fashion following

$$y_a(t) = A\cos(\Omega t) \tag{75}$$

where A is the amplitude of the vertical oscillation and Ω is the frequency of the harmonic motion. The second bar is allowed to rotate at point B thanks to a viscoelastic hinge characterized by a rotating stiffness k and a viscous rotational damping c. The bars themselves are considered massless, and a concentrated mass M lies at the end of the second bar. The motion of the structure is parameterized by the angle $\theta\,(t)$ between the two bars. Such a system is a very simplified model for a heavy post fixed on an oscillating ground. The presence of harmonic excitations forces us to analyze the dynamics of the system.

11.1 Nonlinear equations of motion

We establish the equations of motion of the system of Figure 16. We first write down the kinetic energy of the system. The position of the material point M on the bar is expressed as a function of $\theta\,(t)$ at every time t in the Cartesian frame (O, x, y):

$$\boldsymbol{OM}\,(t) = \left\{ \begin{array}{c} x_a - L\sin\theta(t) \\ y_a(t) + L + L\cos\theta(t) \end{array} \right\} \tag{76}$$

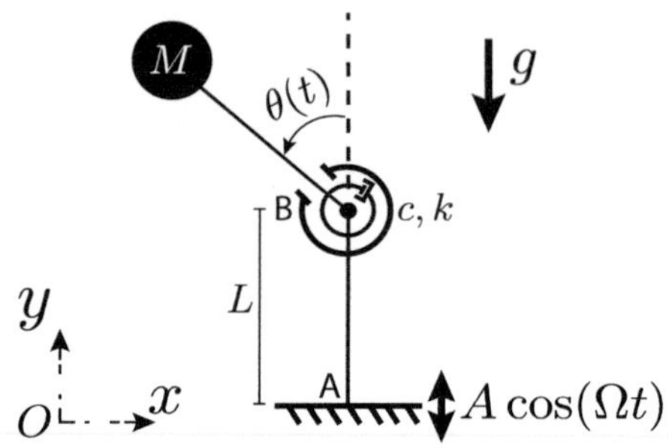

Figure 16. An initially vertical articulated system of two rigid bars, each of length L, supporting a concentrated mass M. The articulation is characterized by a rotational stiffness, k, and a viscous rotational damping, c. The structure is fixed at its base. The base vibrates in a harmonic fashion with an amplitude A and a frequency Ω. At each time t, the position of the moving bar in the $2D$ plane is parameterized by the angle $\theta(t)$ of the second bar with the vertical y-axis.

where x_a is fixed and $y_a(t)$ is given by (75). The velocity of the material point of mass M is therefore

$$\boldsymbol{v}\,(t) = \dot{\boldsymbol{OM}}(t) = \left\{ \begin{array}{c} -L\,\dot{\theta}\,\cos\theta \\ -A\,\Omega\,\sin(\Omega t) - L\,\dot{\theta}\,\sin\theta \end{array} \right\} \tag{77}$$

The kinetic energy then reads

$$T = \frac{1}{2}M\boldsymbol{v}^2(t) = \frac{1}{2}M\left[L^2\dot{\theta}^2 + A^2\Omega^2\sin^2(\Omega t) + 2AL\Omega\dot{\theta}\sin(\Omega t)\sin\theta\right] \tag{78}$$

Virtual power of the external torques The virtual power of the reaction torque due to the stiffness of the circular spring simply writes

$$P_{ek} = -k\theta\,\delta\dot{\theta} \tag{79}$$

where $\delta\dot{\theta}$ is the virtual angular velocity carrying the velocity $\dot{\theta}$. The virtual power of the reaction torque due to the viscous damper reads

$$P_{ec} = -c\dot{\theta}\,\delta\dot{\theta} \tag{80}$$

The virtual power of the external torque due to the weight of the concentrated mass, $\boldsymbol{P} = -Mg\boldsymbol{e}_y$, is calculated considering the power moment of $\boldsymbol{P}$ at point B in the virtual angular velocity $\delta\dot{\theta}$

$$P_{eM} = (\boldsymbol{BM}\times\boldsymbol{P})\cdot\delta\dot{\theta}\,\boldsymbol{e_z} = MgL\,\delta\dot{\theta}\,\sin\theta \tag{81}$$

Note that the spring reaction torque and the torque of external force are conservative torques that derive from the potential energy

$$V = \frac{1}{2}k\theta^2 + MgL\cos\theta. \tag{82}$$

Lagrange formalism From the kinetic energy of the system, Eq. (78), we calculate the virtual power of the quantity of acceleration through Lagrange formula:

$$A_q = \left[\frac{d}{dt}\left(\frac{\partial T}{\partial \dot{q}_i}\right) - \frac{\partial T}{\partial q_i}\right]\delta\dot{q}_i \tag{83}$$

where the q_i are the generalized coordinates. The principle of virtual power, $A_q = P_{ek} + P_{ec} + P_{eM}$, yields the equation of motion of our discrete system

$$\frac{\partial}{\partial t}\left(\frac{\partial T}{\partial \dot{q}_i}\right) - \frac{\partial T}{\partial q_i} + \frac{\partial V}{\partial q_i} = Q_i \tag{84}$$

where Q_i is obtained from the virtual power of the non conservative external forces such that $P_{ec} = Q_i\delta\dot{q}_i$. We apply Eq. (84) to our system and obtain the nonlinear equation of motion for $q_1(t) = \theta(t)$

$$\ddot{\theta} + \frac{c}{ML^2}\dot{\theta} + \frac{k}{ML^2}\theta + \left(\frac{A\Omega^2}{L}\cos(\Omega t) - \frac{g}{L}\right)\sin\theta = 0 \tag{85}$$

We introduce the dimensionless variables $\tau = (\Omega/2)t$ and multiply (85) by $4/\Omega^2$ to rewrite it in dimensionless form

$$\ddot{\theta} + \frac{2}{\beta Q}\dot{\theta} + \frac{4}{\beta^2}\theta + \left(2\varepsilon\cos(2\tau) - \frac{4}{\beta^2}p\right)\sin\theta = 0 \tag{86}$$

where $\dot{(\,)}$ denotes differentiation with respect to τ, $Q = \frac{\sqrt{kML^2}}{c}$ is the quality factor, $\beta = \Omega/\Omega_0$ is the ratio between the excitation and a pseudo natural frequency $\Omega_0 = \sqrt{k/(ML^2)}$, $\varepsilon = 2A/L$ is the dimensionless amplitude of the harmonic motion of the base, and $p = MgL/k$ is a crucial dimensionless loading parameter for the static problem modeling the ratio between weight and stiffness of the system.

For convenience, we write this second order nonlinear ordinary differential equation as a two dimensional dynamical system $\dot{\boldsymbol{x}} = \boldsymbol{f}(\boldsymbol{x},\tau)$:

$$\begin{cases} \dot{\theta} = \phi \\ \dot{\phi} = -\frac{2}{\beta Q}\phi - \frac{4}{\beta^2}\theta - \left(2\varepsilon\cos(2\tau) - \frac{4}{\beta^2}p\right)\sin\theta \end{cases} \tag{87}$$

where $\boldsymbol{x}(\tau) = \{\theta(\tau), \phi(\tau)\}^T$ is the 2-dimensional state vector and $\boldsymbol{f}(\boldsymbol{x},\tau)$ is a nonlinear 2-dimensional vector field.

11.2 Particular case of a standstill base ($\varepsilon = 0$)

In this section we consider the case where the base in not in motion, that is we set $\varepsilon = 0$. We further fix $\beta = 1$ with no loss of generality. The equations of motion take the form $\dot{\boldsymbol{x}} = \boldsymbol{f_1}(\boldsymbol{x}, \tau)$, that is

$$\begin{cases} \dot{\theta} = \phi \\ \dot{\phi} = -\frac{2}{Q}\phi - 4\theta + 4p \sin\theta \end{cases} \tag{88}$$

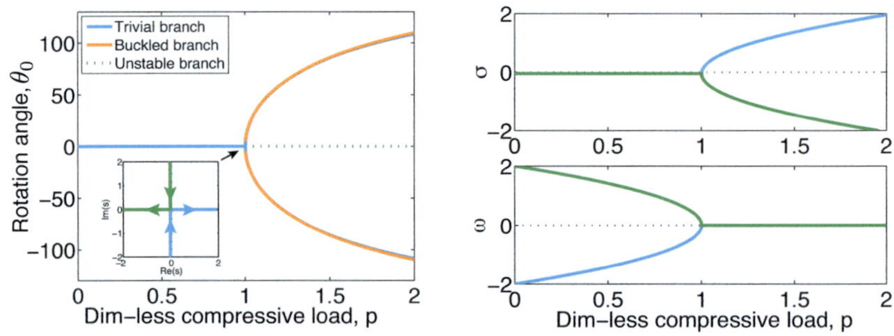

Figure 17. Branches of equilibrium solutions and linear stability analysis. (Left) Bifurcation diagram from the equilibria of system (88): equilibrium angle θ_0 as function of the dimensionless loading parameter p. Inset: Evolution in the Argand plane of the eigenvalues of the associated linearized dynamical equations around the trivial equilibrium point. (Right) Evolution of growth rate σ and frequency ω of the fundamental solutions $\boldsymbol{y}(\tau)$ of Eq. (90) as a function of p.

Bifurcation diagram and stability of equilibrium Equilibrium solutions are defined as solutions to Eq. (88) with $\dot{\phi} = 0 = \dot{\theta}$. We therefore have $\boldsymbol{x}(\tau) = \boldsymbol{x}_0 = \{\theta_0, 0\}^T$ and the equilibrium angle θ_0 solution of

$$\theta_0 - p \sin\theta_0 = 0. \tag{89}$$

For $p \in (0, 2)$, the nonlinear solution to (89) comprises two branches, see Figure 17-Left. The trivial branch is simply $\theta_0 = 0$, i.e. $\boldsymbol{x}_0 = \{0, 0\}^T = \boldsymbol{0}$. To assess the stability of this trivial equilibrium, we study the behavior of Eq. (88) for small $\theta(\tau)$ and $\phi(\tau)$, that is we set $\theta(\tau) = \theta_0 + \theta^*(\tau)$ and $\phi(\tau) = \phi^*(\tau)$ where $\theta^*(\tau)$ and $\phi^*(\tau)$ are small, and $\theta_0 = 0$. Injecting this in Eq. (88) and linearizing to first order, we obtain the so-called linearized equations of motion in the state space

$$\dot{\boldsymbol{y}}(\tau) = \boldsymbol{J_1}\, \boldsymbol{y}(\tau) \tag{90}$$

$$\text{with}\quad \boldsymbol{y}(\tau) = \left\{ \begin{array}{c} \theta^*(\tau) \\ \phi^*(\tau) \end{array} \right\} \quad \text{and} \quad \boldsymbol{J_1} = \left.\frac{\partial \boldsymbol{f_1}}{\partial \boldsymbol{x}}\right|_{\boldsymbol{0}} = \left[\begin{array}{cc} 0 & 1 \\ (4\lambda - 4) & -\frac{2}{Q} \end{array} \right]$$

where $\boldsymbol{J_1}(\boldsymbol{x}_0, \lambda)$ is the Jacobi matrix of the nonlinear vector field $\boldsymbol{f_1}(\boldsymbol{x}, \tau)$, given in Eq. (88), evaluated at the trivial equilibrium point $\boldsymbol{x}_0 = \boldsymbol{0}$. The linear stability of this point is assessed by analyzing the perturbation vector $\boldsymbol{y}(\tau)$ when assuming a perturbation of the form

$$\boldsymbol{y}(\tau) = \boldsymbol{r}\, e^{s\tau} \quad \text{with} \quad \boldsymbol{r} = r \left\{ \begin{array}{c} 1 \\ s \end{array} \right\} \tag{91}$$

where the exponents $s = \sigma + i\omega$ can be complex numbers with σ, the growth rate or Lyapunov exponent and ω, the dimensionless frequency of the pseudo-harmonic eigenmode $\boldsymbol{r}$. Injecting $\boldsymbol{y}\,(\tau)$ in Eq. (90), we obtain the eigenvalue problem

$$[\boldsymbol{J_1} - s\boldsymbol{1}]\, \boldsymbol{r} = \boldsymbol{0} \tag{92}$$

where $\boldsymbol{1}$ is the 2×2 identity matrix. Since the dimension of this eigenvalue problem in the state space is $N = 2$, i.e. twice the number of degree of freedom, Eq.(92) yields 2 eigenmodes $\boldsymbol{r}_n$ and 2 eigenvalues s_n. The evolution of the real, $\Re(s)$, and imaginary, $\Im(s)$, parts of the eigenvalues s_n of the linearized problem Eq. (90) as function of the loading parameter p is drawn in Figure 17-Right. The linear stability of the considered equilibrium

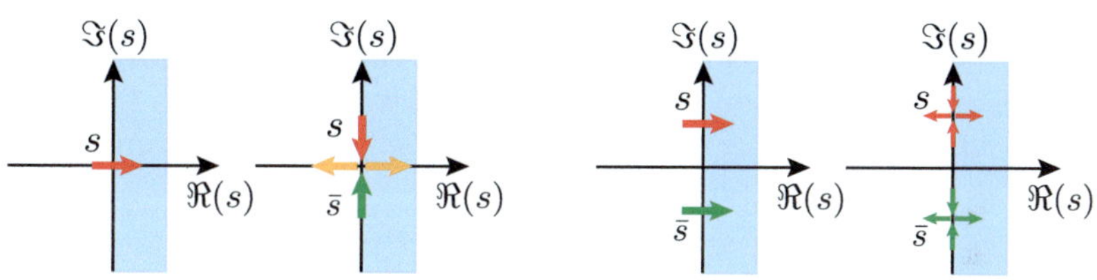

Figure 18. Stability of an equilibrium in the Argand plane. (Left) Static instability or instability by divergence can happen by simple crossing or locking of eigenvalues on the real axis. (Right) Dynamic instability or flutter instability can happen by crossing or locking of eigenvalues away from the real axis.

θ_0 is assessed by analyzing the eigenvalues s_n, i.e. the spectrum of the harmonic eigenmodes $\boldsymbol{y}_n\,(\tau)$ given by Eq. (91). According to Lyapunov theory, here are the following fundamental results when dealing with autonomous dynamical systems such as the one described in Eq. (88):

- If $\Re(s_n) < 0$ for all n, the equilibrium is asymptotically stable.

- If $\Re(s_n) \leq 0$ for all n and there is at least one index k such that $\Re(s_k) = 0$, one cannot conclude with a linearized stability analysis. Higher order considerations are needed.
- If amongst all the eigenvalues there exists one index k such that $\Re(s_k) > 0$, the equilibrium is unstable but two different scenarios have to be considered:
  - If $\Im(s_k) = 0$, the perturbed motion dramatically grows along the direction of the kth mode $\boldsymbol{r}_k$ in an exponential way. This loss of stability is called an instability by divergence or a static instability (buckling presented in the previous sections is such an instability). In the Argand plane, there are two scenarios for such an instability (see Figure 18-Left): *i*) one purely real eigenvalue crosses the imaginary axis, or *ii*) the imaginary parts of two conjugate eigenvalues annihilate to form one purely real eigenvalue crossing the imaginary axis.
  - If $\Im(s_k) \neq 0$, the perturbed motion is an harmonic oscillation that exponentially grows in the direction of the kth mode $\boldsymbol{r}_k$. This loss of stability is refereed to as a flutter instability or more generally as a dynamic instability. The representation in the Argand plane of such instabilities is illustrated in Figure 18-Right where two scenarios have to be considered: *i*) the crossing of the imaginary axis by two conjugate eigenvalues, or *ii*) the locking of two physical eigenvalues that causes the real part of one of them to cross the imaginary axis. The dynamic instability is an indicator of a Hopf bifurcation. In this kind of bifurcation, not only the dynamical system qualitatively bifurcates in the state space but also in the time domain, i.e. the equilibrium bifurcates to a periodic state (limit cycle).

In Figure 17-Right, we represent the evolution of the real and imaginary part of the two eigenvalues s_1 and s_2 of the eigenvalue problem Eq. (92) as function of the loading parameter p when $Q = 20$ (subcritical damping). These curves inform us about the linear stability of the trivial equilibrium branch $\theta_0 = 0$. As expected, above the critical load, $p > 1$, the thermodynamic branch becomes statically unstable and the straight equilibrium configuration $\theta_0 = 0$ is no longer a physical configuration for our articulated system. The inset of Figure 17-Left is the representation of the evolution of eigenvalues s_i in the Argand plane. As illustrated in Figure 18-Left, the system looses stability by the meeting of two purely imaginary and conjugate eigenvalues, which is a particular case of instability by divergence. Above the critical load, $p > 1$, the articulated bar buckles and possible equilibrium solutions with $\theta_0 \neq 0$ exist. To compute those stable bifurcated branches, one

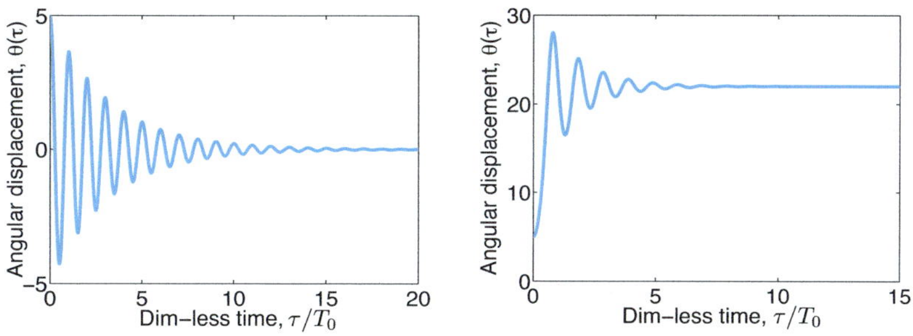

Figure 19. Temporal evolutions $\theta(\tau)$ as a function of dimensionless time τ/T_0 for initial conditions $\boldsymbol{x}(0) = \{5°, 0\}^T$ and $Q = 20$. (Left) For $p = 0.75$. (Right) For $p = 1.025$.

needs to solve the nonlinear algebraic equation given in Eq. (89). Here, we simply perform a Taylor expansion of the sine function and Eq.(89) becomes

$$\theta_0 - p\left(\theta_0 - \frac{\theta_0^3}{6} + \frac{\theta_0^5}{120} - ...\right) = 0. \tag{93}$$

Truncating the series (93) up to the fifth order, we obtain a fourth order polynomial equation with analytical solutions depending on the loading parameter p:

$$\theta_\pm = \sqrt{10 \pm \sqrt{\frac{20(6-p)}{p}}} \tag{94}$$

The solution θ_- is a local minimum of the potential energy and is represented in the bifurcation diagram of Figure 17-Left (we represent also the solution $-\theta_-$, the bifurcation being symmetric). We also compare the approximated solution θ_- with the one obtained by solving Eq. (89) with a classic Newton-Raphson algorithm. As the curves are barely distinguishable, we conclude that a fifth order approximation on the sine function is sufficient to correctly capture the mechanical behavior, up to a loading parameter $p = 2$. Note that the solution θ_+ is associated with a higher potential energy and corresponds to an unstable equilibrium configuration where the articulated rigid bar is rotated by more than a turn. As this solution disappears when the order of the series (93) is changed, we do not consider this 'spurious' solution here, see Domokos and Holmes (1993).

Direct dynamic analysis We recover the qualitative static behavior previously discussed by directly solving the nonlinear system of ordinary differential equations given in Eq. (88) for various initial conditions $\boldsymbol{x}(0) = \{\theta(0), \phi(0)\}^T$. In Figure 19, we represent the evolution of $\theta(\tau)$ as a function of dimensionless time τ/T_0, with $T_0 = 2\pi/\sqrt{4(1-p\cos\theta_0)}$ the period of the small amplitude oscillations around the stable equilibrium $\theta_0(p)$. We set the quality factor to $Q = 20$, use the initial condition $\boldsymbol{x}(0) = \{5°, 0\}^T$, and plot the evolution of $\theta(\tau)$ for $p = 0.75$ in Figure 19-Left and $p = 1.025$ in Figure 19-Right. For $p = 0.75$, a loading parameter below the critical value $p = 1$, the initially perturbed articulated rigid bar undergoes damped oscillations until it eventually converges to the stable equilibrium position $\theta_0 = 0$. When choosing $p = 1.025 > 1$, the motion diverges from its initial position before performing exponentially decreasing oscillations around the equilibrium $\theta \simeq \theta_-$, see Eq. (94). Another representation of the dynam-

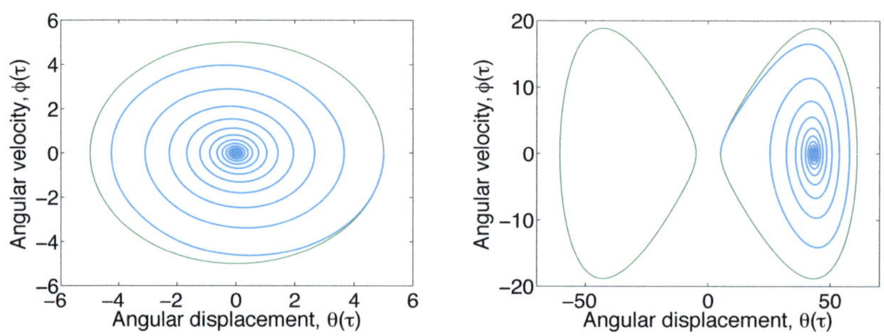

Figure 20. Dynamical response of Figure 19 in the phase plane $(\theta(\tau), \phi(\tau))$. (Left) Trajectory for $p = 0.75$ and $\boldsymbol{x}(0) = \{5°, 0\}^T$. (Right) Trajectories for $p = 1.025$ and $\boldsymbol{x}(0) = \{5°, 0\}^T$ or $\boldsymbol{x}(0) = \{-5°, 0\}^T$.

ical behavior of our articulated system is to draw the phase portrait. In Figure 20-Left and 20-Right, we represent respectively the dynamical response of Figure 19-Left and 19-Right in the phase plane $(\theta(\tau), \phi(\tau))$. We see that the trajectory spirals down to $\boldsymbol{x}(\tau) = \{0, 0\}^T$ when $p < 1$, and to $\{\theta \simeq \theta_-, 0\}^T$ or $\{\theta \simeq -\theta_-, 0\}^T$ for $p > 1$ (whether the trajectory converges to the left or right side of the θ-axis is determined by the initial conditions).

In the absence of dissipation, $Q \to +\infty$ in Eq. (88), the system becomes conservative. The total mechanical E energy is the sum of kinetic, T, and potential energy, V, reads

$$E(\theta, \phi) = T(\phi) + V(\theta) = \frac{1}{2}\phi^2 + \frac{1}{2}4\theta^2 + 4p\cos\theta \tag{95}$$

As $\dot{E} = 0$, the trajectory of a solution in the phase plane of this conservative system is the level curve $E(\theta, \phi) = E_0$ with E_0 determined by the initial condition $\{\theta(0), \phi(0)\}^T$. From Eq. (95), we write

$$\phi = \pm\sqrt{2\left(E_0 - V\left(\theta\right)\right)} \tag{96}$$

and we plot in Figure 20, in green, the trajectories given by (96) for $\boldsymbol{x}(0) = \{5°, 0\}^T$. These trajectories correspond to the motion the articulated bar would have if it was undamped.

11.3 General case of a moving base ($\varepsilon > 0$)

In the general case of a base moving with a harmonic motion $y_a\left(t\right) = A\cos(\Omega t)$, the dynamic response of the articulated rigid bar, $\theta(\tau)$, is determined by Eq. (87). This equation of motion is more complicated than in the previous section since Eq. (87) is a nonlinear ordinary differential equation with a periodic coefficient: the term $2\varepsilon\cos(2\tau)$ with a period $T = \pi$. This type of equation often arises in structural dynamics, notably every time we are interested in characterizing the small oscillations of a structure which is itself in a periodic state, see Bolotin (1964). Here, we are looking for the transverse oscillations of an articulated system resting on a moving foundation. This type of system is called a parametric oscillator, i.e. an oscillator (case of the previous subsection) whose geometrical or mechanical parameters periodically oscillate in time. In our case, it is the quantity of acceleration, A_q, given in Eq. (83), that is changing with time.

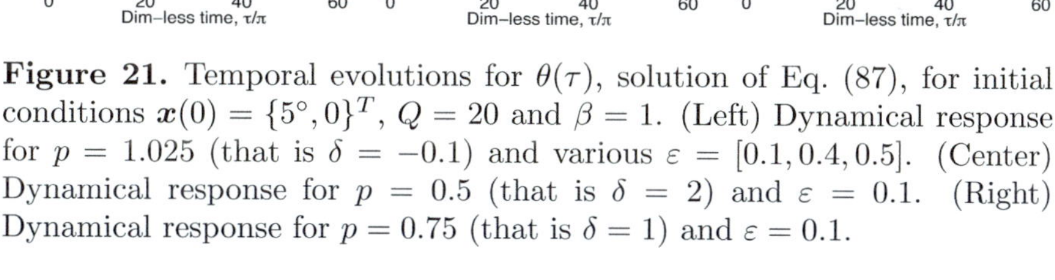

Figure 21. Temporal evolutions for $\theta(\tau)$, solution of Eq. (87), for initial conditions $\boldsymbol{x}(0) = \{5°, 0\}^T$, $Q = 20$ and $\beta = 1$. (Left) Dynamical response for $p = 1.025$ (that is $\delta = -0.1$) and various $\varepsilon = [0.1, 0.4, 0.5]$. (Center) Dynamical response for $p = 0.5$ (that is $\delta = 2$) and $\varepsilon = 0.1$. (Right) Dynamical response for $p = 0.75$ (that is $\delta = 1$) and $\varepsilon = 0.1$.

Direct dynamic analysis To emphasize the complexity of this parametric oscillator, we perform direct computations of Eq. (87) with the operator *ode45* of the Matlab software for initial conditions, $\boldsymbol{x}(0) = \{5°, 0\}^T$, a quality factor $Q = 20$, and a frequency ratio fixed to $\beta = \Omega/\Omega_0 = 1$. The temporal evolution $\theta(\tau)$ as a function of dimensionless time τ/π, with π the dimensionless period of the moving foundation, is given in Figure 21 for different values of the loading parameter p and amplitude of the forcing ε.

In Figure 21-Left, we show results for $p = 1.025$, just above the critical buckling threshold $p = 1$. We have seen in the previous section, Figure 19-Right and Figure 20-Right, that for $\varepsilon = 0$ the dynamical response $\theta(\tau)$ diverges from the unstable trivial equilibrium $\theta_0 = 0$ and settle on the bifurcated equilibrium $\theta_0 \simeq \theta_- = 22°$ in an exponentially damped oscillatory fashion. For small $\varepsilon = 0.1$, we see on Figure 21-Left that the dynamical response of our parametric oscillator also diverges from the unstable configuration $\theta_0 = 0$, but now the solution $\theta(\tau)$ settle to stationary oscillations of period π near the value $\theta_0 \simeq \theta_- = 22°$. This stable limit cycle appears because the moving base is acting as an external harmonic excitation of period π. When increasing ε, the dynamical response still diverges from $\theta_0 = 0$ to converge to a π-periodic limit cycle with larger amplitude, but the mean value of the oscillations becomes closer to $\theta = 0$ as ε increases. Surprisingly, above $\varepsilon \simeq 0.5$, i.e. an amplitude of forcing of the base $A = L/4$, the dynamic response is qualitatively different. The motion of the base stabilizes the equilibrium point $\theta_0 = 0$ and the parametric oscillations asymptotically converge to the straight vertical position.

In Figure 21-Center and 21-Right, we show results for p values below the critical buckling threshold. The amplitude of the forcing is set to $\varepsilon = 0.1$. For $p = 0.5$ the dynamic response, shown in Figure 21-Center, is qualitatively similar to the case with no forcing (see Figure 19-Left): The initially perturbed system undergoes damped oscillations until it eventually exponentially converges to the stable equilibrium position $\theta_0 = 0$. For $p = 0.75$, Figure 21-Right, the temporal evolution $\theta(\tau)$ is qualitatively different from Figure 19-Left and Figure 20-Left: The initially perturbed solution rapidly converges to a limit cycle with oscillations centered around the equilibrium configuration $\theta_0 = 0$, even if we are below the critical buckling threshold. Note that the dynamical response has a period 2π which is twice slower than the motion of the base (we have almost exactly 30 oscillations in 60 periods of the moving base). This phenomenon of self-sustaining oscillations is the well-known parametric instability. For example, a well known parametric instability is a child pumping a swing by periodically standing and squatting to increase the amplitude of the swing's oscillations.

Linear stability analysis In the case of periodic forcing when the base oscillates up and down, the equilibrium solution $\boldsymbol{x}_0 = \mathbf{0}$ nevertheless exists and corresponds to the system being straight and oscillating vertically with the base. We now perform a linear stability analysis of this vertical equilibrium solution to get a better understanding of the physics of the periodic trivial state. We consider the non-autonomous 2-dimensional dynamical systems, Eq. (87), $\dot{\boldsymbol{x}}(\tau) = \boldsymbol{f}(\boldsymbol{x}(\tau), \tau, p)$ where $\boldsymbol{f}$ is a nonlinear 2-dimensional vector field that intrinsically depends on the time τ. The stability of the oscillating vertical equilibrium configuration is studied by considering the dynamics of $\boldsymbol{x}(\tau) = \boldsymbol{x}_0 + \boldsymbol{y}(\tau)$ where $\boldsymbol{y}(\tau)$ is a small disturbance to the equilibrium $\boldsymbol{x}_0$. Injecting $\boldsymbol{x}(\tau)$ into (87), assuming that $\boldsymbol{f}$ is at least twice continuously differentiable, expanding the result in a Taylor series about $\boldsymbol{x}_0$, and retaining only linear terms in the disturbance, we obtain

$$\dot{\boldsymbol{y}}(\tau) = \boldsymbol{J}(\tau)\boldsymbol{y}(\tau) \tag{97}$$

$$\text{with} \quad \boldsymbol{J}(\tau) = \left.\frac{\partial \boldsymbol{f}}{\partial \boldsymbol{x}}\right|_{\boldsymbol{0}} = \begin{bmatrix} 0 & 1 \\ -(\delta + 2\varepsilon\cos 2\tau) & -\frac{2}{\beta Q} \end{bmatrix}$$

and where $\delta = 4(1-p)/\beta^2$ is the ratio between the natural frequency ω_0 of the articulated bar with the concentrated mass M and the frequency of the moving base, Ω. We have $\omega_0 = \sqrt{k/(ML^2) - g/L}$ and hence $\delta = 4\omega_0^2/\Omega^2$. As in the previous section $\boldsymbol{J}(\tau)$ is the 2×2 Jacobi matrix of $\boldsymbol{f}$, evaluated at $\boldsymbol{x}(\tau) = \boldsymbol{x}_0 = \boldsymbol{0}$. The linear system Eq.(97) is a generalization of the autonomous system Eq.(90) with $\varepsilon \neq 0$. The particular form Eq. (97) is called the Mathieu equation. It is more common to express it in the physical space, as a linearization of Eq. (86) about the solution $\theta(\tau) = 0$. It then reads

$$\ddot{\theta}(\tau) + \frac{2}{\beta Q}\dot{\theta}(\tau) + [\delta + 2\varepsilon\cos 2\tau]\,\theta(\tau) = 0 \tag{98}$$

Here the two parameters of interest are on one hand δ, the ratio between the natural frequency of the system and the frequency of the moving base, and on the other hand ε, the dimensionless amplitude of the forcing from the base. Note that for practical purposes δ is tuned either through the compressive loading parameter p, or through the frequency of the moving base Ω.

The linear stability study of the vertical configuration in periodic state consists in computing the solutions of Eq. (97) and determining whether the disturbance $\boldsymbol{y}(\tau)$ fades away or is amplified with time τ. The difficulty though, as compared to the previous section, is that the system Eq. (97) is a linear system *with periodic coefficients* since $\boldsymbol{J}(\tau)$ is π-periodic in time due to the presence of the forcing term $2\varepsilon\cos 2\tau$. The theoretical tools introduced in the previous subsection are therefore no longer applicable. To compute the solutions $\boldsymbol{y}(\tau)$ of linear equations such as the Mathieu equation, we follow the theory introduced by Gaston Floquet at the end of the 19th century which deals with ordinary differential equations with periodic coefficients.

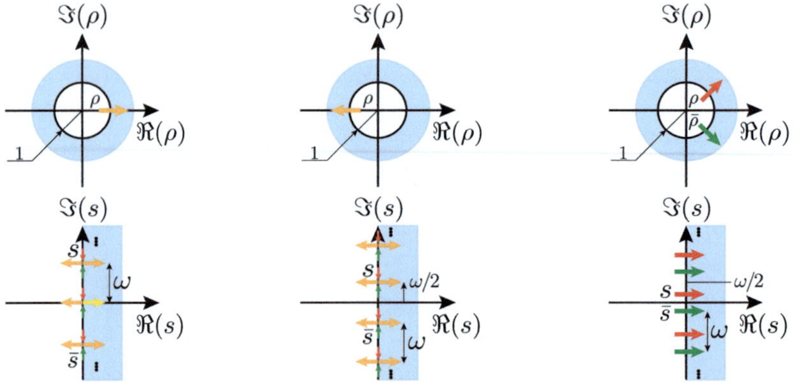

Figure 22. Stability of a periodic state, or limit cycle, in the Argand plane through Floquet multipliers (top row) or Floquet exponents (bottom row). (Left) Steady bifurcation: one purely real Floquet multiplier crosses the unit circle from the right in the time domain or Floquet exponents locked on a multiple of ω in the frequency domain. (Center) Flip or period-doubling bifurcation: one purely real Floquet multiplier crosses the unit circle from the left or Floquet exponents locked on a multiple of $\omega/2$. (Right) Secondary Hopf, or Neimark-Sacker bifurcation: two conjugate Floquet multipliers cross the unit circle at away from the real axis or Floquet exponents cross the imaginary axis at values different from a multiple of $\omega/2$.

Floquet theory (time domain) The following developments are introduced for the particular 2-dimensional system (87), but they are classical and could be generalized to N-dimensional systems without conceptual difficulties. Since the system (97) is linear with respect to the unknowns $\boldsymbol{y}(\tau)$, it has $n = 2$ linearly independent solutions $\boldsymbol{y}_n(\tau)$, so that the general solution $\boldsymbol{y}(\tau)$ of (97) can be written as

$$\boldsymbol{y}(\tau) = \sum_{n=1}^{2} c_n \boldsymbol{y}_n(t) \tag{99}$$

where c_n are 2 constants that depend upon the initial conditions and $\boldsymbol{y}_n(\tau)$ are called fundamental solutions. We gather these into a 2×2 matrix, $\boldsymbol{Y}(\tau) = [\boldsymbol{y}_1(\tau)\ \boldsymbol{y}_2(\tau)]$. As both $\boldsymbol{y}_n(\tau)$ verify (97), the fundamental matrix

$\boldsymbol{Y}(\tau)$ also verifies

$$\dot{\boldsymbol{Y}}(\tau) = \boldsymbol{J}(\tau)\boldsymbol{Y}(\tau) \tag{100}$$

and, as $\boldsymbol{J}(\tau)$ is T-periodic, we have

$$\dot{\boldsymbol{y}}_n(\tau+T) = \boldsymbol{J}(\tau+T)\boldsymbol{y}_n(\tau+T) = \boldsymbol{J}(\tau)\boldsymbol{y}_n(\tau+T) \tag{101}$$

so that the $\boldsymbol{y}_n(\tau+T)$ also verify (97). Consequently the solutions $\boldsymbol{y}_n(\tau+T)$ can be expressed as linear combinations of the 2 independent fundamental solutions $\boldsymbol{y}_n(\tau)$. It thus exists a 2×2 constant matrix $\boldsymbol{\Phi}$, called the monodromy matrix, such that:

$$\boldsymbol{Y}(\tau+T) = \boldsymbol{Y}(\tau)\boldsymbol{\Phi}. \tag{102}$$

This matrix maps a particular set of fundamental solution $\boldsymbol{Y}(\tau)$ at time τ into their values at time $\tau+T$ (this map is called a Poincaré map). We therefore use it to study the stability of the straight equilibrium configuration $\boldsymbol{x}_0 = \boldsymbol{0}$. We compute its eigenvalues ρ_n and eigenvectors. For an eigenvector of $\boldsymbol{\Phi}$ we have, from Eq.(102),

$$\boldsymbol{y}_n(\tau+T) = \boldsymbol{y}_n(\tau)\rho_n \tag{103}$$

The ρ_n are therefore called Floquet multipliers which, interestingly, do not depend upon the choice of the fundamental matrix. As a consequence, it is convenient to use the initial conditions $\boldsymbol{Y}(0) = \boldsymbol{1}$, where $\boldsymbol{1}$ is identity matrix. The monodromy matrix then simply reads $\boldsymbol{\Phi} = \boldsymbol{Y}(T)$, see (102). It is determined by solving the linear system with periodic coefficients (100) over one period $\tau \in (0,T)$ with $\boldsymbol{Y}(0) = \boldsymbol{1}$. Morevover, Floquet theory tells us that we can express the fundamental solutions $\boldsymbol{y}_n(t)$ in the so-called Floquet normal form

$$\boldsymbol{y}_n(\tau) = \boldsymbol{r}_n(\tau)\, e^{s_n\tau} \tag{104}$$

where $\boldsymbol{r}_n(\tau+T) = \boldsymbol{r}_n(\tau)$ is a 2-dimensional complex vector of period T and s_n is a complex number called the Floquet exponent (the Floquet form is a generalization of the eigenmodes of vibration defined in (91) for the autonomous system (90)). Then, it follows from Eq.(104) and from the T-periodicity of $\boldsymbol{r}_n$ that

$$\boldsymbol{y}_n(\tau+T) = \boldsymbol{r}_n(\tau+T)\, e^{s_n(\tau+T)} = \boldsymbol{y}_n(\tau)\, e^{s_nT} \tag{105}$$

We therefore see that the Floquet exponents s_n are linked to the Floquet multipliers ρ_n by:

$$\rho_n = e^{s_nT} \quad \Leftrightarrow \quad s_n = \frac{1}{T}\ln\rho_n + \frac{2\pi m}{T}i, \quad m = 0, \pm 1, \pm 2, \ldots. \tag{106}$$

where $T = \pi$ in our case and i is the unit imaginary number. Whereas the Floquet multipliers are uniquely defined, the above equation shows that the s_n are unique to within an additive integer multiple of $m\,\omega = 2m\,\pi/T$ where the fundamental frequency ω reads $\omega = 2\pi/T$. This last result can also be viewed by replacing s_n by $s_n + 2im\pi/T$ in Eq.(106).

Considering Eqs. (103) or (104), the values of either the moduli $|\rho_n|$ of Floquet multipliers or the Floquet exponents s_n are used to determine the stability of the oscillating straight system characterized by $\boldsymbol{x}_0$. The following statements are very similar to Lyapunov's theory previously described except that, in the present case of Floquet theory, the nature of the instability and bifurcations we encounter are different since the perturbed state is periodic and no more constant in time:

- If $\Re(s_n) < 0$ (or $|\rho_n| < 1$) for all n, all fundamental solutions $\boldsymbol{y}_n(\tau)$ converge toward zero as τ increases, so is any perturbation $\boldsymbol{y}(\tau)$. The periodic state $\boldsymbol{x}_0$ is said to be asymptotically stable;
- If it exists a subscript k such that $\Re(s_k) > 0$ (or $|\rho_k| > 1$), the corresponding fundamental solutions increases exponentially, so are some perturbations $\boldsymbol{y}(t)$. The periodic state is in this case unstable and three different scenarios have to be considered:
  - If $\Im(s_k) = m\,\omega$ for $m = 0, \pm1, \pm2, \ldots$ ($\Im(\rho_k) = 0$ and $\Re(\rho_k) > 0$), the perturbed motion will be a T-periodic oscillation that will exponentially grow in the direction of the kth mode $\boldsymbol{r}_k(\tau)$ (see Figure 22-Left for the representation in the Argand plane). This loss of stability leads to a steady bifurcation since the perturbed state bifurcates away from the periodic state $\boldsymbol{x}_0$ in the state space, but retains its period T.
  - If $\Im(s_k) = \omega/2 + m\,\omega$ for $m = 0, \pm1, \pm2, \ldots$ ($\Im(\rho_k) \neq 0$ and $\Re(\rho_k) < 0$), the perturbed motion will be a $2T$-periodic oscillation that will exponentially grow in the direction of the kth mode $\boldsymbol{r}_k(\tau)$ (see Figure 22-Center for the representation in the Argand plane). This loss of stability leads to a flip or period-doubling bifurcation: the perturbed state bifurcates both in the state space along $\boldsymbol{r}_k(\tau)$ and in the time domain (from a T to $2T$-periodic state). Note that by expressing the periodic state in multiple of $\omega/2$ instead of ω, the flip bifurcation would be transformed in a steady bifurcation. Therefore, flip and steady bifurcation are of same nature.
  - If $\Im(s_k) \neq \omega/2 + m\omega/2$ for $m = 0, \pm1, \pm2, \ldots$ ($\Im(\rho_k) \neq 0$), the perturbed motion is a periodic or quasi-periodic oscillation that exponentially grows in the direction of the kth mode $\boldsymbol{r}_k$ (see Figure 22-Right for the representation in the Argand plane). At

least one of the ratio between the frequencies of the perturbation and the fundamental frequency of the perturbed state is not an integer. As a consequence, this instability leads to a secondary Hopf or Nemark-Sacker bifurcation which introduces one or two new frequencies in the bifurcated stationary state. In that case, not only the limit cycle bifurcates in the state space but also in the time domain.

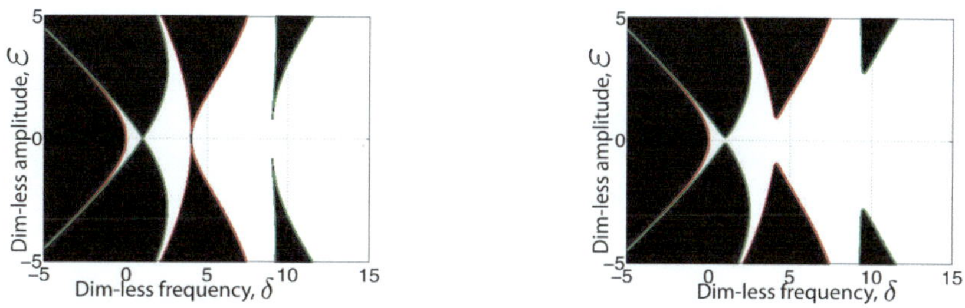

Figure 23. Mathieu's tongues or stability charts of the vertical solution of the system of Figure 16, in the $(\delta,\, \varepsilon)$ plane. Unstable regions are in black. Regions bounded by green and red lines respectively lead to $2T$ and T-periodic steady-state oscillations. (Left) Undamped system with $Q = 10^{10}$, and (Right) Damped system with $Q = 20$.

Numerical applications In this section, we apply Floquet theory to determine the stability of the vertical solution of the system of Figure 16, seen as a periodic state. We compute the Floquet multipliers and exponents of the linearized Eqs.(97)-(98) in the $(\delta,\, \varepsilon)$ space. Figure 23-Left and 23-Right show the stability maps of this state in the $(\delta,\, \varepsilon)$ plane for $Q = 10^{10}$ (undamped system) and $Q = 20$ respectively. Dark regions are such that the modulus of at least one computed Floquet multiplier is superior to unity. These regions of instability are called Mathieu's tongues. Their limits have been drawn with colored lines. Green lines surround regions of instability where $2T$-periodic oscillations develop, while red lines surround regions of instability where T-periodic oscillations develop. Figure 24 gives more details about the shape of Floquet multipliers and exponents for $\varepsilon = 1$ and $Q = 20$ and confirms the alternation between regions of existence of T and $2T$-periodic solutions. In thin dashed line, we also plot the evolution of the natural frequency of the vertical solution when the base is not moving, i.e the eigenvalues of the constant part of the Jacobian, i.e. $\boldsymbol{J}$ in Eq. (97) with $\varepsilon = 0$.

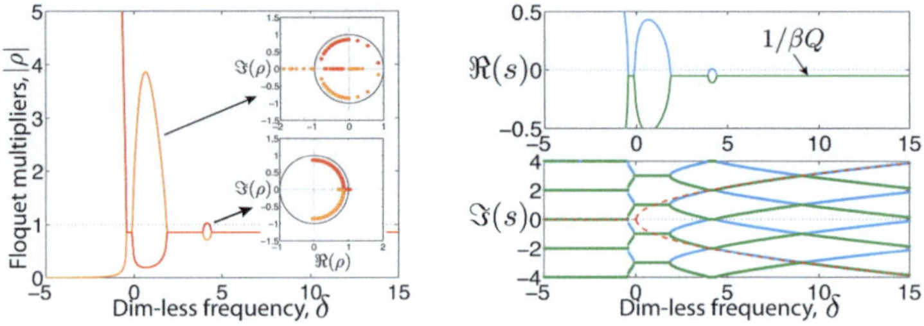

Figure 24. Evolution of Floquet's parameters as a function of dimensionless frequency $\delta = 4\omega_0^2/\Omega^2$ for $Q = 20$ and $\varepsilon = 1$. (Left) Evolution of the modulus of Floquet multipliers. Insets: Evolution of Floquet multipliers in the Argand plane. (Right) Evolution of Floquet exponents. Top: Growth rate. Bottom: Spectrum of the Floquet normal form. Dashed line is the natural frequency of the vertical solution on a standstill base.

Both Figure 23 and Figure 24 give sufficient informations to qualitatively explain the dynamical behavior computed and illustrated in Figure 21. When $\delta < 0$ (i.e. $p > 1$), the vertically oscillating straight configuration is mostly dynamically unstable with the system bifurcating toward a T-periodic solution: in this region, the spectrum of the Floquet exponents shown in Figure 24-Right are locked in the form $\Im(s_k) = m\omega$ for $m = 0, \pm 1, \pm 2, \ldots$. Due to the nonlinear term in Eq.(87), the motion consequently settles of a steady-state that is therefore T-periodic. This is what we observe in Figure 21-Left for $\varepsilon = [0.1, 0.4]$ and $p = 1.025$ which corresponds to $\delta = -0.1$. However, there is an interesting narrow region for $\delta < 0$ illustrated in Figure 23, where the vertical solution is a stable periodic state. This counter-intuitive phenomenon can be observed for the couple of parameters $(\delta, \varepsilon) = (-0.1, 0.5)$ for which the dynamical response is given in Figure 21-Left.

When $\delta > 0$, the system in vertical periodic state is a true parametric oscillator. The transverse oscillations of the moving bar are modulated by the harmonic intrinsic motion. As a consequence, there are alternating windows for the parameter $\delta = 4\omega_0^2/\Omega^2$ for which the vertical solution is unstable, even if it is stable in the case of a immobile base ($\varepsilon = 0$). This qualitative behavior is illustrated by the nonlinear dynamic responses in Figure 21-Center and 21-Right where $(\delta, \varepsilon) = (2, 0.1)$ and $(1, 0.1)$ respectively. The first set of parameters leads to a stable straight configuration when the second corresponds to an unstable one. The regions of instability are often called forbidden regions since in those regions, the natural straight configuration $\theta(\tau) = 0$ is not seen experimentaly. The forbidden regions originate (when $\varepsilon \to 0$) from particular values $\delta \to j^2$ with $j = 0, 1, 2, \ldots$, i.e. $\Omega = (2/j)\omega_0$. This is specific of parametric instabilities: for particular ratios between the frequency of the forcing and the natural frequency of the unforced system, an eventual amplification of any slight perturbation may happen and give rise to a steady-state oscillation. However, we see from Figure 24-Right that parametric instabilities are not due to resonances, as it is often believed, but rather to an intrinsic frequency lock-in of the spectrum of the Floquet normal forms. From a phenomenological point of view, lock-in is a linear phenomenon where the fundamental frequency $\Im(s_n)$ deviates from the expected natural frequency of the unforced system, given by the dashed line in Figure 24-Right, while being close to secondary harmonics $\Im(s_n) + j\omega$. Since quasi-periodic Floquet forms may have multiple harmonics, multiple lock-in may happen for values of δ close to j^2 with $j = 0, 1, 2, \ldots$. Moreover, as the size of the spectrum of the Floquet normal forms decreases with δ (to finally tend to the classic eigenmodes of a structure in equilibrium as $\delta \to +\infty$), lock-in phenomenon happens mostly for small δ, when Ω and ω_0 are of similar scale, and tends to disappear as δ increases. The first region of instability encountered as δ is increased from zero is called the principal region of instability, when the other ones are refereed to as secondary regions of instability. Finally, we note that damping reduces the range of lock-in and therefore the width of forbidden regions for a given set of (δ, ε).

12 Material instabilities

Energetic methods can be applied in some non-conservative cases. We consider here systems of springs with damage. Damage is an irreversible and dissipative phenomenon. Yet, it is possible to formulate the evolution of damage as an energy minimisation problem under a unilateral constraint representing the irreversibility.

12.1 One spring with damage

We consider a linear spring submitted to a tensile force F. The internal deformation energy of this spring is

$$V_{\text{int}} = \frac{1}{2} k \left(x - x_0\right)^2 \tag{107}$$

where k is the stiffness and x the position of the right extremity of the spring. This position is x_0 when no tension ($F = 0$) is applied. In order to model damage happening in the material of the spring, we postulate that

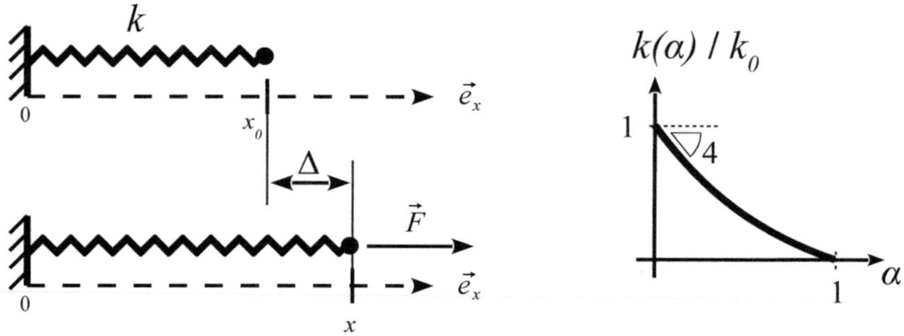

Figure 25. (Left) A damageable spring under tension. During the loading procedure the spring may experience damage and have its stiffness k modified. (Right) The spring stiffness k as function of the damage variable α, see Equation (108).

during a loading experiment the spring stiffness k may decrease

$$k(\alpha) = k_0 \, \frac{1 - \alpha}{1 + 3\alpha} \tag{108}$$

where the internal variable α accounts for the damage status of the material. The material starts with $\alpha = 0$ and α increases as damage occurs. Damage is considered an irreversible phenomenon and therefore modeled by imposing that the damage variable cannot decrease, $\dot{\alpha} = \mathrm{d}\alpha/\mathrm{d}t \geq 0$, with time t. When α reaches $\alpha = 1$ the material is ruined. The damage function is such that $k(0) = k_0$ and $k(1) = 0$. Different damage functions $k(\alpha)$ have been considered in the literature, we chose (108) for simplicity reasons. As damage occurs, some energy is dissipated in the system and it is convenient to consider

$$V(x, \alpha) = V_{\text{int}} + V_{\text{dissip}} \tag{109}$$

as the total energy of the system and treat it as conservative. Here we choose $V_{\text{dissip}} = W \alpha$. The dissipation energy W is related to the energy needed to break the spring: as α reaches $\alpha = 1$ the system no longer opposes any reaction to stretching and can be considered broken. We compute the response of the system in a controlled displacement setup, that is we study (109) under the constraint

$$g(x) = (x - x_0) - \Delta \tag{110}$$

where Δ is the imposed displacement. Introducing the augmented energy

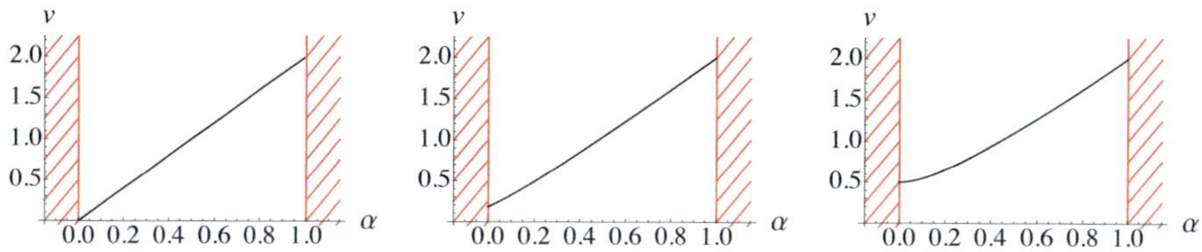

Figure 26. Graphs of the function (113) for (Left) $\epsilon = 0$, (Center) $\epsilon = 0.6$, (Right) $\epsilon = 1$, where $v'(0) = 0$.

$\mathcal{V} = V(x, \alpha) - \lambda g(x)$ and solving for $\partial \mathcal{V}/\partial x = 0$ yields

$$\lambda = k(\alpha)\,(x - x_0) = F \tag{111}$$

where we identify the Lagrange multiplier λ to the applied tension F. Nevertheless, we use the linearity of the constraint (110) to eliminate the variable x and work with a (unconstrained) one-variable energy

$$V(\alpha) = \frac{1}{2} k_0 \, \frac{1-\alpha}{1+3\alpha}\, \Delta^2 + W\,\alpha \tag{112}$$

We use x_0 as unit length and $k_0\,x_0$ as unit force to introduce the dimensionless quantities $\epsilon = \Delta/x_0 = (x - x_0)/x_0$, $w = W/(k_0\,x_0^2)$, $f = F/(k_0\,x_0)$, and $v = V/(k_0\,x_0^2)$. Setting $w = 2$, we obtain

$$v(\alpha) = \frac{1}{2}\,\frac{1-\alpha}{1+3\alpha}\,\epsilon^2 + 2\,\alpha \tag{113}$$

We first consider the loading situation where the extension ϵ is increased quasi-statically from zero, $\epsilon(t) = t$. At each time t we look for the minimum of the energy $v(\alpha)$, under the irreversibility constraint $\dot{\alpha} \geq 0$. We therefore study the graph of $v(\alpha)$ for different values of the imposed parameter ϵ, see Fig. 26. As long as $\epsilon < 1$ the slope at the origin is strictly positive, and the minimum is reached at $\alpha = \alpha^\star = 0$. The damage variable does not evolve, the spring experiences no damage. From (111) we compute the force and obtain that, in this first phase, the force f increases linearly with ϵ, $f = \epsilon$.

As $\epsilon > 1$ a second phase starts and the minimum is now reached for $\alpha = \alpha^\star > 0$, as seen in the graphs of Figure 27. We solve for $v'(\alpha) = 0$ and find $\alpha^\star(t) = (\epsilon - 1)/3 > 1$. The second phase ends at $\epsilon = t = 4$ where the spring is entirely ruined, $\alpha^\star = 1$. In this second phase the force f decreases with ϵ:

$$f = 1 - \alpha^\star = \frac{4-\epsilon}{3} \tag{114}$$

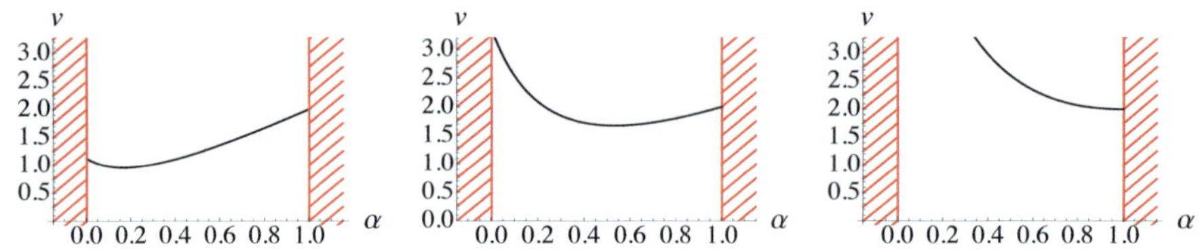

Figure 27. Graphs of the function (113) for (Left) $\epsilon = 1.5$, (Center) $\epsilon = 2.6$, (Right) $\epsilon = 4$, where $v'(1) = 0$.

In summary in the first phase $\dot{\alpha} = 0$ and $v'(\alpha^\star) > 0$, while in the second phase $\dot{\alpha} > 0$ and $v'(\alpha^\star) = 0$. These two phases correspond to the curves P_1 and P_2 in the force-extension diagram, see Figure 28.

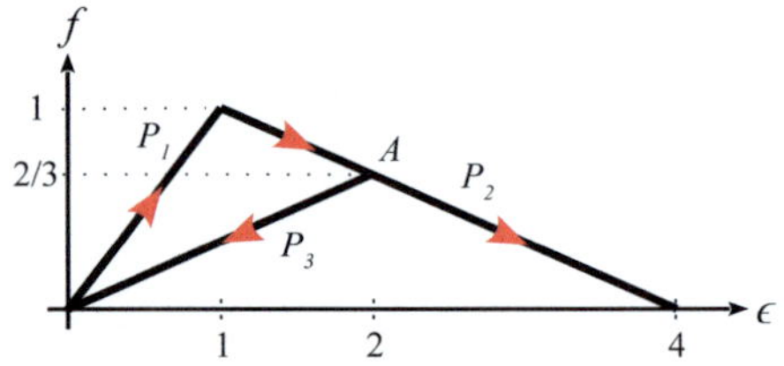

Figure 28. Force-extension diagram for system (113). For paths P_1 and P_2, the loading is such that $\epsilon(t) = t$. For path P_3, the loading is such that $\epsilon(t) = 4 - t$, with $t > 2$.

We consider the point A in Figure 28 where $t = 2$ and $\alpha^\star = 1/3$, and we reverse the loading: we now impose $\epsilon(t) = 4 - t$. The imposed extension ϵ now decreases from $\epsilon = 2$. Consequently we look for the minimum of $v(\alpha)$ with the condition that $\alpha^\star$ cannot decrease under $1/3$. We find that the constraint $\alpha^\star \geq 1/3$ is always active: the minimum is reached at $\alpha^\star = 1/3$, see Figure 29. The force-extension curve P_3 is plotted in Figure 28.

12.2 Two springs with damage

We now turn to the case where two springs are connected in series, see Figure 30. Following the notations of the previous section, we write the energy of the system as

$$V(\Delta_1, \Delta_2, \alpha_1, \alpha_2) = \frac{1}{2}k(\alpha_1)\Delta_1^2 + \frac{1}{2}k(\alpha_2)\Delta_2^2 + V_{\text{dissip}} \qquad (115)$$

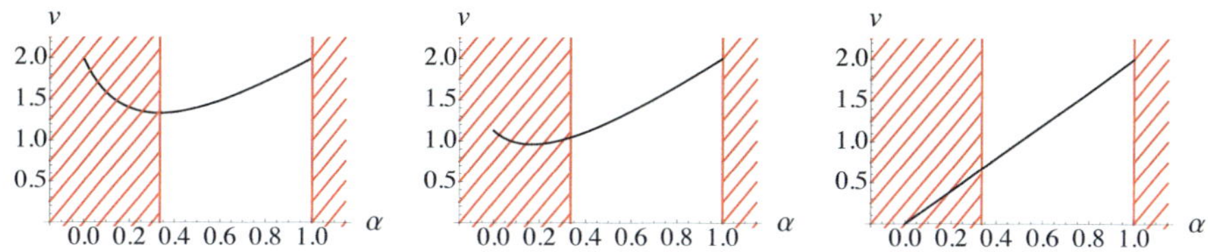

Figure 29. Graph of $v(\alpha)$ during the unloading phase P_3 where $\epsilon = 4 - t$, with $t \in (2, 4)$. According to the past evolution of the system, the minimum of v is searched for $\alpha \geq 1$

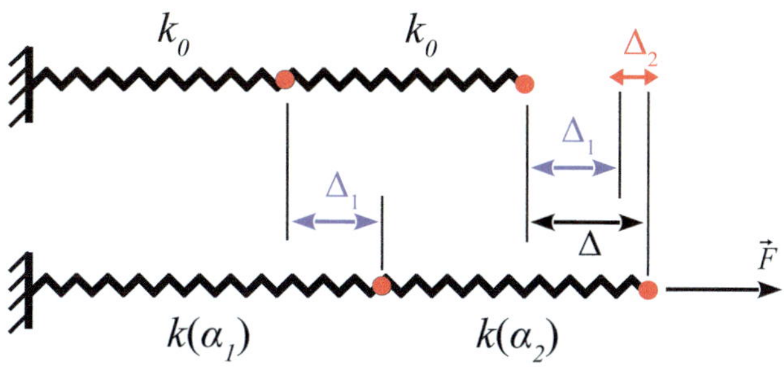

Figure 30. Two damageable springs in series. A displacement Δ is imposed through to application of a tension F. During the loading procedure each spring may experience damage and have its stiffness k modified.

where Δ_1 (respectively Δ_2) is the elongation of spring 1 (resp. spring 2). An internal damage variable is associated to each spring and $V_{\text{dissip}} = W\alpha_1 + W\alpha_2$. We study the behavior of the system under controlled displacement, that is under the constraint $\Delta = \Delta_1 + \Delta_2$. As in the one-spring case, pre-minimisation with regard to Δ_1 and Δ_2 yields:

$$F = k(\alpha_1)\Delta_1 = k(\alpha_2)\Delta_2 = k_{/}(\alpha_1, \alpha_2)\Delta \tag{116}$$

with $k_{/} = \left[k^{-1}(\alpha_1) + k^{-1}(\alpha_2)\right]^{-1}$. We use (116) to eliminate Δ_1 and Δ_2 and work with an unconstrained energy $V(\alpha_1, \alpha_2) = \frac{1}{2}k_{/}(\alpha_1, \alpha_2)\Delta^2 + W\alpha_1 + W\alpha_2$ that we non-dimensionalize as in previous section to finally obtain

$$v(\alpha_1, \alpha_2) = \frac{\epsilon^2/2}{\frac{1+3\alpha_1}{1-\alpha_1} + \frac{1+3\alpha_2}{1-\alpha_2}} + 2\alpha_1 + 2\alpha_2 \tag{117}$$

We study $v(\alpha_1, \alpha_2)$ as the parameter $\epsilon(t) = t$ is increased quasi-statically, under the constraints $\dot{\alpha_1} \geq 0$ and $\dot{\alpha_2} \geq 0$. We consider discrete time step

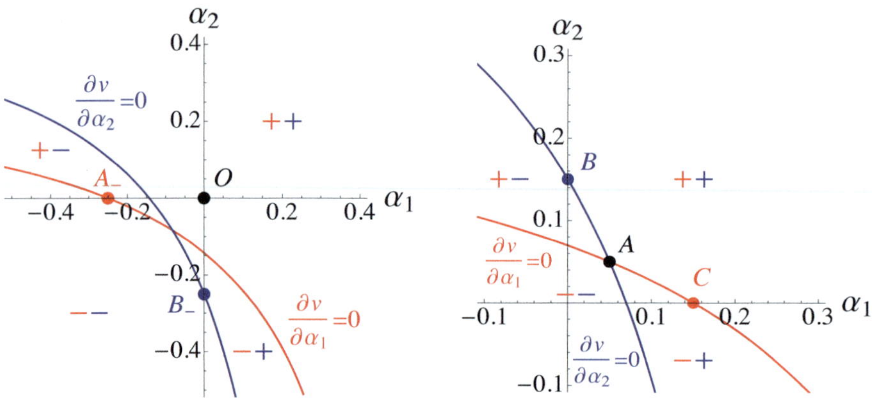

Figure 31. Plane (α_1, α_2) with solutions to (118) with $\epsilon = 1.5$. The curves $\frac{\partial V}{\partial \alpha_i} = 0$ part the plane into regions where the sign of $\frac{\partial V}{\partial \alpha_i}$ is shown, red for $i = 1$, blue for $i = 2$. (Left) Same with $\epsilon = 2.3$

t_i, $i = 0, 1, 2, \ldots$. If at time step t_i, the damage variables α_1 and α_2 have reached values α_1^- and α_2^- respectively, we minimize (117) under the inequality constraints $f_1 = \alpha_1 - \alpha_1^- \geq 0$ and $f_2 = \alpha_2 - \alpha_2^- \geq 0$ that is we solve for the Kuhn-Tucker first order necessary conditions (see Section 8)

$$\frac{\partial V}{\partial \alpha_1} = \mu_1\,, \quad \mu_1\, f_1 = 0 \tag{118a}$$

$$\frac{\partial V}{\partial \alpha_2} = \mu_2\,, \quad \mu_2\, f_2 = 0 \tag{118b}$$

where μ_1 (respectively μ_2) is the Lagrange multiplier associated to the constraint $f_1 \geq 0$ (resp. $f_2 \geq 0$). Eq (118) is a system of 4 equations with 4 unknowns $(\alpha_1, \alpha_2, \mu_1, \mu_2)$. Among the solutions we eliminate those having

$$\mu_1 < 0\,,\ \mu_2 < 0\,,\ f_1 < 0\,,\text{or } f_2 < 0 \tag{119}$$

The stability of each of the remaining solution is then assessed separately.

• For $\epsilon < 2$, the solutions to (118) are (i) the origin $(\alpha_1, \alpha_2) = 0$, or (ii) solutions outside the domain $0 \leq \alpha_{1,2} \leq 1$. We represent in Figure 31-Left two such solutions A_- and B_-. Yet we see that point A_- has $\frac{\partial V}{\partial \alpha_1} = \mu_1 = 0$ but $\frac{\partial V}{\partial \alpha_2} = \mu_2 < 0$, and that point B_- has $\frac{\partial V}{\partial \alpha_2} = \mu_2 = 0$ but $\frac{\partial V}{\partial \alpha_1} = \mu_1 < 0$.

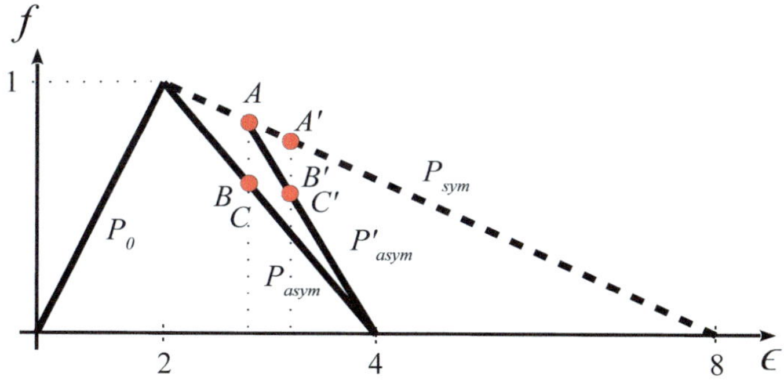

Figure 32. Different paths for the loading of the 2 springs system of Figure (30).

Consequently, in addition to being outside the domain $0 \leq \alpha_{1,2} \leq 1$, points A_- and B_- are eliminated by the test (119). The origin has $\mu_1 > 0$ and $\mu_2 > 0$ and is therefore a local minimum (there is no projected matrix here to test). For this solution the force is $f = 1/2\,\epsilon$, see path P_0 in Figure 32.

• As ϵ is crossing the threshold $\epsilon = 2$, α_1 and α_2 have not yet evolved from 0 and we have $\alpha_1^- = 0$ and $\alpha_2^- = 0$. The solution to (118) are then (i) $\alpha_1 = \alpha_2 = \frac{\epsilon-2}{6}$ (point A in Figure 31-Right), (ii) $\alpha_1 = 0$ and $\alpha_2 = \frac{\epsilon-2}{2}$ (point B in Figure 31-Right), and (iii) $\alpha_2 = 0$ and $\alpha_1 = \frac{\epsilon-2}{2}$ (point C in Figure 31-Right). Stability of point A, where both constraints f_1 and f_2 are inactive (i.e.$f_1 > 0$ and $f_2 > 0$), is assessed with the Hessian $H_{ij} = \frac{\partial^2 V}{\partial\alpha_i \partial\alpha_j}$ which is found to have one positive $\frac{24}{\epsilon}$ and one negative $\frac{24}{\epsilon-8}$ eigenvalues. Point A is therefore an unstable equilibrium[1]. The force is $f = \frac{8-\epsilon}{6}$. Stability of point B, where the constraint f_1 is strongly active ($f_1 = 0$ and $\mu_1(\mathrm{B}) = -6+4\epsilon-\epsilon^2/2 > 0$ for $2 < \epsilon < 6$) is assessed by testing the projected Hessian or equivalently the second derivative $\frac{\partial^2 v}{\partial\alpha_2^2} = \frac{8}{\epsilon} > 0$. Point B is therefore a stable equilibrium point. The force is $f = \frac{4-\epsilon}{2}$. The same conclusion is reached for point C, which is the symmetrical $\alpha_1 \leftrightarrow \alpha_2$ companion to point B. In conclusion we see that the system either moves toward point B or C and therefore evolves to an unsymmetrical ($\alpha_1 \neq \alpha_2$) state. For subsequent time-steps the system follows the path P_{asym} in

---

[1] We note that we have tested the Hessian for all perturbations on α_1 and α_2. Different approaches, with a stricter interpretation of the irreversibility conditions, only consider perturbations that let the α values increase. In this latter case point A would be considered stable (Nguyen, 2000).

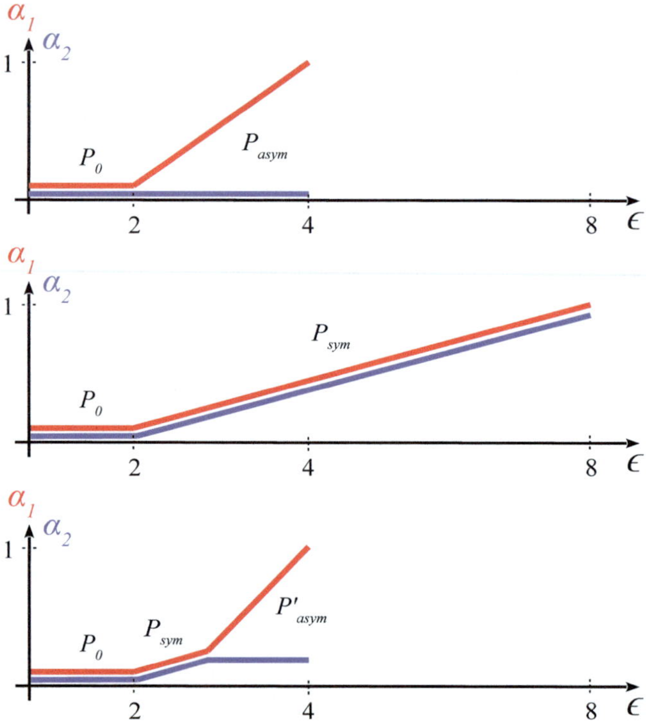

Figure 33. Values of the damage variables along the different paths of Figure (33).

Figure 32. The symmetrical path emerging from point A is labelled P_{sym} in Figure 32.

• For ϵ large than 2, if the system could have evolved along the symmetrical branch P_{sym}, up to say $\alpha_1 = \alpha_2 = \alpha_-$ we would again find 3 solution to (118): (*i*) a symmetrical unstable solution with $\alpha_1 = \alpha_2 = \frac{\epsilon-2}{6}$ (point A' in Figure 32), (*ii*) an unsymmetrical stable solution with $\alpha_1 = \alpha_-$ and $\alpha_2 = \frac{2-\epsilon+\alpha_-(\epsilon+2)}{6\alpha_- -2}$ where the force is $f = \frac{(\epsilon-4)(1-\alpha_-)}{6\alpha_- -2}$ (point B' in Figure 32), and (*iii*) point C' the symmetrical $\alpha_1 \leftrightarrow \alpha_2$ companion to point B'. These last two point are on path P'_{asym}.

We conclude that in a loading experiment where $\epsilon(t) = t$, the system follows path P_0 up to $\epsilon = 2$ and then path P_{asym} up to $\epsilon = 4$. At $\epsilon = 4$ one of the spring is entirely ruined while the other is undamaged.

Acknowledgments

S.N. thanks Patrick Ballard and Marco Rivetti for discussions, comments, and remarks. Financial support from ANR, through grants ANR-09-JCJC-0022-01, ANR-14-CE07-0023-01, and ANR-13-JS09-0009, as well as from 'La Ville de Paris - Programme Émergence' is acknowledged.

Bibliography

Basile Audoly and Yves Pomeau. *Elasticity and Geometry: From hair curls to the non-linear response of shells.* Oxford University Press, 2010.

Davide Bigoni. *Nonlinear Solids Mechanics.* Cambridge University Press, 2012.

V. V. Bolotin. *The dynamic stability of elastic systems.* Holden Day, Inc., 1964.

G. Domokos and P. Holmes. Euler's problem, euler's method, and the standard map; or, the discrete charm of buckling. *Journal of Nonlinear Science*, 3(1):109–151, 1993.

John Guckenheimer and Philip Holmes. *Nonlinear oscillations, dynamical systems and bifurcations of vector fields.* Springer Verlag, New York, 1983.

Gérard Iooss and Daniel D. Joseph. *Elementary Stability and Bifurcation Theory.* Undergraduate Texts in Mathematics. Springer Verlag, 1989.

David G. Luenberger and Yinyu Ye. *Linear and nonlinear programming.* Springer, 3rd edition, 2008.

Quoc Son Nguyen. *Stability and Nonlinear Solid Mechanics.* John Wiley & Sons, 2000.

M. Stein. The phenomenon of change in buckle pattern in elastic structures. Technical Report R39, NASA, 1959.

Steven H. Strogatz. *Nonlinear Dynamics and Chaos, with applications to Physics, Biology, Chemistry and Engineering.* Perseus Publishing, 1994.

Hans Ziegler. *Principles of Structural Stability.* Springer, Basel AG, 2nd edition, 1977.

New phenomena in nonlinear elastic structures: from tensile buckling to configurational forces

D. Bigoni*, F. Bosi*, D. Misseroni*, F. Dal Corso* and G. Noselli†
* DICAM, University of Trento, Trento, IT
† SISSA-International School for Advanced Studies, Trieste, IT

Abstract The theory of the planar elastica is presented in detail and is used to illustrate problems of buckling of a slender structure under tensile dead loading, of buckling as related to constraint's curvature, and of configurational forces. These problems are important tools in the design of compliant mechanisms, in the emergent field of soft robotics and for the understanding of snake and fish locomotion.

1 Introduction

The problem of the planar elastica has a long history, leading back to Jacob Bernoulli (1654-1705), Daniel Bernoulli (1700-1782), Leonhard Euler (1707-1783), and Pieter van Musschenbroek (1692-1761), but is still relevant and rich with applications at times quite unexpected ones. The elastica has attracted a great interest in the past and has involved contributions from first-class scientists, including Kirchhoff, Love, and Born. Research on the elastica marked the initiation of the calculus of variations and promoted the development of the theory of elliptic functions. Nowadays the elastica represents a useful introduction to the theory of nonlinear bifurcation and stability, but is also an important tool in the field of soft robotics and in the design of compliant mechanisms. Moreover, the elastica can be effectively used to explain snake or fish locomotion and to design snake-like robots.

The theory of the Euler elastica is addressed in detail in Section 2 and includes an analysis of the instability of equilibrium configurations. The elastica theory is applied to the problem of tensile buckling, in which an elastic rod bifurcates under tensile dead loading, due to the presence of an internal slider constraint, see Section 3. This problem opens a new perspective on buckling problems, which were usually believed to occur only for compressive loads. The influence of the constraint's curvature on

D. Bigoni (Ed.), *Extremely Deformable Structures*,
CISM International Centre for Mechanical Sciences
DOI 10.1007/ 978-3-7091-1877-1_2 © CISM Udine 2015

buckling and post-critical behaviour is analyzed in Section 4, with reference to elastic systems representing examples of compliant mechanisms. The concept of 'Eshelby-like' or 'configurational' forces in elastic structures is introduced in Section 5, and this represents the key to the understanding of snake locomotion. In fact, roughly speaking, a snake can be idealized as an elastic rod whose propulsion arises from the release of elastic energy that is responsible of an Eshelby-like force. It is shown that this force strongly affects the bifurcation and stability of elastic rods, as explained in Section 6, and that it can be used in the design of a new elastic arm weighing device, Section 7.

2 The Euler elastica

The purpose of this section is to introduce the theory of the Euler's elastica with reference to an elastic, inextensible rod subject to large deflections and with different types of constraints at its ends. When a rectilinear (in its undeformed, reference state) elastic rod is compressed, the determination of the critical load and of the post-critical behaviour provides a beautiful example of linear and nonlinear eigenvalue problems. We will follow and generalize the treatment of Bigoni (2012). Classical references are Timoshenko and Gere (1961), Love (1927), and Reiss (1969), whereas recent work has been presented by Wang (1997), Vaz (2003), Mikata (2007) and by O'Reilly and Peters (2011, 2012). The stability of the equilibrium configurations will also be addressed, which has been previously treated by Maddocks (1984), Manning et al. (1998), Hoffman et al. (2002), Manning (2009, 2014), Kuznetsov and Levyakov (1999, 2002), and by Levyakov and Kuznetsov (2010).

2.1 The kinematics of an inextensible planar rod

We consider an inextensible planar rod of length l, rectilinear in its reference configuration and smoothly deformed into the current configuration as shown in Fig. 1. In the deformed configuration, a generic point can be picked up using a curvilinear coordinate $s \in [0, l]$ (corresponding to $x_0 \in [0, l]$ in the undeformed, straight configuration), such that the inextensibility of the rod implies that $x_0 = s$ and thereby $\mathrm{d}x_0 = \mathrm{d}s$.

If we write $\mathbf{e}_1$ and $\mathbf{e}_2$ as unit vectors, see Fig. 1, then the *displacement* $\mathbf{u}$ of point $\mathbf{x}_0$ from the reference configuration is given by

$$\mathbf{u} = u_1(\mathbf{x}_0)\mathbf{e}_1 + u_2(\mathbf{x}_0)\mathbf{e}_2 = \mathbf{x} - \mathbf{x}_0, \tag{1}$$

which, by introducing the (twice-continuously differentiable) deformation function

$$\mathbf{x} = \mathbf{g}(\mathbf{x}_0), \tag{2}$$

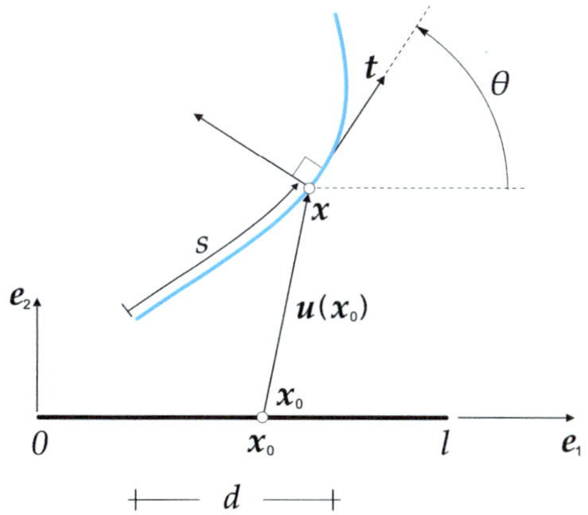

Figure 1. The kinematics of an elastic inextensible rod of length l, rectilinear in the reference configuration. The displacement of a generic point $\mathbf{x}_0$ of coordinate x_0 is $\mathbf{u}(x_0) = \mathbf{x} - \mathbf{x}_0$. Note that inextensibility implies that the curvilinear coordinate s is equal to the coordinate x_0, namely, $s = x_0$.

and noting that the point $\mathbf{x}_0$ has coordinate $x_0\mathbf{e}_1$, becomes

$$\mathbf{u} = \mathbf{g}(x_0\mathbf{e}_1) - x_0\mathbf{e}_1. \tag{3}$$

Since $\mathbf{e}_1$ is fixed, the dependence of function $\mathbf{g}$ on the unit vector can be omitted, so that equation (2) becomes the parametric representation of the curve describing the elastica.

Let us consider now two neighbouring points of the reference configuration at coordinates x_0 and $x_0 + \omega_0$, defining the vector $\mathbf{t}_0 = \omega_0\mathbf{e}_1$. This vector is mapped into

$$\mathbf{g}(\mathbf{x}_0 + \omega_0\mathbf{e}_1) - \mathbf{g}(\mathbf{x}_0), \tag{4}$$

so that, assuming ω_0 to be small, a Taylor series expansion of the deformation around $\omega_0 = 0$ yields the transformed vector (tangent to the deformed line at $\mathbf{x}$)

$$\mathbf{F}(\omega_0\mathbf{e}_1), \tag{5}$$

where

$$\mathbf{F} = \frac{\partial \mathbf{g}}{\partial \mathbf{x}_0} = \left(u_1' + 1\right)\mathbf{e}_1 \otimes \mathbf{e}_1 + u_2'\mathbf{e}_2 \otimes \mathbf{e}_1 + \mathbf{e}_2 \otimes \mathbf{e}_2, \tag{6}$$

in which a prime denotes differentiation with respect to the coordinate $x_0 = s$ and the symbol '$\otimes$' denotes the dyadic product.

Since the elastica is assumed *inextensible*, the length of the transformed vector $\mathbf{F}(\omega_0\mathbf{e}_1)$ must remain equal to the length of the initial vector $\mathbf{t}_0 = \omega_0\mathbf{e}_1$, therefore from equation (5) we obtain

$$|\mathbf{F}\mathbf{e}_1| = 1, \tag{7}$$

which, using equation (6) yields

$$u_1^{'} + 1 = \pm\sqrt{1-(u_2^{'})^2}. \tag{8}$$

Differentiation of equation (8) finally provides the inextensibility constraint in the form

$$u_1^{''} = \mp\frac{u_2^{'}u_2^{''}}{\sqrt{1-(u_2^{'})^2}}. \tag{9}$$

Since the inextensibility constraint is enforced and the tangent to the elastica at $\mathbf{x}$ is given by the unit vector $\mathbf{t}$

$$\mathbf{t} = \left(u_1^{'} + 1\right)\mathbf{e}_1 + u_2^{'}\mathbf{e}_2 = \pm\sqrt{1-(u_2^{'})^2}\,\mathbf{e}_1 + u_2^{'}\mathbf{e}_2, \tag{10}$$

the angle θ of inclination of the tangent $\mathbf{t}$ to the elastica at $\mathbf{x}$ can be introduced to satisfy

$$\sin\theta = x_2^{'} = u_2^{'}, \quad \cos\theta = x_1^{'} = \pm\sqrt{1-(u_2^{'})^2}. \tag{11}$$

Equation $(11)_2$ shows that the '+' sign in equation (8) refers to moderate deflection, for which $\cos\theta$ remains positive. Furthermore, the length d of the projection of the elastica onto the $\mathbf{e}_1$-axis can be written as

$$d = \int_0^l \cos\theta \mathrm{d}s = \pm\int_0^l \sqrt{1-(u_2^{'})^2}\,\mathrm{d}s. \tag{12}$$

The unit vector $\mathbf{n}$ normal to the elastica at $\mathbf{x}$ can be obtained through differentiation (with respect to s) of the scalar product $\mathbf{t}\cdot\mathbf{t}$, so that $\mathbf{t}^{'}$ is found normal to $\mathbf{t}$ in the form

$$\mathbf{t}^{'} = \mp\frac{u_2^{'}u_2^{''}}{\sqrt{1-(u_2^{'})^2}}\mathbf{e}_1 + u_2^{''}\mathbf{e}_2, \tag{13}$$

or, equivalently, in the form

$$\mathbf{t}^{'} = -\theta^{'}\sin\theta\mathbf{e}_1 + \theta^{'}\cos\theta\mathbf{e}_2. \tag{14}$$

The unit normal $\mathbf{n}$ can therefore be obtained from $\mathbf{t}'$ through division by its modulus $|\mathbf{t}'|$, that is, the so-called 'curvature'

$$|\mathbf{t}'| = \frac{|u_2''|}{\sqrt{1-(u_2')^2}} = |\theta'|, \tag{15}$$

thus obtaining

$$\mathbf{n} = \operatorname{sgn}\{u_2''(u_1'+1)\}\left(-u_2'\mathbf{e}_1 + \sqrt{1-(u_2')^2}\,\mathbf{e}_2\right), \tag{16}$$

or equivalently

$$\mathbf{n} = \operatorname{sgn}(\theta')\,(-\sin\theta\mathbf{e}_1 + \cos\theta\mathbf{e}_2)\,. \tag{17}$$

In conclusion, the *signed curvature* χ reads

$$\chi = \operatorname{sgn}(u_1'+1)\,\frac{u_2''}{\sqrt{1-(u_2')^2}}, \quad \text{or} \quad \chi = \theta'. \tag{18}$$

2.2 Constitutive equation, total potential energy and the elastica

The constitutive equation used for the inextensible planar elastica comes from Jacob Bernoulli's celebrated assumption, which, neglecting the effects of normal and shearing forces, defines a linear relation between the bending moment $M(s)$ and the curvature $\theta'(s)$ as

$$M(s) = B\theta'(s), \tag{19}$$

where B is the bending stiffness (considered constant along the rod), defined as the product of the Young's modulus with the principal moment of inertia of the cross section of the rod about the axis normal to the inflection plane.

With reference to the elastic systems reported in Fig. 2, where the rod is loaded at its right end by a longitudinal force P, the total potential energy $\mathcal{V}(\theta(s))$ can be written as

$$\mathcal{V}(\theta(s)) = \int_0^l B\frac{(\theta'(s))^2}{2}\mathrm{d}s - P\left(l - \int_0^l \cos\theta(s)\mathrm{d}s\right) - R\int_0^l \sin\theta(s)\,\mathrm{d}s, \tag{20}$$

where the first integral is the bending strain energy, the second term corresponds to the work done by the external force P, and R is a Lagrange multiplier (representing –as will be shown below– the vertical reaction at the supports) enforcing a global kinematic compatibility condition that corresponds to the vanishing of the relative, vertical displacement of the rod

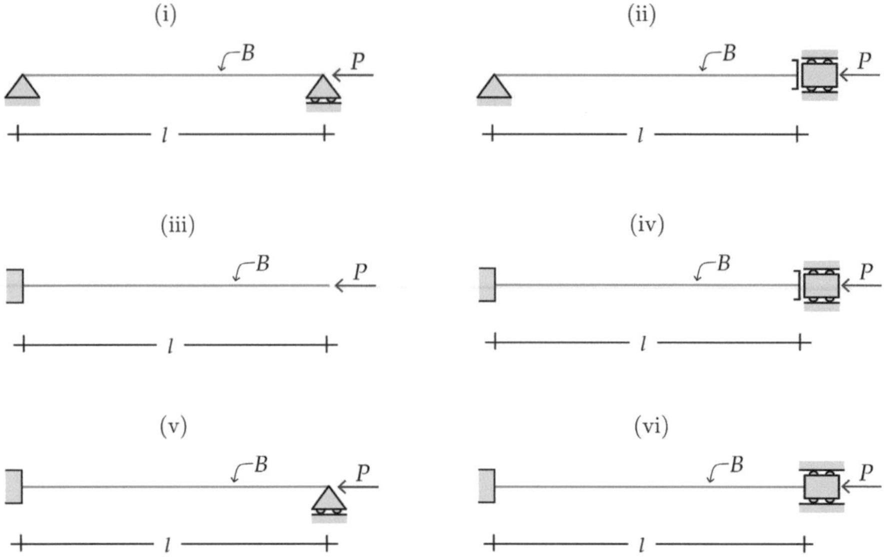

Figure 2. An elastic, inextensible rod loaded by an axial thrust P (positive when compressive) and subject to different constraints at its ends.

extremities. This condition will be enforced while considering the systems (v.) and (vi.) of Fig. 2, as well as in a particular case of system (i.). In contrast, the multiplier R will be set to zero for the elastic structures (ii.)-(iv.).

The functional (20) is defined over the set of kinematically admissible rotation fields, meaning every differentiable rotation field $\theta(s)$ satisfying the boundary conditions introduced by the constraints at the ends of the systems (i.)-(vi.) reported in Fig. 2, namely

$$
\begin{aligned}
&\text{i)} \quad \int_0^l \sin\theta(s)\mathrm{d}s = 0,\\
&\text{ii)} \quad \theta(l) = 0,\\
&\text{iii)} \quad \theta(0) = 0,\\
&\text{iv)} \quad \theta(0) = \theta(l) = 0,\\
&\text{v)} \quad \theta(0) = 0 \quad \text{and} \quad \int_0^l \sin\theta(s)\mathrm{d}s = 0,\\
&\text{vi)} \quad \theta(0) = \theta(l) = 0 \quad \text{and} \quad \int_0^l \sin\theta(s)\mathrm{d}s = 0.
\end{aligned}
\tag{21}
$$

Let us consider now variations $\tilde{\theta}(s)$ of the rotation field $\theta(s)$, satisfying the boundary conditions relevant to each system, namely

$$
\begin{aligned}
&\text{i)} \quad \textstyle\int_0^l \sin\tilde{\theta}(s)\mathrm{d}s = 0,\\
&\text{ii)} \quad \tilde{\theta}(l) = 0,\\
&\text{iii)} \quad \tilde{\theta}(0) = 0,\\
&\text{iv)} \quad \tilde{\theta}(0) = \tilde{\theta}(l) = 0,\\
&\text{v)} \quad \tilde{\theta}(0) = 0 \quad \text{and} \quad \textstyle\int_0^l \sin\tilde{\theta}(s)\mathrm{d}s = 0,\\
&\text{vi)} \quad \tilde{\theta}(0) = \tilde{\theta}(l) = 0 \quad \text{and} \quad \textstyle\int_0^l \sin\tilde{\theta}(s)\mathrm{d}s = 0.
\end{aligned}
\tag{22}
$$

Keeping into account integration by parts

$$
\int_0^l \theta^{'}(s)\tilde{\theta}^{'}(s)\,\mathrm{d}s = -\int_0^l \theta^{''}(s)\tilde{\theta}(s)\,\mathrm{d}s + \theta^{'}(l)\tilde{\theta}(l) - \theta^{'}(0)\tilde{\theta}(0),
$$

the first variation $\delta\mathcal{V}$ of functional $\mathcal{V}$ is obtained as

$$
\delta\mathcal{V} = \int_0^l \left[\theta^{''}(s) + \frac{P}{B}\sin\theta(s) + \frac{R}{B}\cos\theta(s)\right]\tilde{\theta}(s)\,\mathrm{d}s - \theta^{'}(l)\tilde{\theta}(l) + \theta^{'}(0)\tilde{\theta}(0),
\tag{23}
$$

holding for all kinematically admissible variation $\tilde{\theta}(s)$.

Imposing the vanishing of the first variation (23) for every admissible variation $\tilde{\theta}(s)$ yields the nonlinear differential equation for the elastica

$$
\theta^{''}(s) + \frac{P}{B}\sin\theta(s) + \frac{R}{B}\cos\theta(s) = 0,
\tag{24}
$$

and the following natural boundary conditions for the cases (i.)-(iii.) and (v.)

$$
\begin{aligned}
&\text{i)} \quad \theta^{'}(0) = \theta^{'}(l) = 0,\\
&\text{ii)} \quad \theta^{'}(0) = 0,\\
&\text{iii)} \quad \theta^{'}(l) = 0,\\
&\text{v)} \quad \theta^{'}(l) = 0.
\end{aligned}
\tag{25}
$$

Notice that in the equilibrium equation (24) it is apparent that the Lagrange multiplier R corresponds to the vertical reaction at the supports, which are always null except in the following cases:

- ⋄ for the structure (i.), in the special situation when the two supports coincide;
- ⋄ for the structure (v.);
- ⋄ for the structure (vi.), when antisymmetric buckling modes are considered.

Simply supported elastica Let us start by considering the doubly pinned rod (i.), so that, defining $\lambda^2 = P/B$, the equations governing the equilibrium of the elastica are

$$
\begin{array}{ll}
\theta''(s) + \lambda^2 \sin\theta(s) = 0 \quad \forall\, s \in [0, l], & \text{governing diff. equation} \\
\theta'(0) = \theta'(l) = 0, & \text{b.c.: null moment at both supports} \\
u_1(0) = 0, & \text{b.c.: null horiz. displ. at left support} \\
u_2(0) = u_2(l) = 0, & \text{b.c.: null vert. displ. at both supports} \\
u_1'(s) = \cos\theta(s) - 1 \quad \forall\, s \in [0, l], & \text{diff. equation for the horizontal displ.} \\
u_2'(s) = \sin\theta(s) \quad \forall\, s \in [0, l]. & \text{diff. equation for the vertical displ.}
\end{array}
\tag{26}
$$

Equations (26) define a *nonlinear eigenvalue problem*, for which the *trivial solution* $\theta(s) = u_1(s) = u_2(s) = 0$ is always possible, so that the question arises whether nontrival solutions exist or not. Bifurcation corresponds to the situation in which the trivial solution (or possibly a bifurcated solution) of (26) splits into two or more, as λ passes through a critical value λ_{cr}, called 'bifurcation point'.

Before embarking in the solution of the non-linear problem (26), let us consider its linearization about the trivial solution $\theta(s) = 0$, such that the horizontal displacement is null, i.e. $u_1(s) = 0$, and the remaining equations read

$$
\begin{array}{ll}
\theta^{''}(s) + \lambda^2\theta(s) = 0 \quad \forall\, s \in [0, l], & \text{governing diff. equation} \\
\theta^{'}(0) = \theta^{'}(l) = 0, & \text{boundary conditions} \\
u_2(0) = u_2(l) = 0, & \text{boundary conditions} \\
u_2^{'}(s) = \theta(s) \quad \forall\, s \in [0, l]. & \text{diff. equation for the vertical displ.}
\end{array}
\tag{27}
$$

Equations (27) define a *linear eigenvalue problem*, also called 'Sturm-Liouville problem' (Broman, 1970). It admits the infinite solutions

$$
\theta(s) = A_n \cos \frac{n\pi s}{l}, \quad n = 0, 1, 2, ... \tag{28}
$$

and

$$
\underbrace{u_2 = A_0 = 0, \quad n = 0,}_{\text{trivial solution}} \qquad \underbrace{u_2 = \frac{l A_n}{n\pi} \sin \frac{n\pi s}{l}, \quad n = 1, 2, ...}_{\text{nontrivial solutions}} \tag{29}
$$

where the trivial solution holds for every thrust P, while the nontrivial solutions hold if and only if

$$
\lambda = \lambda_n = \frac{n\pi}{l}, \quad \Leftrightarrow \quad P = P_n^{cr} = \frac{n^2\pi^2 B}{l^2}, \quad n = 1, 2, ... \tag{30}
$$

which defines the Euler's critical loads. As a consequence of the linearization, the amplitudes A_n ($n = 1, 2, ...$) of the bifurcation modes remain undetermined, nevertheless the critical loads correctly identify the bifurcation points on the trivial path, λ_n, as will be proven below.

Let us now solve the nonlinear problem (26). First of all, we note that if

$$
\theta(s), \quad u_1(s), \quad u_2(s),
$$

represent a solution corresponding to λ^2, the fields

$$
\pm\theta(s) + 2n\pi, \quad u_1(s), \quad \pm u_2(s), \quad n = ..., -2, -1, 0, 1, 2, ...
$$

also represent solutions (symmetrical with respect to the x_1–axis) and the fields

$$
\pm\theta(s) + (2n+1)\pi, \quad -u_1(s) - 2s, \quad u_2(s), \quad n = ..., -2, -1, 0, 1, 2, ...
$$

are valid for $-\lambda^2$. These are symmetric solutions with respect to the x_1– or the x_2–axis and, without loss of generality, will be ignored in the following.

Therefore, defining $\hat{\theta} = \theta(\hat{s})$, where $\hat{s}$ is the curvilinear coordinate at which the bending moment is null, for the case under consideration

$$\hat{\theta} = \theta(0), \tag{31}$$

and we only address solutions such that $0 \leq \hat{\theta} \leq \pi$. A multiplication of equation $(26)_1$ by $\theta^{'}(s)$ yields

$$\frac{\mathrm{d}}{\mathrm{d}s}\left[\frac{1}{2}(\theta^{'}(s))^2 - \lambda^2 \cos\theta(s)\right] = 0, \tag{32}$$

so that its integration, considering the boundary conditions $(26)_2$ and equation (31), leads to

$$\theta^{'}(s) = \lambda\sqrt{2(\cos\theta(s) - \cos\hat{\theta})}, \tag{33}$$

where we have selected the positive root, since the two solutions merely differ in sign.

An equation formally identical to equation (33) is usually obtained in the analysis of the oscillation of a simple pendulum (Temme, 1996), so that it is a standard expedient to operate the following change of variables

$$\kappa = \sin\frac{\hat{\theta}}{2}, \quad \kappa\sin\phi(s) = \sin\frac{\theta(s)}{2}, \tag{34}$$

leading through trigonometric formulae to the differential problem

$$\frac{\mathrm{d}\,\phi(s)}{\mathrm{d}s} = \lambda\sqrt{1 - \kappa^2\sin^2\phi(s)}. \tag{35}$$

The boundary conditions imply that $\sin\phi(0) = 1$ and $\sin^2\phi(l) = 1$, so that

$$\phi(0) = \frac{4h+1}{2}\pi, \quad \phi(l) = \frac{2j+1}{2}\pi, \quad h, j = 0, \pm1, \pm2, ... \tag{36}$$

and therefore integration of equation (35) by separation of the variables yields

$$s\lambda = \int_{\frac{4h+1}{2}\pi}^{\phi(s)} \frac{\mathrm{d}\phi}{\sqrt{1-\kappa^2\sin^2\phi}}, \quad h = 0, \pm1, \pm2, ... \tag{37}$$

which, for $s = l$, becomes

$$l\lambda = \int_{\frac{4h+1}{2}\pi}^{\frac{2j+1}{2}\pi} \frac{\mathrm{d}\phi}{\sqrt{1-\kappa^2\sin^2\phi}}, \quad h, j = 0, \pm1, \pm2, ... \tag{38}$$

Taken over one period, the integral (38) is equal to $2\mathcal{K}(\kappa)$, where

$$\mathcal{K}(\kappa) = \int_0^{\frac{\pi}{2}} \frac{\mathrm{d}\phi}{\sqrt{1-\kappa^2 \sin^2\phi}}, \tag{39}$$

is the complete elliptic integral of the first kind or the so-called 'real quarter period of the elliptic function' (Byrd and Friedman, 1954; Temme, 1996).

The integral in equation (38) can be rewritten as a function of an integer m as

$$l\lambda = 2m\mathcal{K}(\kappa), \quad \Leftrightarrow \quad P = \frac{B}{l^2} 4m^2 \left[\mathcal{K}\left(\sin\frac{\hat{\theta}}{2}\right)\right]^2, \tag{40}$$

an equation providing the relation between the applied load P and the rotation of the left rod's extremity associated to the m–th bifurcation mode (Reiss, 1969).

For small $\hat{\theta}$, a Taylor series expansion of equation (40) provides exactly equation (30), thus proving that the critical Euler loads (30), calculated from the linearized theory, correctly determine the bifurcation points emanating from the trivial path. Note that this results has a conceptual relevance, since it validates the calculations usually done on the linearized eigenvalue problem.

Let us go back now to equation (37) and note that the integral on the right-hand side can always be written as

$$\int_{\frac{4m+1}{2}\pi}^{\phi(s)} (\cdots)\,\mathrm{d}s = -\int_0^{\frac{4m+1}{2}\pi} (\cdots)\,\mathrm{d}s + \int_0^{\phi(s)} (\cdots)\,\mathrm{d}s, \tag{41}$$

so that, since (Byrd and Friedman, 1954)

$$\int_0^{\frac{4m+1}{2}\pi} \frac{\mathrm{d}\phi}{\sqrt{1-\kappa^2\sin^2\phi}} = (4m+1)\mathcal{K}(\kappa), \quad m = 0, \pm 1, \pm 2, \ldots \tag{42}$$

we obtain

$$s\lambda + (4m+1)\mathcal{K}(\kappa) = \int_0^{\phi(s)} \frac{\mathrm{d}\phi}{\sqrt{1-\kappa^2\sin^2\phi}}, \quad m = 0, \pm 1, \pm 2, \ldots \tag{43}$$

which provides

$$\phi(s) = \mathrm{am}\,(s\lambda + (4m+1)\mathcal{K}(\kappa), \kappa), \quad m = 0, \pm 1, \pm 2, \ldots \tag{44}$$

where 'am' denotes the Jacobi amplitude function of modulus κ. Employing the property (Byrd and Friedman, 1954)

$$\mathrm{am}\,(x \pm 2n\mathcal{K}(\kappa), \kappa) = \mathrm{am}\,(x, \kappa) \pm n\pi, \quad n = 0, \pm 1, \pm 2, ...$$

equation (44) can be simplified to

$$\phi(s) = \mathrm{am}\,(s\lambda + \mathcal{K}(\kappa), \kappa) + 2m\pi, \quad m = 0, \pm 1, \pm 2, ... \tag{45}$$

so that the definition of $\phi(s)$, equation $(34)_2$, yields

$$\sin\frac{\theta(s)}{2} = \kappa\,\mathrm{sn}\,(s\lambda + \mathcal{K}(\kappa), \kappa), \tag{46}$$

where 'sn' is the Jacobi sine amplitude function, defined as

$$\mathrm{sn}\,(x, \kappa) = \sin\,(\mathrm{am}\,(x, \kappa))\,. \tag{47}$$

A substitution of equation (46) into equation (33), where the identity $\cos\theta = 1 - 2\sin^2(\theta/2)$ is employed, yields

$$\theta^{'}(s) = 2\lambda\kappa\,\mathrm{cn}\,(s\lambda + \mathcal{K}(\kappa), \kappa), \tag{48}$$

where 'cn' is the Jacobi cosine amplitude function, defined as

$$\mathrm{cn}\,(x, \kappa) = \cos\,(\mathrm{am}\,(x, \kappa))\,.$$

Note that, due to the properties

$$\mathrm{cn}\,(\mathcal{K}(\kappa), \kappa) = \mathrm{cn}\,(3\mathcal{K}(\kappa), \kappa) = \mathrm{cn}\,[(2m+1)\mathcal{K}(\kappa), \kappa] = 0, \quad m = 0, \pm 1, \pm 2, \pm 3$$

the boundary conditions $(26)_2$, namely $\theta^{'}(0) = \theta^{'}(l) = 0$, are satisfied[1].

According to equations (11) and $(26)_{(5,6)}$, the differential equations determining the points $\mathbf{x}$ of the deformed elastica are

$$x_1^{'}(s) = \cos\theta(s), \quad x_2^{'}(s) = \sin\theta(s), \tag{49}$$

which, since $\cos\theta = 1 - 2\sin^2(\theta/2)$ and $\sin\theta = 2\sin(\theta/2)\sqrt{1 - \sin^2(\theta/2)}$, and using equation (46), provide the two differential equations

$$\begin{aligned} x_1^{'}(s) &= 1 - 2\kappa^2\,\mathrm{sn}^2(s\lambda + \mathcal{K}(\kappa), \kappa), \\ x_2^{'}(s) &= 2\kappa\,\mathrm{sn}\,(s\lambda + \mathcal{K}(\kappa), \kappa)\mathrm{dn}\,(s\lambda + \mathcal{K}(\kappa), \kappa), \end{aligned} \tag{50}$$

[1] Equation (40) has been used in the boundary condition at $\theta^{'}(l)$.

where 'dn' is the Jacobi elliptic function, defined as

$$\mathrm{dn}\,(s\lambda + \mathcal{K}(\kappa), \kappa) = \sqrt{1 - \kappa^2\, \mathrm{sn}^2(s\lambda + \mathcal{K}(\kappa), \kappa)}. \tag{51}$$

Since the following differentiation rules are known (Byrd and Friedman, 1954)

$$\begin{aligned}
&\frac{\partial}{\partial x} E(x, \kappa) = \sqrt{1 - \kappa^2 \sin^2 x}, \\
&\frac{\partial}{\partial x}\, \mathrm{am}\,(x, \kappa) = \mathrm{dn}\,(x, \kappa), \\
&\frac{\partial}{\partial x}\, \mathrm{cn}\,(x, \kappa) = -\,\mathrm{sn}\,(x, \kappa)\, \mathrm{dn}\,(x, \kappa),
\end{aligned} \tag{52}$$

where $E(x, \kappa)$ is the incomplete elliptic integral of the second kind of modulus κ,

$$E(x, \kappa) = \int_0^x \sqrt{1 - \kappa \sin^2 t}\, \mathrm{d}t,$$

taking into account the boundary conditions $(26)_{3,4}$, we integrate equations (50), thus arriving at the equations describing the deformed shape of the elastica[2]

$$\begin{aligned}
x_1(s) &= -s + \frac{2}{\lambda} \left\{ E\left[\mathrm{am}\,(s\lambda + \mathcal{K}(\kappa), \kappa), \kappa\right] - E\left[\mathrm{am}\,(\mathcal{K}(\kappa), \kappa), \kappa\right] \right\}, \\
x_2(s) &= -\frac{2\kappa}{\lambda}\, \mathrm{cn}(s\lambda + \mathcal{K}(\kappa)),
\end{aligned} \tag{53}$$

which are identical with those provided by Love (1927). For compressive loads ($P > 0$) the displacement $u_1(l)$ is negative and, since $|u_1(l)| = l - x_1(l)$, its absolute value can be immediately obtained from equations $(53)_1$ in the form

$$|u_1(l)| = 2l - \frac{2}{\lambda} \left\{ E\left[\mathrm{am}\,(l\lambda + \mathcal{K}(\kappa), \kappa), \kappa\right] - E\left[\mathrm{am}\,(\mathcal{K}(\kappa), \kappa), \kappa\right] \right\}, \tag{54}$$

so that using now equation (40) we obtain

$$\frac{|u_1(l)|}{l} = 2 - \frac{E\left[\mathrm{am}\,((2m+1)\mathcal{K}(\kappa), \kappa), \kappa\right] - E\left[\mathrm{am}\,(\mathcal{K}(\kappa), \kappa), \kappa\right]}{m\mathcal{K}(\kappa)}, \quad m = 1, 2, ... \tag{55}$$

[2]These relations are also valid for the hinged-clamped case (ii), where the value of λ as a function of the rotation at the inflexion point, $\hat{\theta} = \theta(0)$, is obtained through the relation $l\lambda = (2m-1)\mathcal{K}\left(\sin \frac{\hat{\theta}}{2}\right)$.

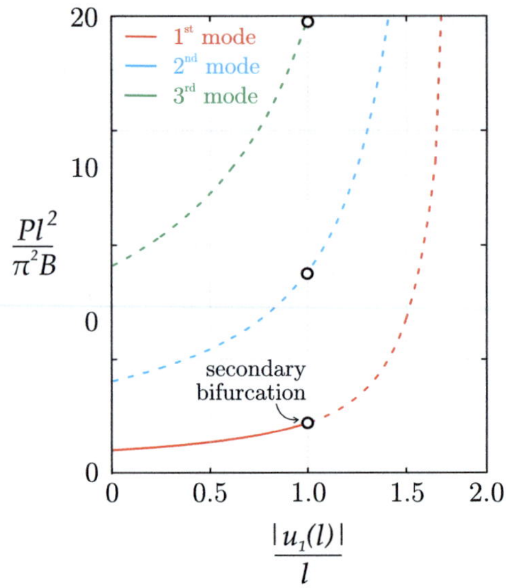

Figure 3. Dimensionless load $Pl^2/(\pi^2 B)$ of a doubly supported rod versus dimensionless displacement $u_1(l)/l$. The first three primary bifurcation points and branches are reported together with the first three secondary bifurcation points. All equilibrium configurations on the second and third branch are unstable (dashed curves). The first branch becomes unstable after the secondary bifurcation point.

which eventually can be simplified to (Reiss, 1969)[3]

$$\frac{|u_1(l)|}{l} = 2 - \frac{2\,E(\kappa)}{\mathcal{K}(\kappa)}, \tag{56}$$

where $E(\pi/2, \kappa) = E(\kappa)$ represents the complete elliptic integral of the second kind. Note that equation (56) is independent of the bifurcation mode m, so that the displacement of the right pin of the rod depends only on $\hat{\theta}$ (through κ).

The mid-span deflection of the rod, null for even values of the mode m, can be evaluated for odd m as

$$\frac{|u_2(l/2)|}{l} = \frac{\kappa}{m\,\mathcal{K}(\kappa)}, \qquad m = 1, 3, 5, ... \tag{57}$$

[3]The following identities turn out to be useful (Byrd and Friedman, 1954):

$$\text{am}\,[\mathcal{K}(\kappa), \kappa] = \pi/2, \quad \text{am}[(2m+1)\mathcal{K}(\kappa), \kappa] = (2m+1)\pi/2, \quad E(n\pi/2, \kappa) = n\,E(\kappa).$$

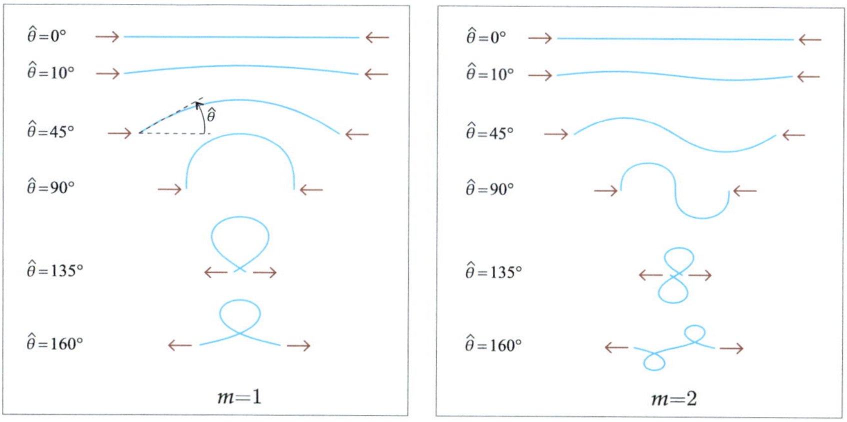

Figure 4. Deformed shapes for the first two modes $m = 1, 2$ of a doubly supported rod at different values of parameters setting the deformation: the initial inflexion angle $\hat{\theta} = \{0, 10°, 45°, 90°, 135°, 160°\}$ and corresponding dimensionless displacement of the end of the rod $u_1(l)/l = \{0, 0.008, 0.149, 0.543, 1.049, 1.340\}$ for both the modes. The deformed shapes of the elastica represent the post-critical behaviour of the structure.

In summary, for a given $\hat{\theta}$ and for a given mode m we can calculate the corresponding λ (using equation (40)) and $u_1(l)$ (using equation (56)) and plot the elastica (using equations (53)). The bifurcation diagram showing the load P (made dimensionless through multiplication by $l^2/(\pi^2 B)$) as a function of the displacement of the right pin of the rod (divided by l) is shown in Fig. 3. In the figure, the first three critical loads and the corresponding three branches are reported. We may note that the branches do not cross each other and the load is continuously increasing during the post-critical behaviour.

The deformed elastic lines have been evaluated and plotted in Fig. 4 for the first two branches at fixed values of $\hat{\theta}$, namely, $\{10°, 45°, 90°, 135°, 160°\}$. These values of rotation correspond to rod end displacements, respectively equal to $\{0.008, 0.149, 0.543, 1.049, 1.340\}\, l$. Note that in Fig. 4 also the undeformed configuration, corresponding to $\hat{\theta} = 0°$, is reported in order to provide the scale bar for the displacement.

It should be noticed that the line of thrust (joining the two forces in Fig. 4) intersects the elastica at points of inflexion (where $\theta' = 0$), separating different so-called 'bays' in the Love's terminology.

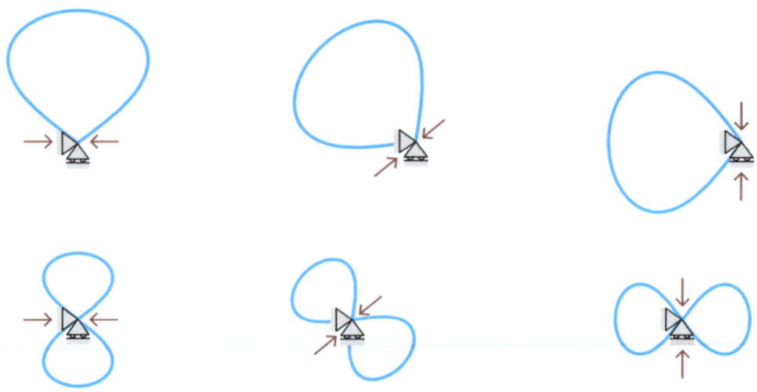

Figure 5. Sketches of the in-plane secondary bifurcation modes of the simply supported elastica, referred to the first (upper part) and second (lower part) mode. When the two supports coincide the structure can suffer a rigid-body rotation. During this rotation, the horizontal load drops to zero (value reached when the elastica is rotated by 90°), so that the force is maintained by the vertical reaction of the support. For rotation angles greater than 90° (not reported) the force changes sign. For an horizontal dead load, the structure becomes unstable when the two supports coincide and snaps to the configuration $u_1 = -2l$, where it is in equilibrium with a tensile load.

In-plane secondary bifurcations of the simply-supported elastica Let us go back to Fig. 3 and note that on each bifurcated branch there is a secondary bifurcation point (marked with a circle). This condition occurs when the two supports of the rod coincide, namely, when $u_1(l) = -l$, corresponding to $\hat{\theta} = 130.7099°$ and different load values: $Pl^2/(\pi B) = 2.1833$ for the first mode, $Pl^2/(\pi B) = 8.7335$ for the second mode, $Pl^2/(\pi B) = 19.6504$ for the third mode, and so on.

These secondary bifurcation modes, which passed unnoticed until Maddocks (1984) (see also Kuznetsov and Levyakov, 2002), have a simple explanation. In fact, *when the two supports of the rod momentarily coincide during deformation along the bifurcation path, the structure can rigidly rotate about the pin.* During the rigid-body rotation, vertical reactions are generated at the two supports, so that the horizontal load P drops and is reduced to zero when the structure is rotated by 90°. Similarly, further rotations imply a negative horizontal force. This situation is sketched in Fig. 5, with reference to the first two modes.

For the case of imposed horizontal load, the situation in which the two supports coincide marks an instability point in the sense that: (i.) at this point the structure rigidly rotates and snaps to the configuration $u_1 = -2l$, where it is subject to a tensile load; (ii.) equilibrium configurations belonging to the post-critical path $m = 1$ and $\hat{\theta} > 130.7099°$ are unstable.

Doubly clamped elastica The equations governing the equilibrium configurations of a doubly clamped elastica are

$$
\begin{aligned}
&\theta''(s) + \frac{P}{B}\sin\theta(s) + \frac{R}{B}\cos\theta(s) = 0 \quad \forall\, s \in [0,l], && \text{governing diff. equation}\\
&\theta(0) = \theta(l) = 0, && \text{b.c.: null rotation}\\
&u_1(0) = 0, && \text{b.c.: null horizontal displ.}\\
&u_2(0) = u_2(l) = 0, && \text{b.c.: null vertical displ.}\\
&u_1'(s) = \cos\theta(s) - 1 \quad \forall\, s \in [0,l], && \text{diff. eq. for horiz. displ.}\\
&u_2'(s) = \sin\theta(s) \quad \forall\, s \in [0,l]. && \text{diff. eq. for vert. displ.}
\end{aligned}
\tag{58}
$$

Similarly to the case of a simply supported rod, equations (58) define a *nonlinear eigenvalue problem*, for which the *trivial solution* $\theta(s) = 0$ is always possible, so that we look for non-trivial solutions.

When symmetric deformed configurations are considered, such that $2m$ inflection points are present, vertical equilibrium requires that $R = 0$ and so the differential equation $(58)_1$ reduces to that governing the previous case of a simply supported rod.

A different solution arises when antisymmetric configurations are sought. In this case, $2m + 1$ inflection points are present and a non-zero vertical reaction R acts at the constraints. A linearization of the problem (58) about the solution $\theta(s) = 0$ leads to

$$
\begin{aligned}
&\theta''(s) + \frac{P}{B}\theta(s) = -\frac{R}{B} \quad \forall\, s \in [0,l], && \text{governing diff. equation}\\
&\theta(0) = \theta(l) = 0, \quad \int_0^l \theta(s)\mathrm{d}s = 0. && \text{boundary conditions}
\end{aligned}
\tag{59}
$$

Equations (59) define a *linear eigenvalue problem* that admits the trivial solution $\theta(s) = 0$ for every thrust P, whereas infinite, non trivial (symmetric and antisymmetric) solutions are only possible when the following characteristic equation is satisfied

$$
2\left(1 - \cos\sqrt{\frac{Pl^2}{B}}\right) = \sqrt{\frac{Pl^2}{B}}\sin\sqrt{\frac{Pl^2}{B}}, \tag{60}
$$

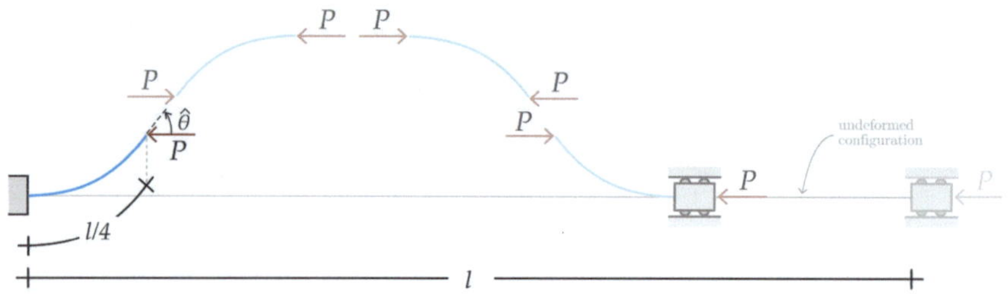

Figure 6. Deformed (and undeformed) configuration for the first (symmetric) mode of the doubly clamped rod. Note that, thanks to symmetry, the problem is reduced to the analysis of a cantilever rod of length $l/4$ loaded with an end thrust P.

defining the Euler's bifurcation loads for the doubly clamped rod as

$$P_1^{cr} = \frac{4\pi^2 B}{l^2}, \quad P_2^{cr} = \frac{8.1830\pi^2 B}{l^2}, \quad P_3^{cr} = \frac{16\pi^2 B}{l^2}, \quad P_4^{cr} = \frac{24.1872\pi^2 B}{l^2}, \ldots \tag{61}$$

where an odd (even) subscript in the definition of the critical loads corresponds to symmetric (antisymmetric) bifurcation paths.

Symmetric buckling modes With reference to the symmetric buckling modes with $2m$ inflection points ($R = 0$), the rotation field $\theta(s)$ is solution of the following differential problem

$$\begin{aligned} &\theta''(s) + \lambda^2 \sin\theta(s) = 0 \quad \forall\, s \in [0, l], \\ &\theta(0) = \theta(l) = 0, \end{aligned} \tag{62}$$

so that, a multiplication of equation $(62)_1$ by $\theta'(s)$ and its integration leads to

$$\theta'(s) = \pm\lambda\sqrt{2(\cos\theta(s) - \cos\hat{\theta})}, \tag{63}$$

where, in this case, $\hat{\theta}$ represents the angle of rotation at the inflection point at the coordinate $s = l/4$.

In order to facilitate the analytical description, the symmetry properties can be exploited, so that four simply clamped rods, of equal length $l/4$, can be identified in the structure, as sketched in Fig. 6.

Therefore, limiting the attention only to one quarter of the rod, $s \in [0, l/4]$, and considering the positive root (representing the curvature) of equation (63) for $s \in [0, l/4]$, the change of variables (34) leads to the differential problem

$$\frac{d\phi(s)}{ds} = \lambda\sqrt{1-\kappa^2\sin^2\phi}. \tag{64}$$

The boundary conditions for the considered problem imply that $\sin\phi(0) = 0$ and $\sin\phi(l/4) = 1$, so that

$$\phi(0) = h\pi, \quad \phi(l/4) = \frac{2j+1}{2}\pi, \quad h, j = 0, \pm 1, \pm 2, ... \tag{65}$$

and therefore integration of equation (64) by separation of the variables yields

$$s\lambda = \int_{h\pi}^{\phi(s)} \frac{d\phi}{\sqrt{1-k^2\sin^2\phi}}, \quad h = 0, \pm 1, \pm 2, ... \tag{66}$$

which, for $s = l/4$, becomes

$$\frac{l}{4}\lambda = \int_{h\pi}^{\frac{2j+1}{2}\pi} \frac{d\phi}{\sqrt{1-k^2\sin^2\phi}}, \quad h, j = 0, \pm 1, \pm 2, ... \tag{67}$$

and can be rewritten as a function of an odd integer m as

$$l\lambda = 2(m+1)\mathcal{K}(\kappa), \quad \Leftrightarrow \quad P = \frac{B}{l^2}4(m+1)^2\left[\mathcal{K}\left(\sin\frac{\hat{\theta}}{2}\right)\right]^2, \quad m = 1, 3, 5, ... \tag{68}$$

This equation provides the relation between the applied load P and the rotation $\hat{\theta}$ at the inflection point ($s = l/4$) associated with the odd (symmetric) m–th buckling mode.

For small $\hat{\theta}$, a Taylor series expansion of equation (68) provides exactly the odd critical Euler's loads calculated through linearization of the governing equations and reported in equation (61).

Going back to equation (66) and employing the definition of $\phi(s)$, equation (34), in addition to the odd nature of Jacobi amplitude function, namely,

$$\text{am}\,(x, \kappa) = -\text{am}\,(-x, \kappa),$$

we can write the rotation field along the rod as[4]

$$\theta(s) = 2\arcsin\left(\kappa\,\text{sn}\,(s\lambda, \kappa)\right) \quad \forall\, s \in [0, l]. \tag{69}$$

[4]We may note that the equations describing the rotational field, equation (69), and the shape of the elastica, equation (70), remain the same also for structure (iii). The only difference lies in the relation between the load P and the angle of the free end $\hat{\theta} = \theta(l)$, associated to the m-th bifurcation mode, which is $l\lambda = (2m-1)\mathcal{K}\left(\sin\frac{\hat{\theta}}{2}\right)$.

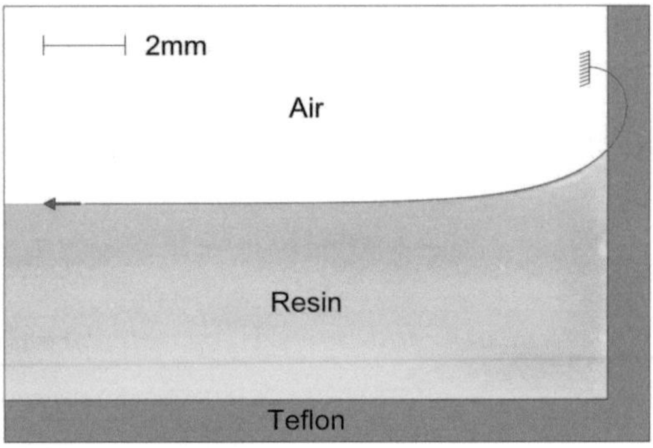

Figure 7. The solution of the elastica superimposed to a photo of fluid (an epoxy resin) forming a meniscus near the boundary of a teflon mould shows that the shape of an elastic cantilever (marked in red) subjected to large displacements (induced by a dead force applied at its free end) is identical to the shape of a fluid meniscus in a capillary channel.

According to equations (11) and (69), and taking into account the boundary conditions $(21)_6$, the analytical formulae describing the shape of the elastica for the entire rod, with $s \in [0, l]$, can be written as[5]

$$
\begin{aligned}
x_1(s) &= -s + \frac{2}{\lambda} \left\{ E \left[\mathrm{am} \left(s\lambda, \kappa \right), \kappa \right] \right\}, \\
x_2(s) &= \frac{2\kappa}{\lambda} \left[1 - \mathrm{cn}(s\lambda) \right].
\end{aligned}
\tag{70}
$$

It might be interesting to notice that a formal analogy exists between the two differential equations governing the equilibrium configurations of an elastic rod and the free surface of a fluid meniscus, see Lamb (1928). Specifically, the term P/B of the elastica is replaced in the case of a fluid meniscus by the ratio between the unit weight and the surface tension of the fluid. This fact is depicted in Fig. 7, where the deformed shape of a clamped elastic rod has been superimposed to a photo of a meniscus formed by an epoxy resin at a boundary of a teflon mould.

Proceeding with our treatment, since $|u_1(l)| = l - x_1(l)$, the displacement of the point of application of the force can be immediately obtained from

[5] Equations (70) describe also the deformed configuration of system (iv), together with the relation (40), providing a connection between the thrust P and the angle $\hat{\theta} = \theta(l/2)$ for different m buckling modes.

equation $(70)_1$ in the form

$$\frac{|u_1(l)|}{l} = -2\left(\frac{E(\kappa)}{\mathcal{K}(\kappa)} - 1\right), \tag{71}$$

which is independent of the bifurcation mode m, so that the displacement of the right end of the rod depends only on $\hat{\theta}$ (through κ). The bifurcation diagram showing the load P (made dimensionless through multiplication by $l^2/(\pi^2 B)$) as a function of the displacement of the right-hand movable end of the rod (divided by l) is shown in Fig. 9 together with the bifurcation diagram valid for antisymmetric buckling modes. In the figure, the first three critical loads and the corresponding three branches (two symmetric and one antisymmetric) are reported. The deformed elastic lines have been evaluated and plotted in Fig. 10 for the first symmetric branch (first buckling mode) at fixed values of $|u_1(l)|/l = \{0, 0.2, 0.6, 1.0, 1.4\}$, together with the first anti-symmetric branch (second buckling mode), described by equations (86), (89), and (95). Note that in Fig. 10 also the undeformed configuration is reported, providing the scale bar for the displacements.

Antisymmetric buckling modes Let us now go back to equations (58) and solve them in the case of antisymmetric (even) buckling modes characterized by $2m+1$ inflection points ($R \neq 0$). Antisymmetry properties allow us to conclude that one inflection point is located at $s = l/2$, so that we may observe that the structure can be regarded as composed of two equal clamped-hinged rods subject to an end thrust P, see Fig. 8. Therefore, we consider only one half of the rod, such that its equilibrium is governed by the following differential problem[6]

$$\begin{aligned}
&\theta''(s) + \frac{P}{B}\sin\theta(s) + \frac{R}{B}\cos\theta(s) = 0 \quad \forall s \in [0, l/2], \\
&\theta(0) = \theta'(l/2) = 0, \\
&\int_0^{l/2} \sin\theta(s)\mathrm{d}s = 0.
\end{aligned} \tag{72}$$

We write $\gamma^2 = \sqrt{P^2 + R^2}/B$ and we introduce the angle $\psi(s) = \theta(s) + \beta$, where β defines the inclination of the resultant of P and R with respect to

[6] We note that the equations for the structure (v.) correspond to those governing the antisymmetric equilibrium configuration of the structure (vi.).

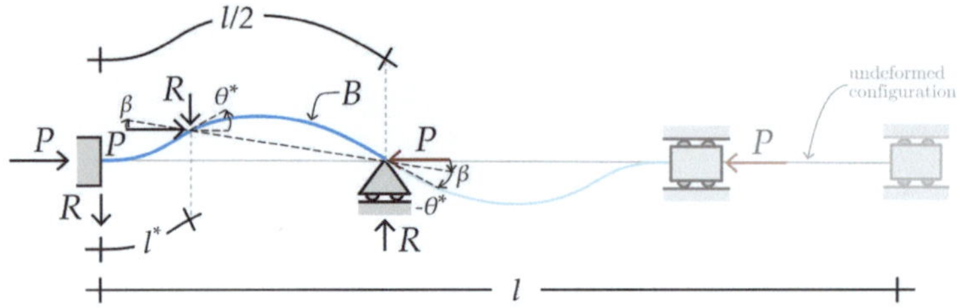

Figure 8. Deformed configuration for the second (antisymmetric) mode of the doubly clamped rod. Note that, thanks to the antisymmetry, one inflection point is located at the mid-span, so that the problem can be reduced to a clamped-guided rod of length $l/2$.

the horizontal axis, so that

$$\cos\beta = \frac{P}{\sqrt{P^2+R^2}}, \qquad \sin\beta = \frac{R}{\sqrt{P^2+R^2}}, \tag{73}$$

and the differential problem (72) can be rewritten as

$$\begin{aligned} &\psi^{''}(s) + \gamma^2 \sin\psi(s) = 0 \quad \forall\, s \in [0, l/2] \\ &\psi(0) = \beta, \quad \psi^{'}(l/2) = 0, \\ &\int_0^{l/2} \sin\left(\psi(s) - \beta\right) \mathrm{d}s = 0. \end{aligned} \tag{74}$$

Before proceeding with the derivation of the non-trivial solutions of the differential problem (74), let us define $\theta(l/2) = -\theta^*$, such that $\psi(l/2) = -\psi^* = -\theta^* + \beta$ and an inflection point exists at $s = l^*$, where the angle of rotation is defined as $\theta(l^*) = \theta^*$, with $\theta^{'}(l^*) = 0$ (note that $\psi(l^*) = \psi^*$ and $\psi^{'}(l^*) = 0$), see Fig. 8. Multiplication of equation $(74)_1$ by $\psi^{'}(s)$ and integration in the variable s yields

$$\frac{\mathrm{d}}{\mathrm{d}s}\left[\frac{1}{2}(\psi^{'}(s))^2 - \gamma^2 \cos\psi(s)\right] = 0, \tag{75}$$

so that imposition of the boundary conditions at the inflection point $s = l^*$ leads to

$$\psi^{'}(s) = \pm\lambda\sqrt{2(\cos\psi(s) - \cos\psi^*)}, \tag{76}$$

where the + (−) sign corresponds to a positive (negative) curvature. Keeping in mind the sketch of Fig. 8, equation (76) provides the following conditions

$$\begin{aligned} \psi'(s) &= +\lambda\sqrt{2\left(\cos\psi(s) - \cos\psi^*\right)} \quad \forall\, s \in (0, l^*)\,, \\ \psi'(s) &= -\lambda\sqrt{2\left(\cos\psi(s) - \cos\psi^*\right)} \quad \forall\, s \in (l^*, l/2)\,. \end{aligned} \tag{77}$$

It is a standard expedient to operate the following change of variables

$$\eta = \sin\frac{\psi^*}{2}, \qquad \eta\sin\omega(s) = \sin\frac{\psi(s)}{2}, \tag{78}$$

leading to the following differential equation

$$\frac{\mathrm{d}\,\omega(s)}{\mathrm{d}s} = \pm\gamma\sqrt{1 - \eta^2\sin^2\omega(s)}. \tag{79}$$

Furthermore, the boundary conditions imply that

$$\omega(0) = \omega_\beta, \quad \omega(l^*) = \frac{\pi}{2}, \quad \omega\left(l/2\right) = -\frac{\pi}{2}, \tag{80}$$

where $\omega_\beta = \arcsin\left(\frac{1}{\eta}\sin\left(\frac{\beta}{2}\right)\right)$.

Therefore, separation of the variables and integration of the positive root of equation (79) between 0 and l^* yields

$$\int_0^{l^*} \gamma \mathrm{d}s = \int_{\omega_\beta}^{\frac{\pi}{2}} \frac{\mathrm{d}\omega}{\sqrt{1 - \eta^2\sin^2\omega}}, \tag{81}$$

which expresses the non-trivial solution for l^*, related to η, γ and β as

$$l^*\gamma = \mathcal{K}(\eta) - \mathcal{K}\left(\omega_\beta, \eta\right), \tag{82}$$

where

$$\mathcal{K}(x, \eta) = \int_0^{x} \frac{\mathrm{d}\omega}{\sqrt{1 - \eta^2\sin^2\omega}}, \tag{83}$$

is the incomplete elliptic integral of the first kind. Now, let us integrate the negative root of equation (79) between l^* and $l/2$. By using the Riemann theorem and the following property for the integral of even functions

$$-\int_{\frac{\pi}{2}}^{-\frac{\pi}{2}} \frac{\mathrm{d}\omega}{\sqrt{1 - \eta^2\sin^2\omega}} = 2\int_0^{\frac{\pi}{2}} \frac{\mathrm{d}\omega}{\sqrt{1 - \eta^2\sin^2\omega}} \tag{84}$$

we obtain

$$\gamma\left(\frac{l}{2}-l^*\right)=2\mathcal{K}(\eta), \tag{85}$$

so that, considering the periodicity of the boundary conditions (80) and using equations (82) and (85), we arrive at

$$\gamma l = 2\left[(m+1)\mathcal{K}(\eta)-\mathcal{K}\left(\omega_\beta,\eta\right)\right] \qquad m=2,4,6,... \tag{86}$$

Note that equation (86) provides the relation between the external load P and the angles θ^* (through η and ψ^*) and β associated with the even, antisymmetric m–th buckled mode. The relation above provides one of the two equations to be solved for the problem under consideration. The other relation that allows for the solution of the problem can be obtained from equation $(72)_3$, which can be rewritten as

$$\int_0^{l^*}\sin\left(\psi(s)-\beta\right)\mathrm{d}s+\int_{l^*}^{l/2}\sin\left(\psi(s)-\beta\right)\mathrm{d}s=0, \tag{87}$$

so that, considering equations (78), we obtain

$$\begin{aligned}&\int_\beta^{\psi^*}\frac{\sin\psi\cos\beta}{\gamma\sqrt{2(\cos\psi-\cos\psi^*)}}\mathrm{d}\psi-\int_\beta^{\psi^*}\frac{\sin\beta\cos\psi}{\gamma\sqrt{2(\cos\psi-\cos\psi^*)}}\mathrm{d}\psi+\\&-\int_{\psi^*}^{-\psi^*}\frac{\sin\psi\cos\beta}{\gamma\sqrt{2(\cos\psi-\cos\psi^*)}}\mathrm{d}\psi+\int_{\psi^*}^{-\psi^*}\frac{\sin\beta\cos\psi}{\gamma\sqrt{2(\cos\psi-\cos\psi^*)}}\mathrm{d}\psi=0.\end{aligned} \tag{88}$$

The third integral in equation (88) is null (in fact the integrand is an odd function of ψ), whereas the other terms can be rewritten, by using equation (78) and by exploiting the following relations

$$\cos\psi=1-2\sin^2(\psi/2),\qquad \sin\psi=2\sin(\psi/2)\sqrt{1-\sin^2(\psi/2)},$$

in the final form (Mikata, 2007)

$$\begin{aligned}&-2\eta\cos\omega_\beta\left(1-2\eta^2\sin^2\omega_\beta\right)+2\eta\sin\omega_\beta\sqrt{1-\eta^2\sin^2\omega_\beta}\Big\{(m+1)\big[2E(\eta)+\\&-\mathcal{K}(\eta)\big]-\left[2E(\omega_\beta,\eta)-\mathcal{K}(\omega_\beta,\eta)\right]\Big\}=0,\qquad m=2,4,6,...\end{aligned} \tag{89}$$

Equations (86) and (89) are highly non-linear and allow for the determination of the nontrivial, antisymmetric equilibrium solution. The relation between β and θ^* (the former contained in ω_β and the latter in η) can be numerically obtained from equation (89) and used into equation (86) to obtain the relation $P - \theta^*$.

Our aim is now to compute the deformed shape of the buckled rod. To this purpose, we first integrate equation (79) from $s = 0$ to a generic point on the left of the inflection $s = l^*$

$$\int_0^s \gamma \mathrm{d}s = \int_{\omega_\beta}^{\omega(s)} \frac{\mathrm{d}\omega}{\sqrt{1 - \eta^2 \sin^2 \omega}}. \tag{90}$$

We then consider the decomposition (41), equation $(78)_2$, and the following properties of the elliptic function 'sn' (Byrd and Friedman, 1954)

$$\mathrm{sn}(-x + 2\mathcal{K}(\eta), \eta) = -\mathrm{sn}(-x, \eta) = \mathrm{sn}(x, \eta), \tag{91}$$

such that we obtain

$$\sin\left(\frac{\omega}{2}\right) = \eta\, \mathrm{sn}\left[\gamma s + \mathcal{K}(\omega_\beta, \eta), \eta\right] \quad \forall s \in [0, l]. \tag{92}$$

Finally, integration of the kinematic fields (11) provides the analytical expressions for the deformed shape of the rod, holding for $s \in (0, l)$, as[7]

[7]Equation (95) holds also for the structure (v.), together with

$$\gamma l = (2m + 1)\mathcal{K}(\eta) - \mathcal{K}\left(\omega_\beta, \eta\right), \qquad m = 1, 2, 3, ... \tag{93}$$

and

$$-2\eta \cos\omega_\beta \left(1 - 2\eta^2 \sin^2 \omega_\beta\right)$$

$$+2\eta \sin\omega_\beta \sqrt{1 - \eta^2 \sin^2 \omega_\beta} \left\{(2m+1)\left[2E(\eta) - K(\eta)\right] - \left[2E(\omega_\beta, \eta) - K(\omega_\beta, \eta)\right]\right\} = 0, \tag{94}$$

defining the relation between P, β and θ^* associated with the m–th bifurcation mode.

$$
\begin{aligned}
x_1(s) = &+\sin\beta\left[-\frac{2\eta}{\gamma}\mathrm{cn}\big(\gamma s+\mathcal{K}(\omega_\beta,\eta),\eta\big)+\frac{2\eta}{\gamma}\mathrm{cn}\big(\mathcal{K}(\omega_\beta,\eta),\eta\big)\right]\\
&\cos\beta\Bigg\{-s+\frac{2}{\gamma}\big[E\left[\mathrm{am}\left(\gamma s+\mathcal{K}(\omega_\beta,\eta),\eta\right),\eta\right]\\
&-E\left[\mathrm{am}\left(\mathcal{K}(\omega_\beta,\eta),\eta\right),\eta\right]\big]\Bigg\},\\
x_2(s) = &\cos\beta\left[-\frac{2\eta}{\gamma}\mathrm{cn}\big(\gamma s+\mathcal{K}(\omega_\beta,\eta),\eta\big)+\frac{2\eta}{\gamma}\mathrm{cn}\big(\mathcal{K}(\omega_\beta,\eta),\eta\big)\right]\\
&-\sin\beta\Bigg\{-s+\frac{2}{\gamma}\big[E\left[\mathrm{am}\left(\gamma s+\mathcal{K}(\omega_\beta,\eta),\eta\right),\eta\right]\\
&-E\left[\mathrm{am}\left(\mathcal{K}(\omega_\beta,\eta),\eta\right),\eta\right]\big]\Bigg\}.
\end{aligned}
\tag{95}
$$

The displacement of the movable, clamped end, where the thrust is applied, can be obtained from equation (95), since $|u_1(l)| = l - x_1(l)$. For a given θ^* and a given even mode m, we can obtain β (through equation (89)), P (using equations (86) and (73)) and $|u_1(l)|$ (through equation $(95)_1$). The bifurcation diagram showing the load P (made dimensionless through multiplication by $l^2/(\pi^2 B)$) as a function of the displacement of the right movable end of the rod (divided by l) is shown in Fig. 9, together with the bifurcation diagram valid for the first and third symmetric buckling mode, see equation (71).

In-plane secondary bifurcation of the doubly clamped elastica It can be noticed from Fig. 9 that the doubly clamped elastica, i.e. the structure (vi.), exhibits a secondary bifurcation (marked with a circle) occurring when the two ends of the rod coincide, namely, when $\hat{\theta} = 130.7099°$ [8]. Under this condition, an '8-shaped' unstable equilibrium configuration is realized by the rod (see the sketches in Fig. 10) and vertical reactions are generated at its ends. By further increasing the external thrust P, the rod snaps onto the stable, 'S-shaped' configuration corresponding to the antisymmetric second mode (Domokos, 1994), see also Fig. 12. The stability of the equilibrium configurations will be addressed in the next section.

[8] The angle $\hat{\theta} = 130.7099°$ corresponds to the value $\kappa = 0.9089$, which is the root of the equation $2E(\kappa) - \mathcal{K}(\kappa) = 0$.

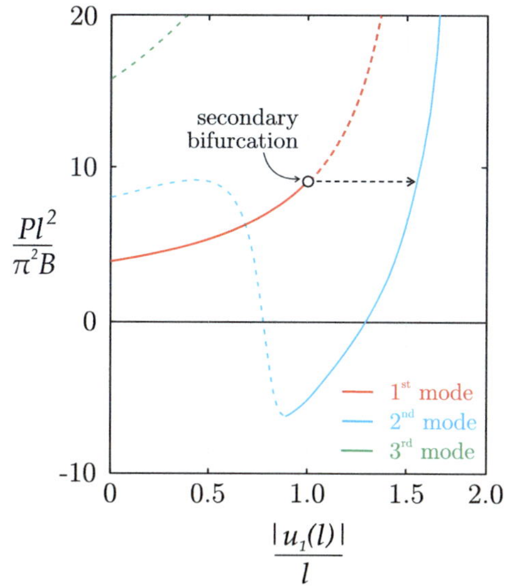

Figure 9. Dimensionless load $Pl^2/(\pi^2 B)$ for a doubly clamped rod versus the dimensionless displacement $|u_1(l)|/l$. The first three bifurcation points and branches are reported together with the secondary bifurcation point. The solid lines represent stable equilibrium configurations, while dashed curves correspond to unstable configurations. Note that the first branch becomes unstable after the secondary bifurcation point.

2.3 Stability of the elastica

The stability of the elastica equilibrium configurations can be judged by analyzing the sign of the second variation of the total potential energy $\mathcal{V}$ with respect to variations $\tilde{\theta}(s)$ compatible with the boundary conditions reported in equation (22), plus the supplementary condition

$$\int_0^l \tilde{\theta}(s) \cos\theta(s)\, \mathrm{d}s = 0, \tag{96}$$

following from the vanishing of the integral constraint in equations $(22)_1$, $(22)_5$, and $(22)_6$ and holding for the systems (i.), (v.) and (vi.), in which both the ends of the rod are constrained to null vertical displacements.

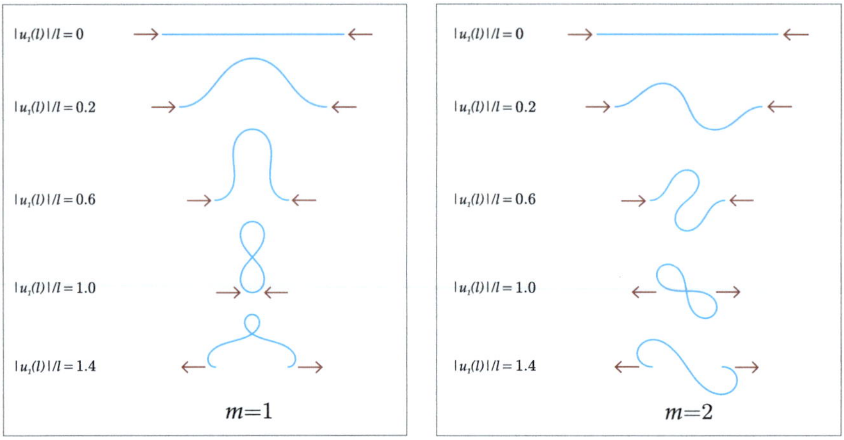

Figure 10. Deformed elastic lines for the first two modes $m = 1, 2$ (symmetric and antisymmetric) at different values of dimensionless displacements $|u_1(l)|/l = \{0, 0.2, 0.6, 1.0, 1.4\}$. The deformed shapes of the elastica represent the post-critical behaviour of the structure.

The second variation of the functional $\mathcal{V}$ can be computed as

$$\delta^2\mathcal{V} = \int_0^l \left[\left(\tilde{\theta}'(s)\right)^2 - \frac{P}{B}\tilde{\theta}^2(s)\cos\theta(s) + \frac{R}{B}\tilde{\theta}^2(s)\sin\theta(s)\right] \mathrm{d}s, \tag{97}$$

an equation that, using integration by parts

$$\int_0^l \left(\tilde{\theta}'(s)\right)^2 \mathrm{d}s = -\int_0^l \tilde{\theta}''(s)\tilde{\theta}(s)\,\mathrm{d}s,$$

yields

$$\delta^2\mathcal{V} = -\int_0^l \left[\tilde{\theta}''(s) + \frac{P}{B}\tilde{\theta}(s)\cos\theta(s) - \frac{R}{B}\tilde{\theta}(s)\sin\theta(s)\right]\tilde{\theta}(s)\,\mathrm{d}s, \tag{98}$$

for all kinematically admissible rotation fields $\tilde{\theta}(s)$, so that the *stability criterion* reads

$$\delta^2\mathcal{V}\begin{cases} > 0 & \text{stability,} \\ < 0 & \text{instability.}\end{cases} \tag{99}$$

In order to treat condition (98), let us denote with $\phi_n(s)$ (where $s \in [0, l]$) the non-trivial solutions of the following Sturm-Liouville problem[9]

$$\phi_n''(s) + \delta_n \left(\frac{P}{B} \cos\theta(s) - \frac{R}{B} \sin\theta(s) \right) \phi_n(s) = C_{Rn} \cos\theta(s), \qquad (101)$$

subject to the following constraints representing the boundary conditions (21)

$$\begin{array}{ll} \text{i)} & \phi_n'(0) = \phi_n'(l) = 0 \quad \text{and} \quad \int_0^l \cos\theta(s)\phi_n(s)\mathrm{d}s = 0, \\ \text{ii)} & \phi_n'(0) = \phi_n(l) = 0, \\ \text{iii)} & \phi_n(0) = \phi_n'(l) = 0, \\ \text{iv)} & \phi_n(0) = \phi_n(l) = 0, \\ \text{v)} & \phi_n(0) = \phi_n'(l) = 0 \quad \text{and} \quad \int_0^l \cos\theta(s)\phi_n(s)\mathrm{d}s = 0, \\ \text{vi)} & \phi_n(0) = \phi_n(l) = 0 \quad \text{and} \quad \int_0^l \cos\theta(s)\phi_n(s)\mathrm{d}s = 0. \end{array} \qquad (102)$$

In the equations above, $\phi_n(s)$ are the eigenfunctions associated to the eigenvalues δ_n with weight function

$$\frac{P}{B} \cos\theta(s) - \frac{R}{B} \sin\theta(s), \qquad (103)$$

and C_{Rn} is a constant that represents the isoperimetric constraint (96) and is 'linked' to R, so that for the unconstrained systems (for the systems (ii.), (iii.), (iv.), and (i.), but the last except when the two supports coincide) $C_{Rn} = 0$.

It is known (Broman, 1970) that: (i.) problem (101) admits a countably infinite set of eigenvalues δ_n and these can be arranged in an increasing sequence ($\delta_n < \delta_{n+1}$ for each integer n)[10], (ii.) $\delta_n \longrightarrow \infty$ when $n \longrightarrow \infty$,

[9] When isoperimetric constraints are not present, namely condition (96) is absent, stability can also be addressed by evaluating the eigenvalues δ_n of the following Sturm-Liouville problem

$$-\phi_n''(s) - \left(\frac{P}{B} \cos\theta(s) - \frac{R}{B} \sin\theta(s) \right) \phi_n(s) = \delta_n \phi_n(s). \qquad (100)$$

When at least one eigenvalue is negative, the associated equilibrium configuration is *unstable*, while, if all eigenvalues are positive, the equilibrium is *stable*, see Manning et al. (1998) and Hoffman et al. (2002).

[10] Note in addition that $\delta_n \geq 0$, see Broman (1970), pag. 41 – 44.

(iii.) the system $\phi_n(s)$ is an orthogonal system with the weight function (103). Multiplication of the differential equation (101) by ϕ_n and integration between 0 and l yields (keeping into account the boundary conditions (102) and using integration by parts)

$$\delta_n \int_0^l \left(\frac{P}{B}\cos\theta(s) - \frac{R}{B}\sin\theta(s)\right)\phi_n^2(s)\,\mathrm{d}s = \int_0^l (\phi_n^{'}(s))^2\,\mathrm{d}s, \tag{104}$$

while writing equation (101) for the eigenfunction ϕ_m, that is

$$\phi_m^{''}(s) + \delta_m\left(\frac{P}{B}\cos\theta(s) - \frac{R}{B}\sin\theta(s)\right)\phi_m(s) = C_{Rm}\cos\theta(s), \tag{105}$$

and combining equations (101) and (105) multiplied by ϕ_m and ϕ_n, respectively, integration between 0 and l (taking into account the boundary conditions) provides the following orthogonality condition

$$\int_0^l \phi_n(s)\phi_m(s)\left(\frac{P}{B}\cos\theta(s) - \frac{R}{B}\sin\theta(s)\right)\mathrm{d}s = 0, \quad n \neq m. \tag{106}$$

Condition (104) defines a norm and equation (106) a weighted orthogonality condition for the functions $\phi_n(s)$ with weight function (103).

Therefore, the system $\phi_n(s)$ with weight function (103) can be used to provide a Fourier series representation (converging in the mean) to the square-integrable function $\tilde{\theta}(s)$,

$$\tilde{\theta}(s) = \sum_{n=1}^{\infty} c_n\phi_n(s), \tag{107}$$

where c_n are the Fourier coefficients.

We do not need to specify the coefficients c_n, rather we can simply substitute the Fourier representation (107) into condition (98) and keep into consideration equation (101) to obtain

$$\begin{aligned}\delta^2\mathcal{V} = \int_0^l &\left[\sum_{n=1}^{\infty}(\delta_n - 1)c_n\phi_n(s)\left(\frac{P}{B}\cos\theta(s) - \frac{R}{B}\sin\theta(s)\right)\right] \\ &\times\left[\sum_{m=1}^{\infty} c_m\phi_m(s)\right]\mathrm{d}s,\end{aligned} \tag{108}$$

which, employing conditions (104) and (106) finally becomes

$$\delta^2\mathcal{V} = \sum_{n=1}^{\infty}(1-\frac{1}{\delta_n})c_n^2 \int_0^l (\phi_n'(s))^2\,\mathrm{d}s \quad \begin{cases} >0 & \text{stability,} \\ <0 & \text{instability,} \end{cases} \tag{109}$$

so that we arrive at the following stability requirement

$$\begin{cases} \delta_n \notin [0,1] & \text{stability,} \\ \delta_n \in [0,1] & \text{instability,} \end{cases} \tag{110}$$

where δ_n are solutions of the Sturm-Liouville problem (101). The values $\delta_n = 0$ or $\delta_n = 1$ represent 'transition' points and thus are called 'critical'.

Stability of the simply supported elastica Let us consider first the stability of the straight configuration of a simply supported rod, such that $\cos\theta(s) = 1$. In this case, $R = C_{Rn} = 0$ and the Sturm-Liouville problem, equation (101), becomes

$$\phi_n''(s) + \delta_n \frac{P}{B}\phi_n(s) = 0, \quad \phi_n'(0) = \phi_n'(l) = 0, \tag{111}$$

which has the nontrivial solutions

$$\phi_n(s) = \cos\frac{n\pi s}{l}, \quad \delta_n = \frac{P_n^{cr}}{P}, \tag{112}$$

where P_n^{cr} is the Euler's critical load corresponding to the n-th buckling mode, so that when $\delta_1 < 0$ or $\delta_1 > 1$ ($0 < \delta_1 < 1$) the straight configuration is stable (unstable) and $P < P_1^{cr}$ ($P > P_1^{cr}$)[11].

To judge the stability of the deformed configurations, we can substitute equation (40) into equation (50) to obtain

$$\cos\theta(s) = 1 - 2\kappa^2\,\mathrm{sn}^2\left[\left(\frac{s}{l}2m+1\right)\mathcal{K}(\kappa),\kappa\right], \tag{113}$$

[11] This statement hold also true for other structures when $R = 0$, namely systems (ii.), (iii.) and (iv.), whose nontrivial solutions are

ii) $\phi_n(s) = \cos\dfrac{\pi + 2n\pi}{2}\dfrac{s}{l}$,

iii) $\phi_n(s) = \sin\dfrac{\pi + 2n\pi}{2}\dfrac{s}{l}$,

iv) $\phi_n(s) = \sin\dfrac{n\pi s}{l}$,

with $\delta_n = \dfrac{P_n^{cr}}{P}$.

and rewrite the Sturm-Liouville problem (101) using the dimensionless variable $\tilde{s} = s/l \in [0, 1]$, so that, for a given mode m and rotation of the rod end $\kappa = \sin(\hat{\theta}/2)$, the *smallest eigenvalue* δ_m has to be determined as solution of

$$\phi_m''(\tilde{s}) + \delta_m \, 4m^2 \mathcal{K}^2(\kappa) \left\{1 - 2\kappa^2 \, \mathrm{sn}^2 \left[(2m \, \tilde{s} + 1) \, \mathcal{K}(\kappa), \kappa\right]\right\} \, \phi_m(\tilde{s}) = 0, \quad (114)$$

(where now a prime denotes differentiation with respect to $\tilde{s}$) subject to the boundary conditions $(101)_2$, namely, $\phi_m'(0) = \phi_m'(1) = 0$.

The solutions of equation (114) can be easily computed by means of a numerical routine. As noticed by Kuznetsov and Levyakov (2002), a numerical procedure can be easily set, solving the differential equation (114) with the boundary conditions

$$\phi_m(0) = 1, \quad \phi_m'(0) = 0, \quad (115)$$

so that the trivial solution is always eliminated and iterations on δ_m can be performed to match the condition $\phi_m'(1) = 0$[12]. A numerical integration and iterations performed on the basis of a bisection method yield the graphs reported in Fig. 11.

The smallest eigenvalues δ_m for the first three modes $m = 1, 2, 3$ are reported versus the inclination $\hat{\theta}$ (in degrees) of the ends of the deformed rod. It is clear that the first mode, $m = 1$, is stable (the eigenvalues range between 1 and 10) until the two supports coincide for $\hat{\theta} = 130.7099°$, at which point the eigenvalues become discontinuous and fall to values within $[0, 1]$ (in particular, $\delta_1 = 9.9228$ at $\hat{\theta} = 130.7°$ and $\delta_1 = 0.0059$ at $\hat{\theta} = 130.8°$). All modes higher than the first ($m = 2, 3$) are unstable with eigenvalues belonging to $[0, 1]$. Note that for the considered modes, all the eigenvalues coincide when $\hat{\theta} \geq 130.7099°$.

With respect to the unstable configuration that occurs at $\hat{\theta} > 130.7099°$ for $m = 1$ (Maddocks, 1984), the instability mode and the corresponding 'self-intersecting' elastica is illustrated through an experiment in Bigoni (2012).

[12] The nontrivial solution of the Sturm-Liouville problem (101) for the systems (ii.), (iii.) and (iv.) can be numerically found using the following boundary conditions

ii) $\phi_m'(0) = 0, \quad \phi_m(l) = 0, \quad \phi_m(0) = 1,$

iii) $\phi_m(0) = 0, \quad \phi_m'(l) = 0, \quad \phi_m(l) = 1,$

iv) $\phi_m(0) = 0, \quad \phi_m(l) = 0, \quad \phi_m'(0) = 1.$

In all these cases, the first mode, $m = 1$, is stable (all eigenvalues are external to $[0, 1]$), while the higher modes $m = 2, 3$ are both unstable with at least one eigenvalue belonging to the interval [0,1].

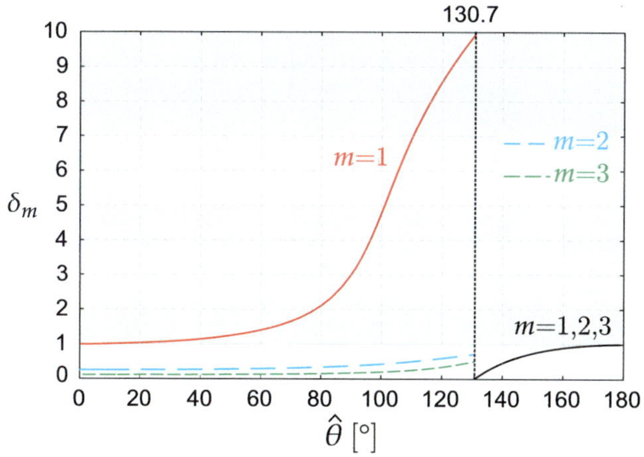

Figure 11. The smallest eigenvalues δ_m of the Sturm-Liouville problem (114) as functions of the rotation $\hat{\theta}$ of the ends of the simply supported rod. These determine the stability of the different modes of the elastica ($m = 1, 2, 3$ are investigated). The light gray region corresponds to stability, so that only the first mode $m = 1$ is stable and only until the two supports of the rod coincide, a situation corresponding to $\hat{\theta} = 130.7099°$.

We finally note that, with the proposed procedure to check stability, it is not directly possible to conclude that *all* the modes with $m > 1$ are unstable, though physical intuition suggests that this might be the case. In fact, our stability study for the modes $m = 2, 3$ substantiates the Love's (Love, 1927) statement according to which 'the instability of forms of the elastica with more than the smallest possible number of inflexions between the ends is well known as an experimental fact'.

Stability of the doubly clamped elastica We move now to the analysis of a doubly clamped rod: system (vi.) in Fig. 2. Stability of the straight configuration can be analyzed by noting that $\sin\theta(s) = 0$ and $\cos\theta(s) = 1$, so that the Sturm-Liouville problem (101) becomes

$$\phi_n''(s) + \delta_n \frac{P}{B}\phi_n(s) = C_{Rn}, \quad \phi_n(0) = \phi_n(l) = 0, \quad \int_0^l \phi_n(s)\mathrm{d}s = 0, \quad (116)$$

and admits solutions in the following form

$$\phi_n(s) = A_1 \cos\left(\sqrt{\frac{\delta_n P}{B}}\, s\right) + A_2 \sin\left(\sqrt{\frac{\delta_n P}{B}}\, s\right) + \frac{A_3 B}{\delta_n P}, \qquad (117)$$

where A_1, A_2, and A_3 are constants. By substitution of equation (117) into the boundary conditions of the Sturm-Liouville problem (116), we arrive at a system of homogeneous equations for A_1, A_2 and A_3 which can be written in the following matrix form

$$\begin{bmatrix} \dfrac{\delta_n P}{B} & 0 & 1 \\ \cos\left(\sqrt{\dfrac{\delta_n P}{B}}\,l\right) & \sin\left(\sqrt{\dfrac{\delta_n P}{B}}\,l\right) & \dfrac{B}{\delta_n P} \\ \dfrac{\delta_n P}{B}\sin\left(\sqrt{\dfrac{\delta_n P}{B}}\,l\right) & \dfrac{\delta_n P}{B}\cos\left(\sqrt{\dfrac{\delta_n P}{B}}\,l\right) & l \end{bmatrix} \begin{bmatrix} A_1 \\ A_2 \\ A_3 \end{bmatrix} = \begin{bmatrix} 0 \\ 0 \\ 0 \end{bmatrix}. \tag{118}$$

Therefore, nontrivial solutions of the system (117) can be obtained by imposing the vanishing of the determinant of its coefficient matrix. This leads to the following characteristic equation

$$2\left[\cos\left(\sqrt{\frac{\delta_n P}{B}}\,l\right) - 1\right] + \sqrt{\frac{\delta_n P}{B}}\,l \sin\left(\sqrt{\frac{\delta_n P}{B}}\,l\right) = 0, \tag{119}$$

which can be numerically solved for δ_n to show that the straight, natural configuration is stable only for $P < P_1^{cr}$. In fact, when $P > P_1^{cr}$ at least one eigenvalue δ_n belongs to the interval $[0, 1]$, and therefore the trivial configuration becomes unstable.

We explore in what follows the stability of the nontrivial equilibrium configurations and, to that purpose, we make use of the numerical procedure proposed by Levyakov and Kuznetsov (2010) while retaining as much as possible the notation employed by those authors[13]. We introduce the dimensionless arc-length $\tilde{s} = s/l \in [0, 1]$, such that the Sturm-Liouville problem (101) for a given mode m can be rewritten as

$$\phi_m''(\tilde{s}) + \delta_m\, L(\tilde{s})\, \phi_m(\tilde{s}) = C_{Rm}\, N(\tilde{s}), \tag{120}$$

where

$$L(\tilde{s}) = \begin{cases} \dfrac{P}{B}\cos\theta(\tilde{s}), & \text{for odd modes} \\ \dfrac{P}{B}\cos\theta(\tilde{s}) - \dfrac{R}{B}\sin\theta(\tilde{s}), & \text{for even modes} \end{cases} \tag{121}$$

[13] The analysis of stability with isoperimetric constraints can be also performed following Manning et al. (1998), Hoffman et al. (2002) and Manning (2009, 2014) through the conjugate point theory.

and $N(\tilde{s}) = \cos\theta(\tilde{s})$. Notice that, for the odd buckling modes with $2m$ inflections, equation $(68)_1$ can be substituted into the definition of $\cos\theta(s)$ to obtain

$$\cos\theta(\tilde{s}) = 1 - 2\kappa^2 \operatorname{sn}^2\left[(2(m+1)\tilde{s})\,\mathcal{K}(\kappa), \kappa\right], \qquad m = 1, 3, 5, \tag{122}$$

which is a function of the rotation $\hat{\theta}$ of the rod at the inflection point.

For the even buckling modes with $2m+1$ inflections, we can write

$$\begin{aligned}\cos\theta(\tilde{s}) &= \sin\beta\left[2\eta \operatorname{sn}\left[\gamma\tilde{s} + \mathcal{K}(\omega_\beta, \eta), \eta\right] \operatorname{dn}\left[\gamma\tilde{s} + \mathcal{K}(\omega_\beta, \eta), \eta\right]\right] \\ &\quad + \cos\beta\left[1 - 2\eta^2 \operatorname{sn}^2\left[\gamma\tilde{s} + \mathcal{K}(\omega_\beta, \eta), \eta\right]\right], \\ \sin\theta(\tilde{s}) &= \cos\beta\left[2\eta \operatorname{sn}\left[\gamma\tilde{s} + \mathcal{K}(\omega_\beta, \eta), \eta\right] \operatorname{dn}\left[\gamma\tilde{s} + \mathcal{K}(\omega_\beta, \eta), \eta\right]\right] \\ &\quad - \sin\beta\left[1 - 2\eta^2 \operatorname{sn}^2\left[\gamma\tilde{s} + \mathcal{K}(\omega_\beta, \eta), \eta\right]\right],\end{aligned} \tag{123}$$

where β, γ, P and R can be computed through equations (89), (86) and (73) as functions of the rotation θ^* of the rod at inflection point. Notice also that equations (120) are subjected to the boundary conditions $\phi_m(0) = 0$, $\phi_m(1) = 0$[14] and to the additional constraint

$$\int_0^1 \phi_m(\tilde{s}) N(\tilde{s}) \mathrm{d}\tilde{s} = 0. \tag{124}$$

The numerical procedure to compute the eigenvalues δ_m consists in partitioning the interval $0 \leq \tilde{s} \leq 1$ into n segments of equal length $h = n^{-1}$, such that the coordinates of the starting points of the segments can be denoted by $s_{i-1} = h(i-1)$, with $i = 1, ..., n$. For the i-th segment, the functions $L(\tilde{s})$ and $N(\tilde{s})$ are approximated by their average values L_i and N_i as computed at the midpoint of the segment, so that equation (120) becomes

$$\phi''_m(\tilde{s}) + \delta_m \, L_i \, \phi_m(\tilde{s}) = C_{Rm} \, N_i, \tag{125}$$

an ordinary differential equation with constant coefficients, which solution is

$$\phi_m(\tilde{s}) = A_{1i} F_{1i}(\tilde{s} - \tilde{s}_{i-1}) + A_{2i} F_{2i}(\tilde{s} - \tilde{s}_{i-1}) + C_{Rm} \frac{N_i}{\delta_n \, L_i}. \tag{126}$$

In equation (126), A_{1i} and A_{2i} are constants, whereas the coefficients F_{1i} and F_{2i} are defined for $\delta_m L_i > 0$ as

$$F_{1i}(\tilde{s} - \tilde{s}_{i-1}) = \cos\, a_i(\tilde{s} - \tilde{s}_{i-1}), \quad F_{2i}(\tilde{s} - \tilde{s}_{i-1}) = \sin\, a_i(\tilde{s} - \tilde{s}_{i-1}), \tag{127}$$

[14]For the system (v.) the boundary conditions are $\phi_m(0) = \phi'_m(1) = 0$.

while for $\delta_m L_i < 0$ they are

$$F_{1i}(\tilde{s}-\tilde{s}_{i-1}) = \cosh a_i(\tilde{s}-\tilde{s}_{i-1}), \quad F_{2i}(\tilde{s}-\tilde{s}_{i-1}) = \sinh a_i(\tilde{s}-\tilde{s}_{i-1}), \tag{128}$$

with $a_i = \sqrt{|\delta_m L_i|}$. Furthermore, the constants A_{1i} and A_{2i} can be expressed by means of $\phi_m(\tilde{s}_{i-1}) = \phi_{m,i-1}$ and by means of $\phi_m^{'}(\tilde{s}_{i-1}) = \phi_{m,i-1}^{'}$ as

$$A_{1i} = \phi_{m,i-1} - C_{Rm}\frac{N_i}{\delta_m L_i}, \qquad A_{2i} = \frac{\phi_{m,i-1}^{'}}{a_i}, \tag{129}$$

so that the quantity $\phi_m(\tilde{s}_i) = \phi_{m,i}$ at the right end of the segment is computed as

$$\phi_{m,i} = \phi_{m,i-1}\, F_{1i}(h) + \phi_{m,i-1}^{'}\frac{F_{2i}(h)}{a_i} + C_{Rm}N_i\frac{1-F_{1i}(h)}{\delta_m L_i}, \tag{130}$$

and its derivative reads

$$\phi_{m,i}^{'} = \phi_{m,i-1}\, F_{1i}^{'}(h) + \phi_{m,i-1}^{'}\frac{F_{2i}^{'}(h)}{a_i} - C_{Rm}N_i\frac{F_{1i}^{'}(h)}{\delta_m L_i}. \tag{131}$$

The general solution of equation (125) can now be constructed by using equations (130)-(131) and the condition of continuity of ϕ_m and $\phi_m^{'}$ at the extremities of every integration segment. Since equation (125) is linear, its general solution can be written as a combination of three particular solutions, that is

$$\phi_m(\tilde{s}) = c_1\varphi_1(\tilde{s}) + c_2\varphi_2(\tilde{s}) + C_{Rm}\varphi_3(\tilde{s}), \tag{132}$$

where c_1 and c_2 are constants. Using the following initial data

$$\begin{array}{lll} \varphi_1(0) = 1, & \varphi_1^{'}(0) = 0, & C_{Rm} = 0, \\ \varphi_2(0) = 0, & \varphi_2^{'}(0) = 1, & C_{Rm} = 0, \\ \varphi_3(0) = 0, & \varphi_3^{'}(0) = 0, & C_{Rm} = 1, \end{array} \tag{133}$$

the functions $\varphi_i(\tilde{s})$ $(i = 1, 2, 3)$ can be constructed separately using the recurrence relations (130) and (131). Note that the particular solutions $\varphi_i(\tilde{s})$ can also be obtained by numerically solving the differential equation (120) –where $\phi_m(\tilde{s})$ is replaced by $\varphi_i(\tilde{s})$– through the function NDSolve of Mathematica© and imposing the boundary conditions (133).

By substituting the general solution (132) into the relation (124), the constraint can be rewritten as

$$\int_0^1 \left[c_1\varphi_1(\tilde{s}) + c_2\varphi_2(\tilde{s}) + C_{Rm}\varphi_3(\tilde{s})\right] N(\tilde{s})\mathrm{d}\tilde{s} = 0, \tag{134}$$

while the boundary conditions $\phi_m(0) = 0$ and $\phi_m(1) = 0$ provide the following relations

$$\begin{aligned} &c_1 = 0, \\ &c_1\varphi_1(1) + c_2\varphi_2(1) + C_{Rm}\varphi_3(1) = 0. \end{aligned} \tag{135}$$

Equations (134) and $(135)_2$ represent an homogeneous system of algebraic equations for c_2 and C_{Rm}, which can be rewritten in matrix form as

$$\begin{bmatrix} \varphi_2(1) & \varphi_3(1) \\ \int_0^1 \varphi_2(\tilde{s})N(\tilde{s})\mathrm{d}\tilde{s} & \int_0^1 \varphi_3(\tilde{s})N(\tilde{s})\mathrm{d}\tilde{s} \end{bmatrix} \begin{bmatrix} c_2 \\ C_{Rm} \end{bmatrix} = \begin{bmatrix} 0 \\ 0 \end{bmatrix}. \tag{136}$$

Nontrivial solutions of the system (136) can be obtained by imposing the vanishing of the coefficient matrix's determinant Δ, so that, by varying the eigenvalue δ_m from 0 to 1, we can study the behaviour of Δ. When at least one value of $\delta_m \in [0, 1]$ exists such that the determinant vanishes, the non trivial solution of (132) satisfies the boundary conditions, so that the corresponding equilibrium configuration is *unstable*. Therefore, the stability of the equilibrium configurations of the doubly clamped elastica can be judged by analyzing the determinant Δ, which becomes[15]

$$\Delta = \varphi_2(1)\int_0^1 \varphi_3(\tilde{s})N(\tilde{s})\mathrm{d}\tilde{s} - \varphi_3(1)\int_0^1 \varphi_2(\tilde{s})N(\tilde{s})\mathrm{d}\tilde{s}, \tag{137}$$

where the integrals have to be numerically computed. The above mentioned numerical procedure arrives at the results reported in Fig. 9, where a stable (unstable) equilibrium path is depicted by a solid (dashed) curve. Note the secondary bifurcation point that is highlighted in the figure by a circle along the red path: once the critical configuration is attained such that the two ends of the rod touch, the elastic system snaps from the fist into the second mode.

In order to better understand the instability mode that takes place for $m = 1$ and $\hat{\theta} > 130.7099°$, a qualitative experiment was performed with a beam model made up of a PMMA strip of overall dimensions 490 mm ×

[15] For the elastic system (v.), the determinant of the coefficient matrix is $\Delta = \varphi_2'(1)\int_0^1 \varphi_3(\tilde{s})N(\tilde{s})\mathrm{d}\tilde{s} - \varphi_3'(1)\int_0^1 \varphi_2(\tilde{s})N(\tilde{s})\mathrm{d}\tilde{s}$.

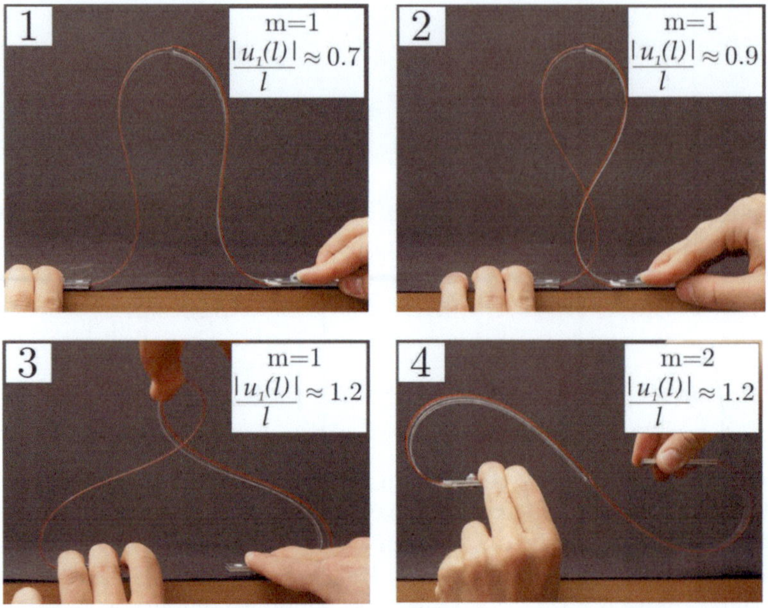

Figure 12. An experiment documenting the stability predictions obtained for the doubly clamped elastica. Configurations (1) and (2) of the first mode are stable for $|u_1(l)|/l < 1$; the photo (3) shows an unstable self-intersecting configuration of the first symmetric mode beyond the secondary bifurcation point (note that this configuration is held in position by hand, otherwise the structure would snap to configuration (4)). The configuration realized in (4) for the second 'S-shaped' antisymmetric mode is stable when $|u_1(l)|/l > 1$.

25 mm × 1.5 mm. The realization of the corresponding 'self-intersecting' configurations of the elastica was made possible by providing the model with a longitudinal 13 mm wide cut, dividing the strip into two parts (one 12 mm wide and the other '∩-shaped' with each of the two legs 6 mm wide). Four configurations of the model are reported in Fig. 12, fully confirming the theoretical expectations.

3 Buckling under tensile dead loading

Buckling of elastic structures is known from the ancient times[16] and is typically associated to compressive forces[17]. Examples have been provided of elastic systems exhibiting buckling under a tensile load, such as those proposed by Ziegler (1977), though instability of these structures is clearly linked to the presence of compressed elements, which are responsible for buckling[18]. Another example of tensile buckling is that reported by Gajewski and Palej (1974), but in that case the applied load is not dead, rather it is provided by a container partially filled with a liquid, i.e., an example of live load.

The first example of an elastic system exhibiting buckling under tensile dead loading has been provided by Zaccaria et al. (2011) and is reported in Fig. 13. The one-degree-of-freedom structure comprises two rigid bars, hinged at their extremities and internally connected through a 'slider', a mechanical, frictionless device that keeps the two bars parallel, leaving the possibility of relative transversal sliding. Notice that the two bars composing the elastic system are subject to tension.

The buckling load and the equilibrium paths of the structure sketched in Fig. 13 can be easily computed with reference to its deformed shape, that is defined by the rotation ϕ. The elongation of the system reads

$$\Delta = 2l\left(\frac{1}{\cos\phi} - 1\right), \tag{138}$$

so that the total potential energy is

$$\mathcal{V}(\phi) = \frac{1}{2}k\phi^2 - 2Fl\left(\frac{1}{\cos\phi} - 1\right), \tag{139}$$

and the solutions of the equilibrium problem are

$$F = \frac{k}{2l}\frac{\phi\cos^2\phi}{\sin\phi}, \tag{140}$$

[16] In fact, it has been experimentally investigated by Pieter van Musschenbrok (1692-1761) and mathematically solved by Leonhard Euler (1707-1783).

[17] We refer here only to manifestations of structural instability and not of material instability such as necking of cylindrical bars, wrinkling of membranes and shear bands formation.

[18] For instance, one of these examples can be exemplified as a system of three rigid bars of equal length and connected by elastic hinges, whose natural configuration is the flat, folded configuration AB-BA-AB (like after flattening the Z letter). When the endpoints A and B are pulled apart, this system of 3 bars unfolds by an instability, but buckling here is clearly due to the element subject to compression.

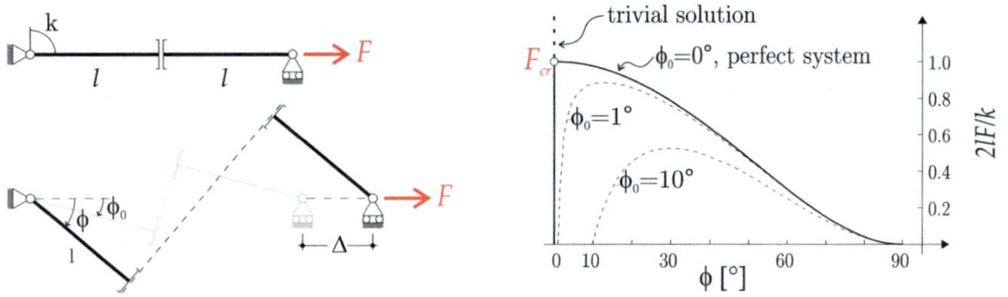

Figure 13. Bifurcation of an elastic one-degree-of-freedom system under *tensile dead loading*. The system comprises two rigid bars of length l that are internally jointed by a slider, whereas a rotational spring of stiffness k provides the elasticity to the structure. An imperfect system is also shown on the left, and its imperfection is given by the initial inclination ϕ_0 of the bars. Note that bifurcation is 'purely geometrical' and is induced by the internal constraint which transmits rotation, but not shear. The bifurcation diagram is reported on the right for $\phi_0 = \{0°, 1°, 10°\}$ and shows that the structure suffers softening once the critical load is attained.

plus the trivial solution $\phi = 0$, $\forall F$. The stability of the equilibrium solutions can be judged by studying the sign of the higher order derivatives of the total potential energy $\mathcal{V}(\phi)$. For the system being considered, we find that the trivial solution is stable up to the critical load

$$F_{cr} = \frac{k}{2l}, \tag{141}$$

while the nontrivial path, *evidencing softening*, is always unstable. The imperfect system, characterized by an initial rotation ϕ_0, can be analyzed in a similar manner. In fact, its elongation reads now as

$$\Delta = 2l\left(\frac{1}{\cos\phi} - \frac{1}{\cos\phi_0}\right), \tag{142}$$

and hence the total potential energy becomes a function of both the rotation ϕ and of the imperfection ϕ_0, that is

$$\mathcal{V}(\phi, \phi_0) = \frac{1}{2}k\,(\phi - \phi_0)^2 - 2Fl\left(\frac{1}{\cos\phi} - \frac{1}{\cos\phi_0}\right), \tag{143}$$

so that the equilibrium configurations are obtained as

$$F = \frac{k}{2l}\frac{(\phi - \phi_0)\cos^2\phi}{\sin\phi}. \tag{144}$$

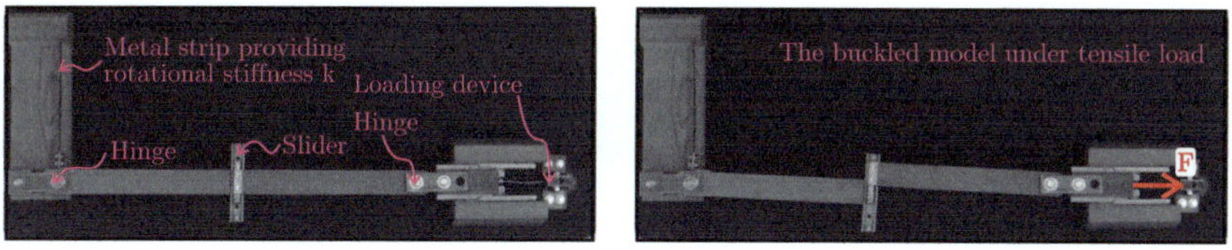

Figure 14. The practical realization of the one-degree-of-freedom system sketched in Fig. 13 (left). Two wooden rods have been employed and joined by means of a linear bearing, whereas elasticity has been provided to the system by clamping a thin metal strip to the hinge at the left end of the structure. The buckled configuration, as obtained by pulling with a dead load at the right end of the structure, is reported on the right.

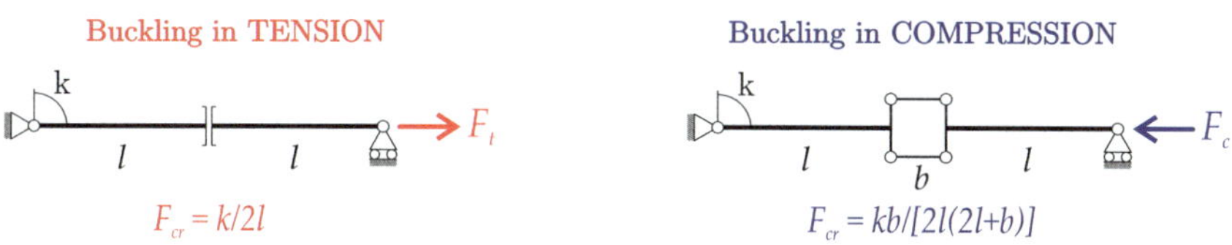

Figure 15. Two one-degree-of-freedom elastic systems composed by two rigid bars internally joined either with a slider (left) or with a parallelogram linkage (right). Although the linearized behaviour of the two examples is the same, it turns out that buckling under tensile loading is only feasible when the internal constraint is realized with a slider.

In order to provide experimental evidence of buckling under tensile dead loading, a physical model was built that resembles the structure sketched in Fig. 13. Specifically, two wooden rods were joined by means of two linear bearings (type Easy Rail SN22-80-500-610 from Rollon®), whereas elasticity was provided to the system by clamping a thin metal strip to its left end. The prototype is shown in Fig. 14, together with its buckled configuration as obtained by hanging a dead load at its right end.

Notice that bifurcation of the system in Fig. 13 is 'purely geometrical' and due to the presence of the internal constraint that connects the rigid bars. The central role of the slider in promoting tensile buckling can be further emphasized by comparing the two structural systems sketched in Fig. 15. Although the linearized behaviour of the two examples is the same, it turns out that by replacing the slider (left) with a parallelogram linkage (right) tensile buckling is suppressed and the structure can only buckle for a compressive force.

Once we have understood the crucial role played by the slider in favoring tensile buckling, we can exploit the concept to invent other structures with

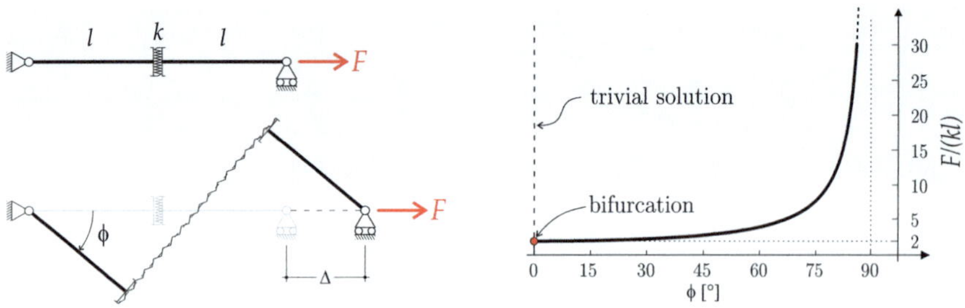

Figure 16. A one-degree-of-freedom elastic system (left) subject to a tensile dead load and having a linear elastic spring of stiffness k acting within the slider. Buckling of this structure still occurs by tensile loading, but now the post-critical behavior (right) exhibits hardening and hence is stable.

specific elastic responses.

A first example is that of Fig. 16, where the rotational spring has been replaced by a linear spring of stiffness k acting within the slider. It is worth noting that buckling still occurs for a tensile dead load, but now the post-critical behavior shows hardening instead of softening and consequently is stable.

Another example of structures buckling under tensile loading is shown in Fig. 17. A 'T-shaped' frame comprises three sliders and is simultaneously loaded by a horizontal force F and a vertical force αF, whereas elasticity is again provided by a rotational spring of stiffness k. Buckling and post-critical behaviour of this system are highly sensitive to the ratio α between the vertical and the horizontal force as shown on the right-hand side of Fig. 17.

A further generalization of the concepts presented above to elastic structures is possible by replacing the rigid bars of the system of Fig. 13 by two flexible elastic rods of equal length l and bending stiffness B, clamped at their outer extremities and internally connected by a slider, see Fig. 18. This elastic structure was first proposed and analyzed in detail by Zaccaria et al. (2011).

With reference to the sketch of Fig. 18, the determination of the critical loads of the system under consideration requires integration of the linearized equation of the elastica,

$$\frac{d^4 v(z)}{dz^4} - \alpha^2 \operatorname{sgn}(F) \frac{d^2 v(z)}{dz^2} = 0, \tag{145}$$

where F is assumed positive when tensile and $\alpha^2 = |F|/B$. Equation (145)

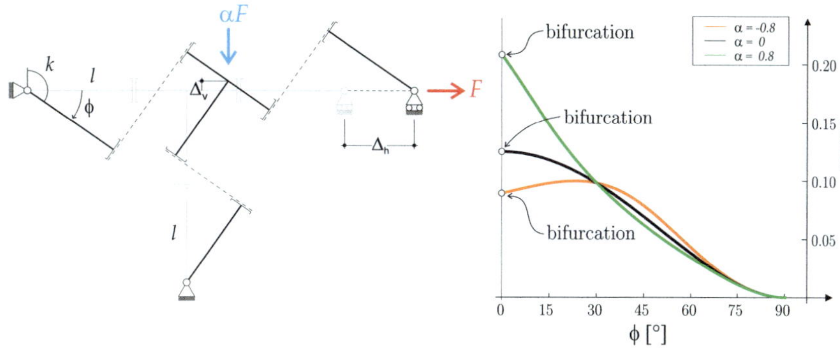

Figure 17. A 'T-shaped' frame exhibiting tensile buckling. The structure (left) comprises three sliders and is simultaneously subject to a horizontal force F and a vertical force αF; elasticity is provided by a rotational spring of stiffness k. Note that both buckling and post-critical behaviour (right) are significantly influenced by the ratio α between the two applied forces.

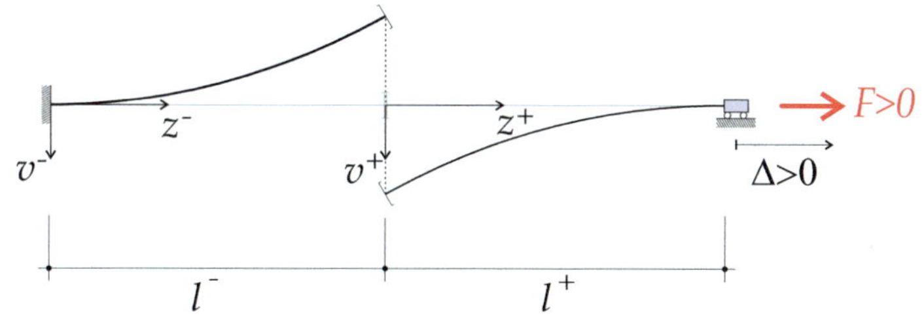

Figure 18. An elastic structure exhibiting buckling both under tensile ($F > 0$) and compressive ($F < 0$) loading. The structure has been designed by replacing the rigid bars of the one-degree-of-freedom system with elastic, flexible rods of length l and bending stiffness B.

holds for both the left and right rod, denoted in the following by '−' and '+', respectively, and its integration requires the following boundary conditions to be imposed (a prime denotes differentiation with respect to z),

$$\begin{aligned}
&v_-(0) = v_-^{'}(0) = v_+(l) = v_+^{'}(l) = 0, && \text{zero displ. and rot. at clamps} \\
&v_-^{'}(l) = v_+^{'}(0), && \text{continuity of rot. at the slider} \\
&v_-^{'''}(l) = v_+^{'''}(0) = 0, && \text{zero shear force at the slider} \\
&v_-^{''}(l) - v_+^{''}(0) = \alpha^2 \,\text{sgn}(F)[v_-(l) - v_+(0)]. && \text{equilibrium of the slider}
\end{aligned} \tag{146}$$

For the case of elastic rods of equal length[19], we obtain the following characteristic equations for the critical loads

$$\begin{cases} \sinh(\alpha\, l)\,[1 - \alpha\, l\, \tanh(\alpha\, l)] = 0, & \text{for } F > 0, \\ \sin(\alpha\, l)\,[1 + \alpha\, l\, \tan(\alpha\, l)] = 0, & \text{for } F < 0. \end{cases} \tag{147}$$

showing that there is only one bifurcation load in tension, whereas there are infinite bifurcation loads in compression. Specifically, buckling of the elastic structure depicted in Fig. 18 occurs when

$$\frac{F_{cr}(2l)^2}{\pi^2 B} = \quad +0.58, \quad -3.17, \quad -4, \quad -15.19, \quad -16, \quad \ldots \tag{148}$$

A more detailed analysis of the structure, including the determination of its post-critical behaviour, can be developed through integration of the nonlinear equation of the elastica. To attack this problem we start by employing the local reference systems shown in Fig. 19, and then we impose one global kinematic compatibility condition and three equilibrium conditions. These are listed in the following.

- ⋄ Kinematic compatibility condition: the jump in displacement across the slider (measured orthogonally to the line of the elastica), i.e. Δ_s, can be related to the angle of rotation of the slider Φ_s as

$$\left[x_1^-(l^-) + x_1^+(l^+)\right] \tan\Phi_s + x_2^-(l^-) + x_2^+(l^+) + \Delta_s = 0, \tag{149}$$

where $x_1^\pm(l^\pm)$ and $x_2^\pm(l^\pm)$ are the coordinates of the elastica evaluated at $s = l^\pm$. Note that Φ_s is assumed positive when anticlockwise and Δ_s is not restricted in sign (negative in the case of Fig. 19).

[19] The treatment of the more general case of two rods of different length is reported in Zaccaria et al. (2011).

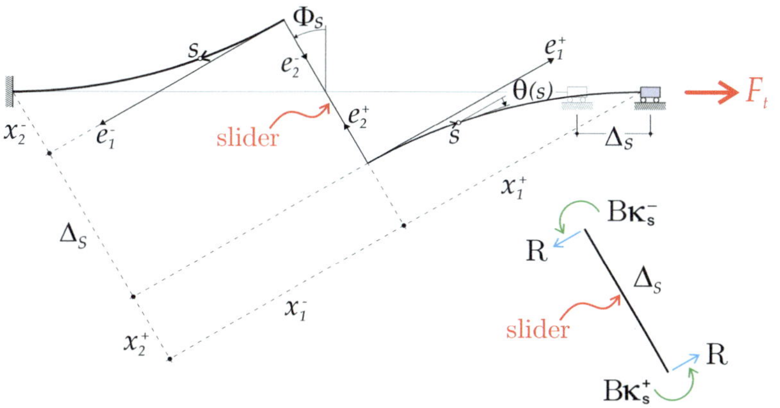

Figure 19. A sketch of the planar problem of the elastica at large displacements under tensile axial loading. Note the local reference systems employed to carry out the analysis.

◇ Global equilibrium of the structure: since the slider can only transmit a moment and a force R orthogonal to it, equilibrium of the structure requires that (see the inset of Fig. 19)

$$R = \frac{F}{\cos\Phi_s}, \tag{150}$$

where F is the external, axial force applied to the system, assumed positive (negative) when tensile (compressive). Since $\Phi_s \in [-\pi/2, \pi/2]$, R is positive (negative) for tensile (compressive) loading. Notice also that with the above definitions we have

$$\theta^+(0) = \theta^-(0) = 0, \quad \theta^+(l^+) = \theta^-(l^-) = -\Phi_s. \tag{151}$$

◇ Rotational equilibrium of the slider:

$$\kappa_s^- + \kappa_s^+ = \frac{R}{B}\Delta_s, \tag{152}$$

where B is the bending stiffness of the rods and $\kappa_s^\pm$ denotes their curvature evaluated at $s = l^\pm$. Note that B is positive by definition, but R, $\kappa_s^\pm$ and Δ_s can take any sign.

◇ Equation of the elastica for both the rods:

$$\theta''(s) - \frac{R}{B}\sin\theta(s) = 0, \tag{153}$$

where $\theta(s)$ is the rotation of the normal to the elastica at each point s, assumed positive when anticlockwise.

Equation (153) is usually[20] written with a sign '+' replacing the sign '−', meaning that R is assumed positive when compressive.

Integration of the elastica (details can be found in Zaccaria et al., 2011) leads to the coordinates x_1 and x_2 of the deformed elastic rod expressed in terms of the co-ordinate $u = s/\xi\sqrt{|R|/B}$, that is

$$\begin{cases} x_1 = \dfrac{1}{\xi\,\tilde{\alpha}}\left[\left(2-\xi^2\right)u - 2\,\mathrm{E}\left[\mathrm{am}\left[u,\xi\right],\xi\right] + 2\xi^2\mathrm{sn}\left[u,\xi\right]\mathrm{cn}\left[u,\xi\right]\right], \\ x_2 = \dfrac{2}{\xi\,\tilde{\alpha}}\sqrt{1-\xi^2}\left(\dfrac{1-\mathrm{dn}\left[u,\xi\right]}{\mathrm{dn}\left[u,\xi\right]}\right), \end{cases} \tag{154}$$

for a tensile ($R > 0$) axial load, whereas for a compressive ($R < 0$) axial load

$$\begin{cases} x_1 = \dfrac{1}{\xi\,\tilde{\alpha}}\left[\left(\xi^2-2\right)u + 2\,\mathrm{E}\left[\mathrm{am}\left[u,\xi\right],\xi\right]\right], \\ x_2 = \dfrac{2}{\xi\,\tilde{\alpha}}\left(1-\mathrm{dn}\left[u,\xi\right]\right). \end{cases} \tag{155}$$

To derive equations (154)-(155), the constants of integration have been chosen such that x_1 and x_2 vanish at $s = 0$, whereas $\tilde{\alpha} = \sqrt{|R|/B}$. Moreover, am, dn, sn, cn and E are elliptic functions (Byrd and Friedman, 1971) of modulus $\xi = 2\tilde{\alpha}/\sqrt{\kappa_s^2 + 4\tilde{\alpha}^2 H(R)}$, in which H denotes the Heaviside step function.

With reference to Fig. 19, we may also note that the horizontal displacement Δ_c of the right clamp can be written in the form

$$\Delta_c = \frac{x_1^-(l^-) + x_1^+(l^+)}{\cos\Phi_s} - \left(l^+ + l^-\right). \tag{156}$$

Finding the axial load F as a function of the slider rotation Φ_s, or similarly as a function of the end displacement Δ_c, is not particularly difficult, although the procedure involves the numerical solution of the nonlinear equation that arises from the global kinematic compatibility condition (149), see Zaccaria et al. (2011) for details. For reference, computed deformed elastica in tension and compression (the first mode in tension and the first mode in compression) are reported in Fig. 20.

[20] See equation (24), where R has be taken null and P plays the role of R in equation (153).

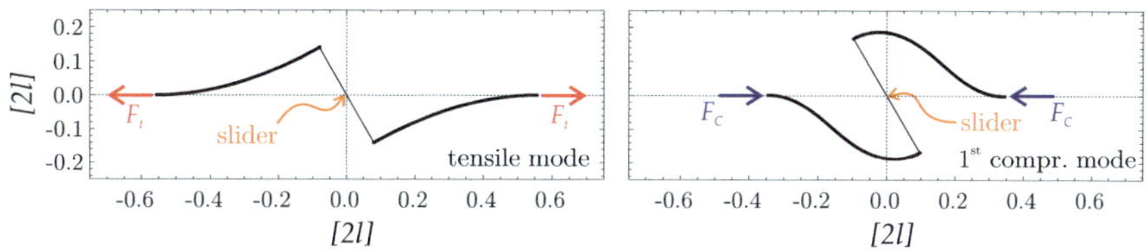

Figure 20. Deformed shapes in tension (left) and in compression (right) of the structure shown in Fig. 19 (with rods of equal length) at a slider rotation of 30°. Note that the scale of the axes is $2l$.

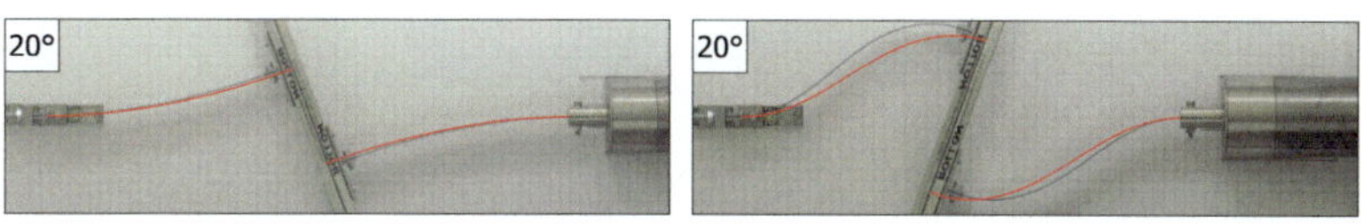

Figure 21. Computed deformed shapes (red lines) superimposed on photos taken during experimentation. Note the remarkable agreement between theory and experiments both in the case of a test under tensile (left) and compressive (right) loading.

The theoretical predictions have been compared with experimental results obtained by means of the structural model shown in the inset of Fig. 22. A first comparison is reported in Fig. 21, where computed deformed shapes (red, solid curves) have been superimposed on photos taken during experimentation. Note the remarkable agreement between theory and experiments both in the case of a test under tensile and compressive loading. An additional comparison between experiments and theory is reported in Fig. 22 in terms of buckling load and post-critical response of the elastic structure shown in the inset.

As a final remark, we emphasize the analogy between the deformed shape of an elastic rod buckled in tension and the shape of a water meniscus in a capillary channel (already pointed out, see Fig. 7). This can be appreciated from the comparison reported in Fig. 23 and is a consequence of the fact that both the physical phenomena are governed by the same differential equation.

4 Influence of constraint's curvature on buckling

The influence of the constraints upon the stability of elastic, structural systems has been pointed out by Zaccaria et al. (2011), showing that a

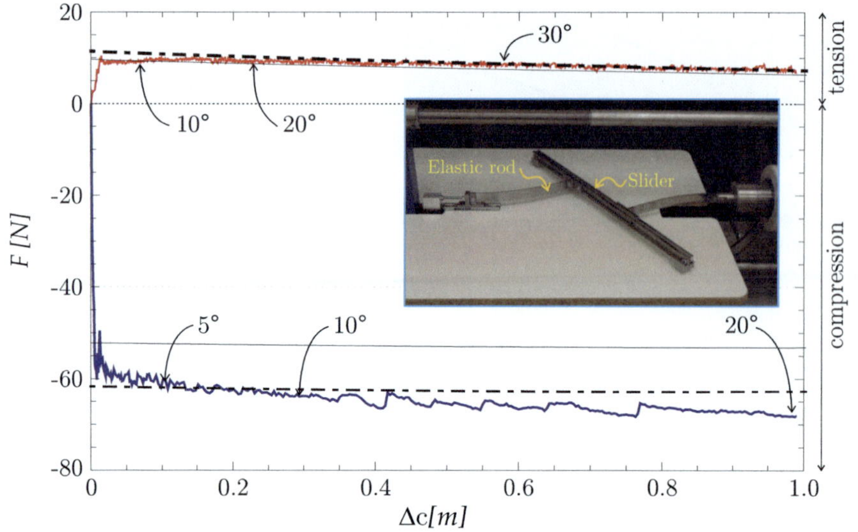

Figure 22. Buckling and post-critical response of the elastic system: theoretical prediction (black dashed line) versus experiments reported in red (tension) and blue (compression) lines. Numbers in degrees refer to slider rotations.

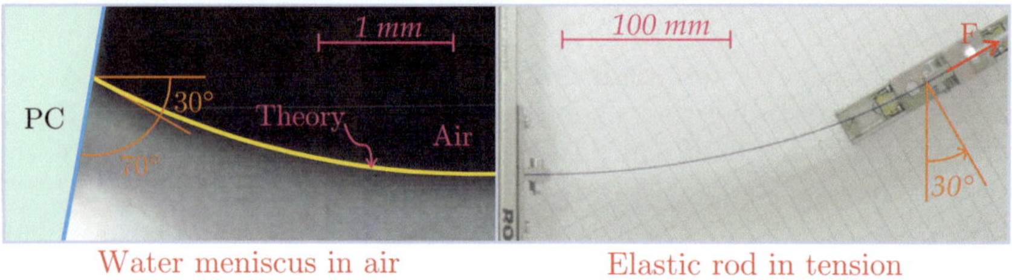

Figure 23. The shape of a water meniscus in a capillary channel (left: photo near a water-air-Polycarbonate contact superimposed to the solution of the elastica, highlighted with a yellow line) versus the elastic line of a rod buckled under tensile loading (right: photo taken during an experiment).

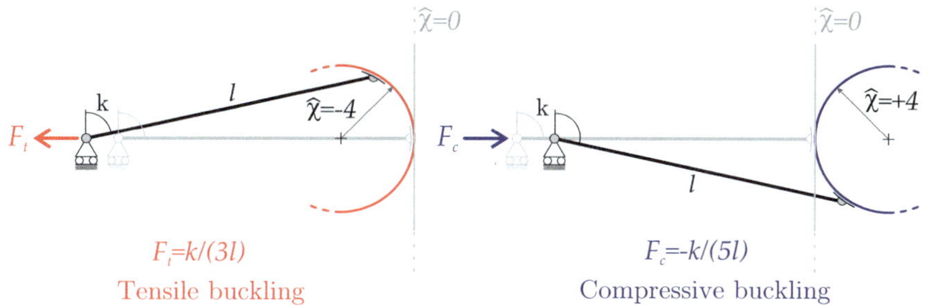

Figure 24. A one-degree-of-freedom structure (with a rotational elastic spring at its left end) evidencing compressive (right) or tensile (left) buckling as a function of the curvature of the constraint (a circular profile with constant curvature, $\hat{\chi} = \pm 4$) along which the pin on the right of the structure has to slide.

slider can induce tensile buckling of elastic structures, and by Bigoni et al. (2012, 2014a), demonstrating the deep influence of the constraint curvature on buckling and post-critical behaviour.

In fact, the curvature of a constraint plays a fundamental role in a bifurcation problem, as highlighted by the one-degree-of-freedom structure shown in Fig. 24, where a rigid bar is connected to a rotational spring on its left end and to a roller on its right end, which is constrained to move along a circle of radius R_c centered on the bar's axis. The structure is subject to a horizontal force F, so that when this load is compressive and the circle degenerates to a line (null curvature), the structure buckles at the compressive force $F = -k/l$. But the curvature of the circle strongly affects the critical load, which results to be *tensile* in the case of negative curvature ($F_t = k/(3l)$, for $\hat{\chi} = l/R_c = -4$, see Fig. 24 on the left) and compressive in the case of positive curvature ($F_c = -k/(5l)$, for $\hat{\chi} = l/R_c = 4$, see Fig. 24 on the right).

To extend this analysis to systems with diffuse elasticity, the rigid bar can be replaced by an elastic rod, as shown in Fig. 25, and the fully nonlinear problem can be solved through integration of the elastica, see Bigoni et al. (2012) for the treatment. The analytical solution to this problem is reported in Fig. 26 and is compared with results of experiments shown as orange lines. A remarkable feature of the system is the force reversal found along the post-critical behaviour both in tension and in compression (points '2' and '5' in the figure).

Since the constraint's curvature deeply influences both buckling and post-critical behaviour, one may think to exploit this feature in an alter-

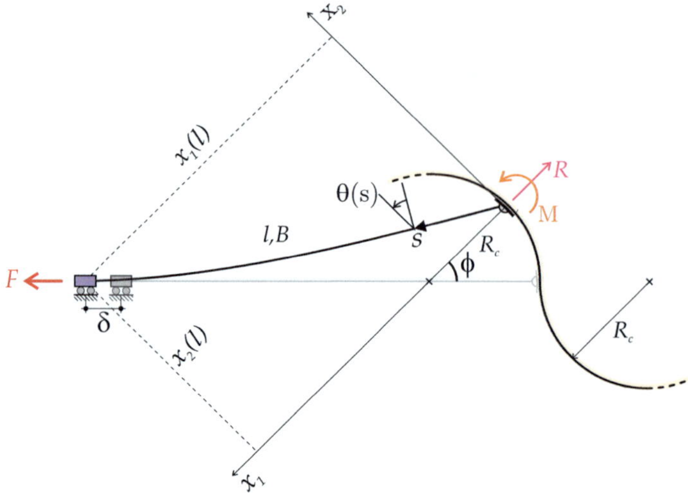

Figure 25. The elastica problem of a rod clamped at the left end and constrained to slide with a roller along a circle at its right end. Note the reference system employed in the analysis.

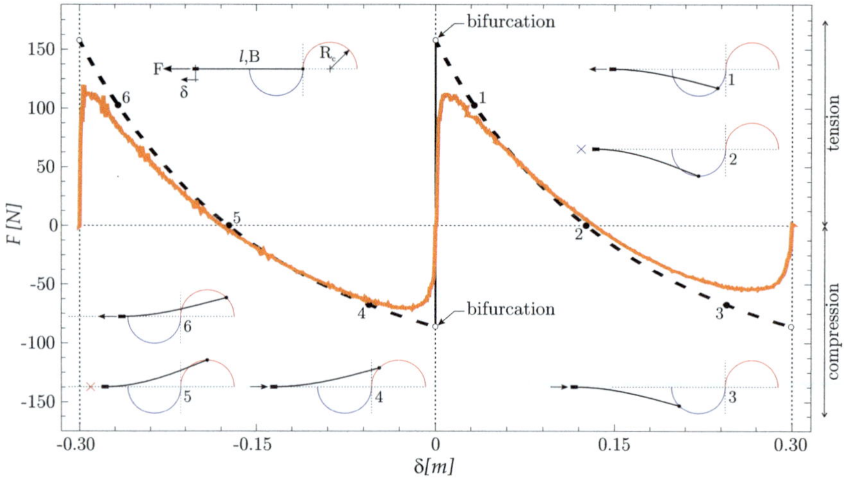

Figure 26. Load/displacement experimental results (orange, solid line) versus theoretical prediction (black, dashed line) for an elastic rod constrained to slide along a 'S-shaped', bi-circular profile. Deformed shapes are also reported in the insets.

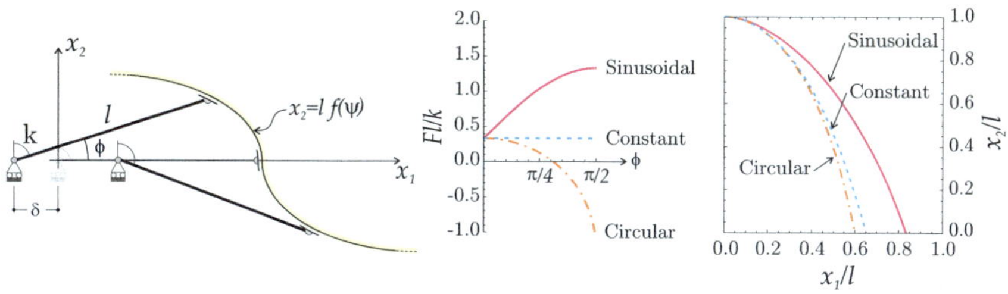

Figure 27. A one-degree-of-freedom structure (left) with one end constrained to slide along a generic, curved profile. Three distinct post-critical behaviours are shown (centre) together with the shape of the constraint (right) that was designed to realize them. Specifically, the three force-versus-rotation relations correspond to a sinusoidal, a circular and a constant (or 'neutral') elastic behaviour.

native way. In other words, one could think to design the shape of the constraint's profile (on which one end of the elastic system is prescribed to move) to obtain a 'desired post-critical behaviour' (for details see Bigoni et al., 2012). Fig. 27 reports on three examples of elastic structures with curved profiles designed to obtain a sinusoidal, a circular, and a constant post-critical response in terms of force F versus rotation ϕ. Furthermore, experiments have been performed to confirm the relation between the geometry of the constraint's profile and post-critical response of elastic systems. As an example, Fig. 28 shows the setup that was used to carry out experiments on the one-degree-of freedom structure exhibiting a 'neutral' post-critical behaviour shown in Fig. 27 (for further details see Bigoni et al., 2012, 2014a).

5 Eshelby-like forces in elastic structures with variable length

Eshelby (1956) introduced the concept of configurational force, acting on inhomogeneities or defects in solids to move them until the total potential energy is minimized. In other words, considering a defect in an elastic body characterized by a configurational parameter ν, so that the total potential energy of the body is $\mathcal{V}(\nu)$, the configurational force is defined as

$$-\frac{\partial \mathcal{V}(\nu)}{\partial \nu}.$$

The Peach-Koehler force acting on a dislocation, the crack-extension force of fracture mechanics, and the material force developing on a phase boundary

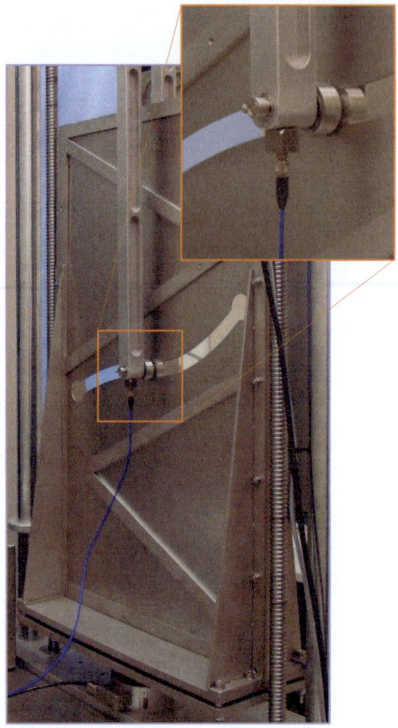

Figure 28. The experimental setup employed to test the one-degree-of-freedom system prescribed to move along a profile that was designed to provide a constant post-critical response.

in a solid under loading are all known examples of Eshelby forces, so that they are considered nowadays the cornerstone of a well-developed theory (see for instance Gurtin, 2000; Kienzler and Herrmann, 2000; Maugin, 1993, 2011; Dascalu et al., 2010; Bigoni and Deseri, 2011). The purpose of this section is to show that Eshelby-like forces can develop in elastic structures when the configuration of the structure changes. The simplest example of evidence of a configurational force is shown in Fig. 29, where an elastic inextensible rod of total length $\bar{l}$, bending stiffness B, and straight in its undeformed configuration is constrained with a sliding sleeve on the left end and loaded with a transversal dead load P at the other end. The sliding sleeve fully constrains rotation and vertical displacement for the part of the rod currently contained in it, but leaves the possibility of free

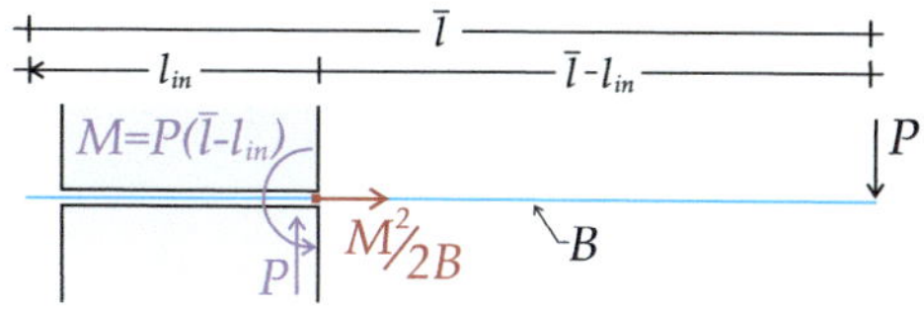

Figure 29. An elastic rod of total length $\bar{l}$, constrained with a sliding sleeve on its left, is subject to a transversal load P on its right end. Equilibrium of this configuration is impossible, because a horizontal Eshelby-like force is generated in the sliding sleeve. The Eshelby-like force equals the square of the bending moment at the end of the sliding sleeve, $P^2(\bar{l} - l_{in})^2$, divided by twice the flexural bending stiffness of the rod, $2B$.

horizontal sliding (without friction). Therefore, the elastic system has a variable length $\bar{l} - l_{in}$ and the total potential energy $\mathcal{V}(l_{in})$ can be evaluated as a function of the free configurational parameter l_{in}, defining the amount of the rod constrained by the sliding sleeve. Restricting for the moment attention to small displacements of the rod, the linear elastic solution for this structure yields a transverse displacement at the point of application of the force equal to $P(\bar{l} - l_{in})^3/(3B)$, so that the total potential energy is

$$\mathcal{V}(l_{in}) = \frac{P^2(\bar{l} - l_{in})^3}{6B}. \tag{157}$$

In analogy with the Eshelby concept of configurational force, the derivative of the potential energy (157) with respect to the configurational parameter l_{in} yields the following non-null horizontal Eshelby-like force

$$-\frac{\partial \mathcal{V}(l_{in})}{\partial l_{in}} = \frac{P^2(\bar{l} - l_{in})^2}{2B} = \frac{M^2}{2B}. \tag{158}$$

Note that the equilibrium for the elastic system (Fig. 29) is impossible since the Eshelby-like force (158) is not null except in the trivial cases of $P = 0$ or $l_{in} = \bar{l}$. The Eshelby-like force can also be expressed in terms of the transverse displacement at the loaded end of the structure $v_{\bar{l}} = P(\bar{l} - l_{in})^3/(3B)$ as

$$-\frac{\partial \mathcal{V}(l_{in})}{\partial l_{in}} = \frac{9B}{2(\bar{l} - l_{in})^2} \left(\frac{v_{\bar{l}}}{\bar{l} - l_{in}} \right)^2, \tag{159}$$

showing that for small deflections the force is small, but, as we will see later, this force grows and becomes dominant when displacements are large.

Following Bigoni et al. (2015), an inextensible elastic rod (straight in its unloaded configuration, with bending stiffness B and total length $\bar{l}$) is considered now in a more general setting. In particular, the rod is constrained

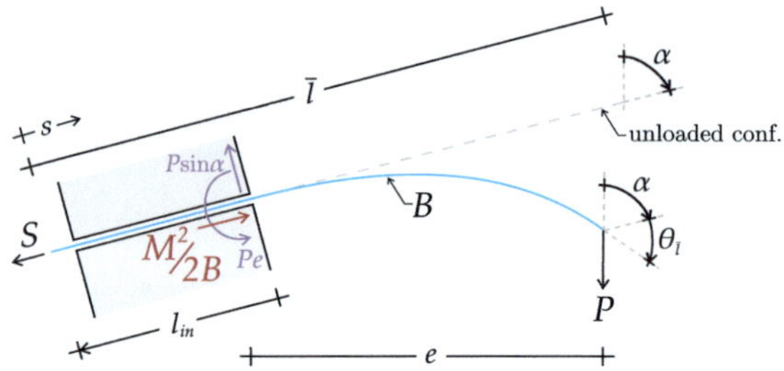

Figure 30. Structural scheme of the elastic system used to demonstrate an Eshelby-like force. The elastic rod of total length $\bar{l}$ is subject to a dead vertical load P on its right end, is constrained with a sliding sleeve inclined at an angle α (with respect to the vertical) and has a axial dead force S applied at its left end. The presence of the Eshelby-like force $M^2/(2B)$ influences the force S at equilibrium, which results different from $P\cos\alpha$. The other reactions at the end of the sliding sleeve are the force $P\sin\alpha$ perpendicular to the axis of the sleeve and the counterclockwise reaction moment Pe.

with a sliding sleeve as in Fig. 29, but is now inclined with respect to the horizontal direction and is subject at one end to an axial (dead) force S, while the other end is subject to a dead load P (inclined at an angle α), see Fig. 30.

Introducing the curvilinear coordinate $s \in [0, \bar{l}]$, the length l_{in} of the segment of the rod inside the sliding sleeve, and the rotation $\theta(s)$ of the rod's axis, the constraint imposes the condition

$$\theta(s) = 0 \text{ for } s \in [0, l_{in}].$$

Denoting by a prime the derivative with respect to s, the bending moment along the elastic rod is $M(s) = B\theta'(s)$, so that at the loaded end of the rod, the zero-moment boundary condition $\theta'(\bar{l}) = 0$ has to hold.

The total potential energy of the system is

$$\mathcal{V}(\theta(s), l_{in}) = B \int_{l_{in}}^{\bar{l}} \frac{\left[\theta'(s)\right]^2}{2} \mathrm{d}s - P\left[\bar{l}\cos\alpha - \cos\alpha \int_{l_{in}}^{\bar{l}} \cos\theta(s)\mathrm{d}s \right. \\ \left. + \sin\alpha \int_{l_{in}}^{\bar{l}} \sin\theta(s)\mathrm{d}s\right] - S\, l_{in}, \tag{160}$$

which at equilibrium becomes

$$\mathcal{V}(\theta_{eq}(s,l_{eq}),l_{eq}) = B\int\limits_{l_{eq}}^{\bar{l}} \frac{\left[\theta'_{eq}(s,l_{eq})\right]^2}{2}\mathrm{d}s - S\,l_{eq} - P\Bigg[\bar{l}\cos\alpha - \cos\alpha\int\limits_{l_{eq}}^{\bar{l}}\cos\theta_{eq}(s,l_{eq})\mathrm{d}s + \sin\alpha\int\limits_{l_{eq}}^{\bar{l}}\sin\theta_{eq}(s,l_{eq})\mathrm{d}s\Bigg], \tag{161}$$

where l_{eq} is the length of the amount of the elastic rod inside the sliding sleeve and θ_{eq} is the rotation of the rod's axis at the equilibrium.

The Eshelbian force related to the presence of the sliding sleeve can be calculated by differentiating with respect to l_{eq} the total potential energy at equilibrium, equation (161), which, considering the Leibniz's rule[21], becomes

$$\frac{\partial\mathcal{V}(l_{eq})}{\partial l_{eq}} = B\int\limits_{l_{eq}}^{\bar{l}} \theta'_{eq}(s)\frac{\partial\theta'_{eq}(s)}{\partial l_{eq}}\mathrm{d}s + P\Bigg[\cos\alpha\int\limits_{l_{eq}}^{\bar{l}}\sin\theta_{eq}(s)\frac{\partial\theta_{eq}(s)}{\partial l_{eq}}\mathrm{d}s - \sin\alpha\int\limits_{l_{eq}}^{\bar{l}}\cos\theta_{eq}(s)\frac{\partial\theta_{eq}(s)}{\partial l_{eq}}\mathrm{d}s\Bigg] - P\cos\alpha - B\frac{[\theta'_{eq}(l_{eq})]^2}{2} - S. \tag{163}$$

From equation (163), keeping into account integration by parts

$$\theta'_{eq}\frac{\partial\theta'_{eq}}{\partial l_{eq}} = \left(\theta'_{eq}\frac{\partial\theta_{eq}}{\partial l_{eq}}\right)' - \theta''_{eq}\frac{\partial\theta_{eq}}{\partial l_{eq}}, \tag{164}$$

the elastica

$$B\theta''_{eq}(s) + P\left[\cos\alpha\sin\theta_{eq}(s) + \sin\alpha\cos\theta_{eq}(s)\right] = 0, \qquad s\in[l_{eq},\bar{l}] \tag{165}$$

[21] The Leibniz rule of differentiation is

$$\frac{\mathrm{d}}{\mathrm{d}\zeta}\int\limits_{\alpha(\zeta)}^{\beta(\zeta)} f(x,\zeta)\mathrm{d}x = f(\beta,\zeta)\frac{\mathrm{d}\beta}{\mathrm{d}\zeta} - f(\alpha,\zeta)\frac{\mathrm{d}\alpha}{\mathrm{d}\zeta} + \int\limits_{\alpha(\zeta)}^{\beta(\zeta)}\frac{\partial f(x,\zeta)}{\partial\zeta}\,\mathrm{d}x. \tag{162}$$

and the boundary condition $\theta_{eq}'(\bar{l}) = 0$, the following expression for the Eshelby-like force is obtained

$$-\frac{\partial \mathcal{V}(l_{eq})}{\partial l_{eq}} = B\frac{[\theta_{eq}'(l_{eq})]^2}{2} + B\theta_{eq}'(l_{eq}) \left.\frac{\partial \theta_{eq}}{\partial l_{eq}}\right|_{s=l_{eq}} + P\cos\alpha + S. \quad (166)$$

Considering that θ_{eq} is a function of $s - l_{eq}$ and of the angle of rotation of the rod at the loaded end $\theta_{\bar{l}}$ (function itself of l_{eq}) one arrives at the following condition

$$\left.\frac{\partial \theta_{eq}}{\partial l_{eq}}\right|_{s=l_{eq}} = -\left.\frac{\partial \theta_{eq}}{\partial s}\right|_{s=l_{eq}} + \left.\frac{\partial \theta_{eq}}{\partial \theta_{\bar{l}}}\frac{\partial \theta_{\bar{l}}}{\partial l_{eq}}\right|_{s=l_{eq}}. \quad (167)$$

Since θ_{eq} is always zero at $s = l_{eq}$ for all $\theta_{\bar{l}}$, the second term in the right-hand side of equation (167) is null, so that equation (167) becomes

$$\left.\frac{\partial \theta_{eq}}{\partial l_{eq}}\right|_{s=l_{eq}} = -\theta_{eq}'(l_{eq}). \quad (168)$$

Therefore, the vanishing of the Eshelby-like force, namely, of the derivative of the total potential energy $\mathcal{V}(l_{eq})$ with respect to l_{eq}, eqn (166), represents the global axial equilibrium for the structure shown in Fig. 30

$$\underbrace{\frac{M^2}{2B}}_{\text{Eshelby-like force}} = S + P\cos\alpha, \quad (169)$$

where $M = B\theta_{eq}'(l_{eq})$ is the reaction moment, equal to Pe, where e is the load eccentricity (to the sliding sleeve).

The term $M^2/(2B)$ is a 'counterintuitive term' which depends on the configurational parameter l_{eq} (and would be absent if the elastic rod were constrained with a movable clamp instead than a sliding sleeve) and is for this reason indicated as the 'Eshelby-like force'. This term has wrongly been neglected by a number of authors who have considered sliding sleeve constraints. On the other hand, a term $M^2/(2B)$ was previously correctly considered in a context different from that addressed here, namely, adhesion mechanics, in which the term is equated to an 'adhesion energy' (Majidi, 2007; Majidi et al., 2012) .

In the following the existence of the Eshelby-like force (169) will be theoretically demonstrated via two independent approaches, namely an asymptotic method and a variational technique.

5.1 Asymptotic technique

The Eshelby-like force (169) can be obtained via an asymptotic approach (inspired by the work by Balabukh et al. (1970)) where an imperfect sliding sleeve having a small gap Δ (the distance between the two rigid, frictionless and parallel surfaces making the sliding device) is considered, Fig. 31. Within this gap, the elastic rod is deflected, so that $\vartheta(\Delta)$ denotes the angle at its right contact point, where the forces H, V, M are applied. The length of the rod detached from the two surfaces representing the imperfect sliding sleeve is denoted with $a(\Delta)$. The frictionless contact generates the reaction forces R and Q, in equilibrium with the axial dead force S at the other end. For small Δ, the equilibrium is given by

$$Q = \frac{M}{a(\Delta)}, \qquad R = V + \frac{M}{a(\Delta)}, \qquad S = \left(V + \frac{M}{a(\Delta)}\right)\vartheta(\Delta) - H. \quad (170)$$

On application of the virtual work for a linear elastic inextensible rod yields the geometric quantities $a(\Delta)$ and $\vartheta(\Delta)$

$$a(\Delta) = \sqrt{\frac{6B\Delta}{M}}, \qquad \vartheta(\Delta) = \frac{1}{2}\sqrt{\frac{6M\Delta}{B}}, \quad (171)$$

so that forces Q, R and S can be rewritten as

$$Q = M\sqrt{\frac{M}{6B\Delta}}, \qquad R = V + M\sqrt{\frac{M}{6B\Delta}}, \qquad S = \frac{M^2}{2B} + \frac{V}{2}\sqrt{\frac{6M\Delta}{B}} - H. \quad (172)$$

In the limit of perfect (zero-thickness) sliding sleeve, $\Delta \to 0$, the horizontal component of the reaction R *does not vanish*, but becomes the Eshelby-like force (169)

$$\lim_{\Delta \to 0} R(\Delta)\vartheta(\Delta) = \frac{M^2}{2B}. \quad (173)$$

5.2 Variational technique

The total potential energy (160) has a movable boundary l_{in}, so that it is expedient (Courant and Hilbert, 1962; Majidi et al., 2012) to introduce a small parameter ϵ and take variations (subscript '*var*') of an equilibrium configuration (subscript '*eq*') in the form

$$\theta(s,\epsilon) = \theta_{eq}(s) + \epsilon\theta_{var}(s), \qquad l_{in}(\epsilon) = l_{eq} + \epsilon l_{var}\ , \quad (174)$$

with the boundary conditions

$$\theta_{eq}(l_{eq}) = 0, \quad \theta(l_{eq} + \epsilon l_{var}) = 0, \quad \theta_{eq}'(\bar{l}) = 0. \quad (175)$$

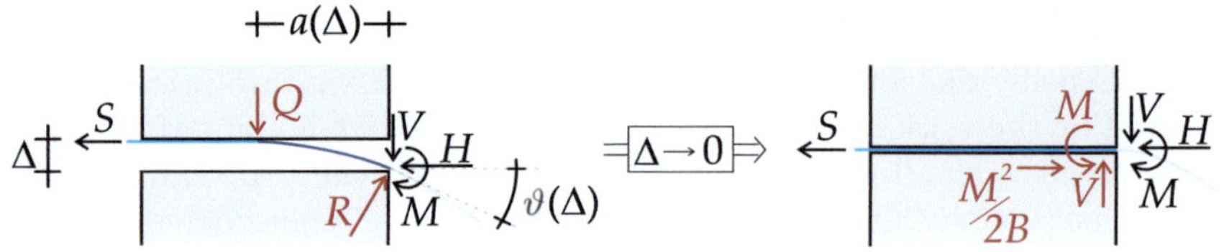

Figure 31. Deformed configuration of an elastic rod within an imperfect sliding sleeve made up of two smooth, rigid and frictionless planes placed at a distance Δ. Applied and reaction forces (left) and values obtained in the limit $\Delta \to 0$, revealing the Eshelbian force (right).

A Taylor series expansion of $\theta(l_{in})$ for small ϵ yields

$$\theta(l_{eq} + \epsilon l_{var}, \epsilon) = \theta_{eq}(l_{eq}) + \epsilon\left(\theta_{var}(l_{eq}) + \theta'_{eq}(l_{eq})l_{var}\right) + \frac{\epsilon^2}{2} l_{var}\left(2\theta'_{var}(l_{eq}) + \theta''_{eq}(l_{eq})l_{var}\right) + \mathcal{O}\left(\epsilon^3\right), \tag{176}$$

so that the boundary conditions (175) lead to the following compatibility equations

$$\theta_{var}(l_{eq}) + \theta'_{eq}(l_{eq})l_{var} = 0, \quad 2\theta'_{var}(l_{eq}) + \theta''_{eq}(l_{eq})l_{var} = 0. \tag{177}$$

Taking into account the Leibniz rule of differentiation (162) and the boundary (175) and compatibility (177) conditions, through integration by parts, the first variation of the functional $\mathcal{V}$ is obtained in the following form

$$\begin{aligned}\delta_\epsilon \mathcal{V} &= -\int_{l_{eq}}^{\bar{l}} \left[B\theta''_{eq}(s) + P\left(\cos\alpha \sin\theta_{eq}(s) + \sin\alpha\cos\theta_{eq}(s)\right)\right]\theta_{var}(s)\mathrm{d}s \\ &\quad + \left[B\frac{\theta'_{eq}(l_{eq})^2}{2} - P\cos\alpha - S\right] l_{var},\end{aligned} \tag{178}$$

so that the equilibrium equations (165) and (169) are derived, the latter of which, representing the so-called 'transversality condition' of Courant and Hilbert (1962), provides the Eshelby-like force.

5.3 Analytical expression of the Eshelby-like force

The equilibrium configuration of the elastic rod satisfies the elastica equation (165) (see Love, 1927; Bigoni, 2012), that introducing the auxiliary angle $\varphi(s) = \theta_{eq}(s) + \alpha$ and the dimensionless load $\lambda^2 = P/B$ writes as

$$\varphi^{''}(s) + \lambda^2 \sin\varphi(s) = 0, \qquad s \in [l_{eq}, \bar{l}] \tag{179}$$

subject to the boundary conditions $\varphi(l_{eq}) = \alpha$ and $\varphi^{'}(\bar{l}) = 0$. Integration of equation (179) yields

$$\varphi^{'}(s) = \pm\lambda\sqrt{2\left[\cos\varphi(s) - \cos(\theta_{\bar{l}} + \alpha)\right]}, \tag{180}$$

taken in the following with the '+' sign, so that introducing the change of variable

$$\eta = \sin\frac{\theta_{\bar{l}} + \alpha}{2}, \qquad \eta \sin\phi(s) = \sin\frac{\varphi(s)}{2}, \tag{181}$$

where $\theta_{\bar{l}} = \theta_{eq}(\bar{l})$ represents the rotation measured at the free end of the rod, we end up with the following differential problem

$$\phi^{'}(s) = \lambda\sqrt{1 - \eta^2 \sin^2\phi(s)}, \tag{182}$$

subject to $\phi(l_{eq}) = m = \arcsin\left[\sin(\alpha/2)/\,\eta\right]$ and $\phi(\bar{l}) = \pi/2$.

Restricting the attention to the first (stable) mode of deformation, the integration of equation (182) leads to the relation between the rotation measured at the free end of the rod $\theta_{\bar{l}}$ and the applied vertical load

$$P = \frac{B}{(\bar{l} - l_{eq})^2}\left[\mathcal{K}(\eta) - \mathcal{K}(m, \eta)\right]^2, \tag{183}$$

where $\mathcal{K}(\eta)$ is the complete elliptic integral of the first kind (39), whereas $\mathcal{K}(m, \eta)$ is the incomplete elliptic integral of the first kind (83). Moreover, through the integration of equation (182) and the implemented change of variable, the rotation field (for the first mode) can be obtained as

$$\theta_{eq}(s) = 2\arcsin\left[\eta\,\mathrm{sn}\left((s - l_{eq})\sqrt{\frac{P}{B}} + \mathcal{K}(m, \eta), \eta\right)\right] - \alpha, \tag{184}$$

where sn is the Jacobi sine amplitude function (47).

The Eshelby-like force (169) can be expressed, taking into account equation (180), as

$$\frac{M^2}{2B} = 2P\left(\eta^2 - \sin^2\frac{\alpha}{2}\right), \tag{185}$$

so that the axial force S at the end of the sliding sleeve (which has been experimentally measured through a load cell by Bigoni et al. (2015)) is given by

$$S = -P\cos(\alpha + \theta_{\bar{l}}) = -P\cos\alpha + \underbrace{2P\left(\sin^2\frac{\theta_{\bar{l}} + \alpha}{2} - \sin^2\frac{\alpha}{2}\right)}_{\text{Eshelby-like force}}. \tag{186}$$

From equation (186) it can be noted that the load S is (in modulus) bounded by P and that it tends to P only in the 'membrane limit', when B tends to zero and $\theta_{\bar{l}} + \alpha$ to π.

The following three different cases may arise, explaining the experiments shown in Fig. 33:

(i.) the elastic rod within the sliding sleeve is in compression, or 'pushed in', if $\theta_{\bar{l}} + \alpha < \pi/2$;

(ii.) the elastic rod within the sliding sleeve is unloaded if $\theta_{\bar{l}} + \alpha = \pi/2$;

(iii.) the elastic rod within the sliding sleeve is in tension, or 'pulled out', if $\theta_{\bar{l}} + \alpha > \pi/2$.

The case of null axial force, $S = 0$, occurs when $M^2/(2B)$ equals the axial component of the dead load, $P \cos\alpha$, and corresponds to deformed configurations having the tangent at the free end orthogonal to the direction of the load P, as in Fig. 33 (center).

Finally, it can be noted that the Eshelby-like force $M^2/(2B)$ is greater than the applied load P when

$$\cos\alpha - 2\cos^2\left(\frac{\theta_{\bar{l}} + \alpha}{2}\right) > 0. \tag{187}$$

Regions in the $\theta_{\bar{l}} - \alpha$ plane where the axial force S is positive/negative and where $M^2/(2B) > P$ are shown in Fig. 32. From the figure it can be concluded that $M^2/(2B) > P$ is possible only for positive axial load, $S > 0$, and high deflections of the rod (for rotations at the free end $\theta_{\bar{l}}$ greater than $\pi/3$).

5.4 Experimental evidence of configurational force

Fig. 33 gives a qualitative (but indisputable) proof of the existence of the Eshelby-like force in the structure shown in Fig. 30, in fact for this system equilibrium is only possible when equation (169) is satisfied, so that the presence of the Eshelby-like force (parallel to the direction of sliding) explains the reason why the configuration shown in the central photo in Fig. 33 is in equilibrium and why the rod is 'expelled' from the sliding sleeve in the lower photo.

'Expulsion' of the elastic rod is consequent to a value of the Eshelby-like force exceeding the axial component of P, namely, $P \cos\alpha$. Quantitative experiments are reported in Bigoni et al. (2015), whereas movies of the experiments can be found at http://ssmg.unitn.it/eshelbylikeforce.html. The purpose is now to analyze the effects of Eshelby-like force on bifurcation and instability of structures.

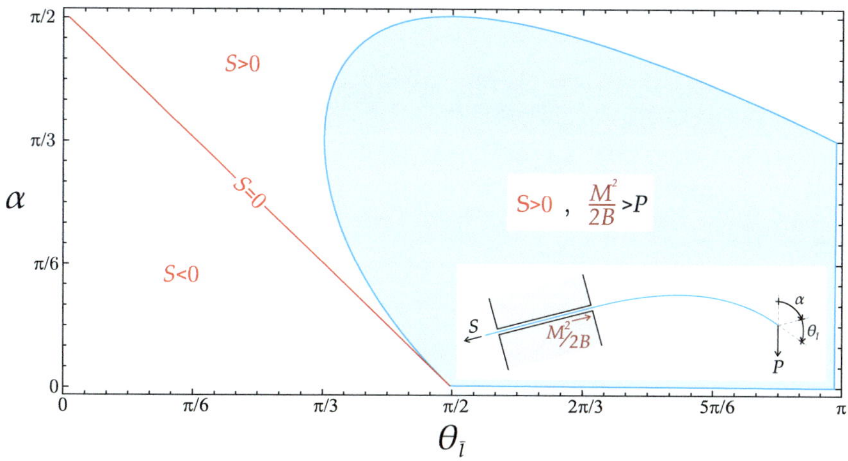

Figure 32. Regions in the plane $\theta_{\bar{l}} - \alpha$ where $S > 0$, $S < 0$ and $M^2/(2B) > P$.

6 Configurational forces and buckling

To highlight the effects of configurational forces on elastic structures, we consider the problem in which a blade (an elastic planar rod) is forced to penetrate into an elastic compliant sliding sleeve clamp (a frictionless sliding sleeve with a final linear spring of stiffness k) through the application of a dead compressive load P at the other end.

A structural system with a free end loaded with an axial dead force was shown to exhibit several surprising and counterintuitive mechanical behaviours by Bigoni et al. (2014b) and will not be repeated here. In the following, the governing equations will be obtained for the structures shown in Fig. 34, differing only in the boundary conditions at the right end. These structures are similar to those analyzed in Section 2, except for the presence of the Eshelby-like force, arising from the sliding sleeve and which strongly affects the post-critical behaviour and its stability (see Bigoni et al., 2014b).

6.1 Finite number of critical loads

The Euler formula for an inextensible elastic planar rod (of length l) provides the n-th critical load (associated to the n-th instability mode) as

$$P_{cr,n} = \frac{\pi^2 B}{(\rho_n l)^2}, \qquad n \in \mathbb{N}^+, \tag{188}$$

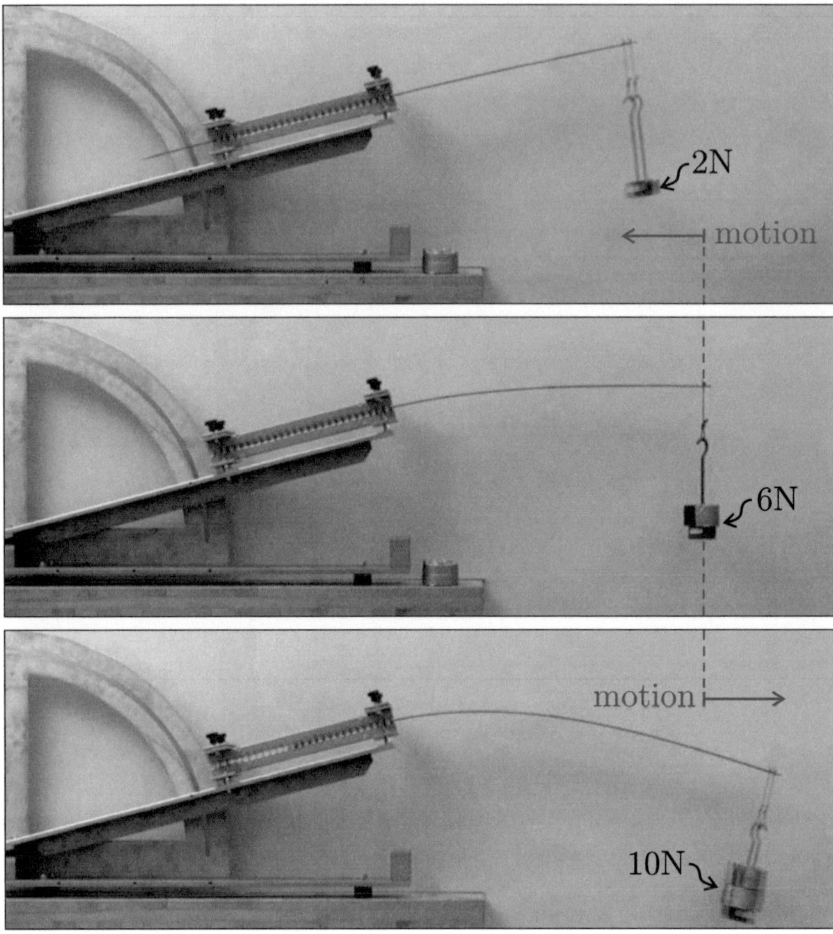

Figure 33. The practical realization of the elastic structure shown in Fig. 30 reveals an axial Eshelby-like force, so that, while at low vertical force (2 N) the elastic rod tends, as expected, to slip inside the sliding sleeve (upper photo), at 6 N the equilibrium is surprisingly possible (note that the tangent at the loaded end of the elastic rod is horizontal, see the photo in the centre) and at 10 N the elastic rod is expelled from the sliding sleeve (lower photo), even if the system is inclined at 15° with respect to the horizontal direction ($\alpha = 75°$).

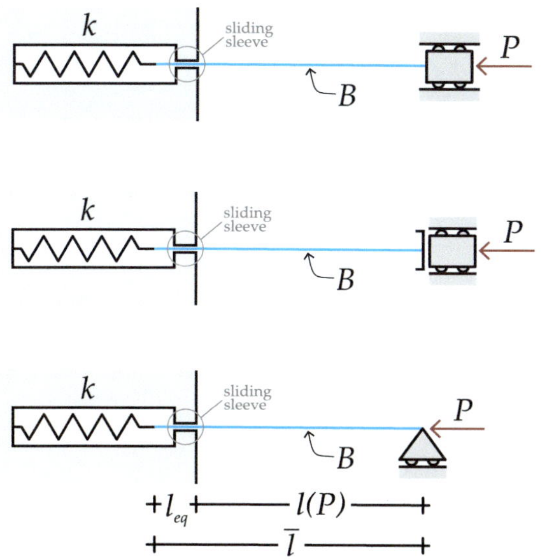

Figure 34. The penetrating blade is an elastic planar rod whose free length l is a function of the amount of the applied axial dead load P. The blade slides into a frictionless sleeve and is restrained by an axial linear spring, while at the other loaded end three types of constraints are considered: clamped (upper), constrained rotation (center) and simply supported (lower).

where ρ_n is a slenderness factor (depending on the constraints applied to the elastic system[22]). Equation (188) shows that the n-th critical load P_{cr} is unique whenever the rod length l is fixed, but this uniqueness may be lost when the length becomes a function of the applied axial load, $l = l(P)$.

If an axial spring is introduced (as shown in Fig. 34), the elastic rod (or 'blade') can penetrate the constraint (a sliding sleeve) by a length l_{eq}, so

[22] With reference to a structure with the left end constrained by a sliding sleeve, the factor ρ_n is

- $\rho_n = 1/(n+1)$ for $n = 1, 3, 5, ...$, $\rho_2 \approx 0.350$, $\rho_4 \approx 0.203$, and $\rho_n \approx 1/(n+1)$ for $n = 6, 8, 10, ...$ for the structure with the clamped right end, Fig. 34 (upper part);
- $\rho_n = 1/n$, for the structure with constrained rotation at the right end, Fig. 34 (central part);
- $\rho_1 = 0.699$, $\rho_2 = 0.407$ $\rho_n \approx \dfrac{2}{2n+1}$ for $n \geq 3$, for the structure with simply supported right end, Fig. 34 (lower part).

that

$$l(P) = \bar{l} - l_{eq}(P), \tag{189}$$

where $\bar{l}$ is the outer length of the blade at null axial load P, $l(P = 0) = \bar{l}$. In the particular case of a linear spring with stiffness k and considering the blade in the straight configuration[23] the equilibrium equation in the axial direction is given by

$$P = kl_{eq}, \tag{190}$$

so that the length of the outer part of the blade is $l(P) = \bar{l} - P/k$ and the Euler formula (188) becomes

$$P_{cr,n} = \frac{\pi^2 B}{\rho_n^2 \left(\bar{l} - \dfrac{P_{cr,n}}{k}\right)^2}, \qquad n \in \mathbb{N}^+. \tag{191}$$

The determination of the critical load $P_{cr,n}$ defined by equation (191) leads to the following cubic equation,

$$p_{cr,n}^3 - 2p_{cr,n}^2 + p_{cr,n} - \frac{4}{27q_n} = 0, \qquad n \in \mathbb{N}^+, \tag{192}$$

where $p_{cr,n}$ and q_n are respectively the dimensionless n-th critical load and dimensionless relative stiffness (spring stiffness multiplied by the rod's length and divided by a critical load) of the elastic system given as,

$$p = \frac{P}{k\bar{l}}, \qquad q_n = \frac{4\rho_n^2 k\bar{l}^3}{27\pi^2 B}, \qquad n \in \mathbb{N}^+. \tag{193}$$

Note that the dimensionless stiffness parameter is positive, $q_n > 0$, and that the dimensionless critical load has to satisfy the following inequality,[24]

$$p_{cr,n} \le 1, \qquad n \in \mathbb{N}^+, \tag{194}$$

to avoid the the full penetration of the blade, occurring when $l(P) < 0$.

From the cubic equation (192) it can be concluded that:

---

[23] If the rod is in a deformed configuration, an Eshelby-like force affects the axial equilibrium, see section 5.

[24] This restriction holds only for the calculation of the critical loads. Indeed, equilibrium configurations with $p \ge 1$ are possible for non-trivial deformation paths (see Bigoni et al., 2014b).

i) since all the coefficients of the cubic equation (192) are real, the following infinite sequence always exists of real roots

$$p_{cr,n}^{\mathcal{C}} = \frac{1}{3}\left[2 + \sqrt[3]{\frac{q_n}{2 - q_n + 2\sqrt{1 - q_n}}} + \sqrt[3]{\frac{2 - q_n + 2\sqrt{1 - q_n}}{q_n}}\right] > 1, \quad n \in \mathbb{N}^+, \tag{195}$$

all violating the constraint (194) and thus representing mathematical solutions which are meaningless from mechanical point of view;

ii) in the case when for a given $m \in \mathbb{N}^+$ the inequality

$$q_{m+1} < 1 < q_m, \qquad \text{or equivalently} \qquad \left(\frac{\rho_1}{\rho_m}\right)^2 < q_1 < \left(\frac{\rho_1}{\rho_{m+1}}\right)^2, \tag{196}$$

is satisfied, in addition to the sequence of real roots (195) other $2m$ real roots exist for the cubic equation (192),

$$\left.\begin{array}{c} p_{cr,n}^{\mathcal{A}} \\ \\ p_{cr,n}^{\mathcal{B}} \end{array}\right\} = \frac{1}{3}\left[2 - \frac{1 \pm i\sqrt{3}}{2}\sqrt[3]{\frac{q_n}{2 - q_n + 2\sqrt{1 - q_n}}}\right. \\ \left. - \frac{1 \mp i\sqrt{3}}{2}\sqrt[3]{\frac{2 - q_n + 2\sqrt{1 - q_n}}{q_n}}\right], \qquad n \in \mathbb{N}^+ \tag{197}$$

satisfying

$$0 < p_{cr,n}^{\mathcal{A}} \leq p_{cr,m}^{\mathcal{A}} \leq p_{cr,m}^{\mathcal{B}} \leq p_{cr,n}^{\mathcal{B}} < 1, \qquad n \leq m \qquad n, m \in \mathbb{N}^+ \tag{198}$$

so that $2m$ critical loads are obtained, corresponding to two critical loads referred to the same n-th instability mode;

iii) in the particular case when $q_m = 1$ (or equivalently $q_1 = \rho_1^2/\rho_m^2$), two real roots associated to the m-th mode (197) are coincident,

$$p_{cr,m}^{\mathcal{A}} = p_{cr,m}^{\mathcal{B}} = \frac{1}{3}, \qquad m \in \mathbb{N}^+, \tag{199}$$

therefore $2m - 1$ critical loads are obtained in this case.

In Bigoni et al. (2014b) it is shown that $p_{cr}^{\mathcal{A}}$ corresponds to a critical *buckling* load, while $p_{cr}^{\mathcal{B}}$ to a critical *restabilization* load for which the trivial path returns to be stable. In a similar vein, it can be concluded for the structures now under consideration that:

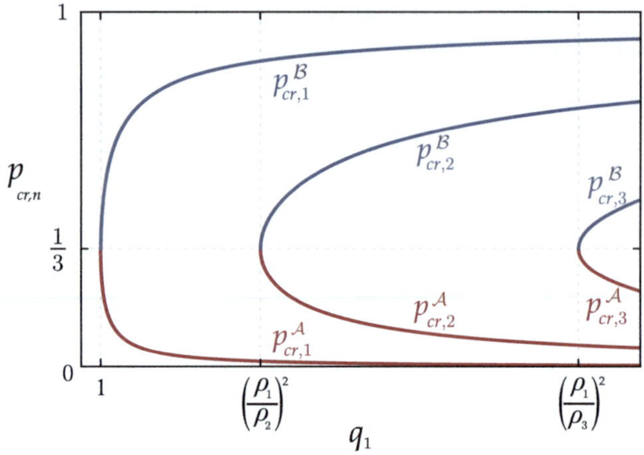

Figure 35. Dimensionless bifurcation loads $p_{cr,n}$ (*buckling* load $p^{\mathcal{A}}_{cr,n}$ and *restabilization* load $p^{\mathcal{B}}_{cr,n}$) as a function of the dimensionless relative stiffness q_1. Note that if the stiffness ratio q_1 decreases then the n-th buckling load increases while the n-th restabilization load decreases, and the number of bifurcation modes can even reduce to zero in the case of 'highly compliant systems' ($q_1 < 1$), where bifurcation does not exist.

- ◇ for fixed properties of the rod, a reduction of the spring stiffness k leads to an increase of the *buckling* load $p^{\mathcal{A}}_{cr}$ (a result agreeing with Tarnai, 1980) and to a reduction of the *restabilization* load $p^{\mathcal{B}}_{cr}$;
- ◇ differently from the behaviour observed in usual structures, the elastic systems under consideration can have a countable (finite) number of critical loads;
- ◇ 'highly compliant systems' ($q_1 < 1$) do not show any bifurcation.

6.2 Non-trivial configurations for the blade

Clamped blade Let us consider the system shown in Fig. 34 (upper part), where an inextensible elastic planar rod is forced to penetrate a sliding sleeve, restrained by an axial linear spring of stiffness k, through the application of a dead compressive load P at the other end, which is constrained by a movable clamp.

Considering only the outer part of the rod at an equilibrium configuration with $\theta(s) = 0$ for $s \in [0, l_{eq}]$ and introducing the dimensionless axial load $\lambda^2 = P/B$, the rotation field at equilibrium $\theta_{eq}(s)$ (unknown for the outer part, $s \in [l_{eq}, \bar{l}]$) is a solution of the following differential problem

$$\begin{cases} \dfrac{\mathrm{d}^2\theta_{eq}(s)}{\mathrm{d}s^2} + \lambda^2 \sin\theta_{eq}(s) = 0, & s \in \left(l_{eq}, \bar{l}\right) \\ \theta_{eq}\left(l_{eq}\right) = \theta_{eq}\left(\bar{l}\right) = 0, & \\ \left.\dfrac{\mathrm{d}\theta_{eq}(s)}{\mathrm{d}s}\right|_{s=\frac{\bar{l}+3l_{eq}}{4}} = 0, & \\ \lambda^2 = \dfrac{k}{B} l_{eq} + \dfrac{1}{2}\left[\theta'_{eq}(l_{eq})\right]^2, & \end{cases} \tag{200}$$

Equation $(200)_4$ represents the equilibrium in the axial direction, revealing the presence of the configurational force $B\left[\theta'_{eq}(l_{eq})\right]^2/2$ (Bigoni et al., 2014b). For conciseness, only the first bifurcation mode will be analyzed, so that the symmetry of the problem allows us to consider only the first quarter of the outer blade, similarly to Section 2.2. The rotation at the inflection point is $\theta_{eq}\left((\bar{l}+3l_{eq})/4\right) = \theta_q$ so that, through a change of variable and the Riemann theorem, an integration of the differential problem (200) leads to the relation between the load parameter λ and the angle θ_q as follows

$$4\mathcal{K}\left(\sin\frac{\theta_q}{2}\right) = \lambda\left[\bar{l} - \lambda^2\frac{B}{k}\left(1 - 2\sin^2\frac{\theta_q}{2}\right)\right]. \tag{201}$$

The coordinates $x_1(s)$ and $x_2(s)$ of the rod's axis in the deformed configuration for $s \in \left(l_{eq}, \bar{l}\right)$ can be computed from rotation field $\theta_{eq}(s)$ by integration of equation (11) as

$$\begin{aligned} x_1(s) &= -s + \frac{2}{\lambda} E\left[\mathrm{am}\left(\lambda\left(s - l_{eq}\right), \sin\frac{\theta_q}{2}\right), \sin\frac{\theta_q}{2}\right] - l_{eq}, \\ x_2(s) &= \frac{2}{\lambda}\sin\frac{\theta_q}{2}\left[1 - \mathrm{cn}\left(\lambda\left(s - l_{eq}\right), \sin\frac{\theta_q}{2}\right)\right]. \end{aligned} \tag{202}$$

Rotationally constrained blade The differential problem governing the equilibrium of the system reported in Fig. 34 (center), where the elastic blade is constrained in rotation at its loaded end, is

$$\begin{cases} \dfrac{\mathrm{d}^2\theta_{eq}(s)}{\mathrm{d}s^2} + \lambda^2 \sin\theta_{eq}(s) = 0, \qquad s \in \left(l_{eq}, \bar{l}\right) \\ \theta_{eq}\left(l_{eq}\right) = \theta_{eq}\left(\bar{l}\right) = 0, \\ \left.\dfrac{\mathrm{d}\theta_{eq}(s)}{\mathrm{d}s}\right|_{s=\frac{\bar{l}+l_{eq}}{2}} = 0, \\ \lambda^2 = \dfrac{k}{B} l_{eq} + \dfrac{1}{2}\left[\theta'_{eq}(l_{eq})\right]^2, \end{cases} \tag{203}$$

where the rotation at the inflection point (mid-span of the outer part of the elastic rod, $s = (\bar{l} + l_{eq})/2$) is denoted with θ_m. For this system, the relation between the dimensionless load parameter λ and the rotation at the rod's mid-span θ_m is written for the first buckling mode as

$$2\mathcal{K}\left(\sin\frac{\theta_m}{2}\right) = \lambda\left[\bar{l} - \lambda^2\frac{B}{k}\left(1 - 2\sin^2\frac{\theta_m}{2}\right)\right]. \tag{204}$$

The coordinates $x_1(s)$ and $x_2(s)$ of the rod's axis in the deformed configuration can be computed from the rotation field $\theta_{eq}(s)$ through integration of equation (11) and result to be expressed by equation (202), already obtained in the previous case of clamped end.

Simply supported blade Considering the system shown in Fig. 34 (lower part), where the elastic planar rod is simply supported at its loaded end and may slide into the frictionless sliding sleeve, the rotation field $\theta(s)$ of the outer part of the rod is obtained as the solution of the following differential problem

$$\begin{cases} \dfrac{\mathrm{d}^2\theta_{eq}(s)}{\mathrm{d}s^2} + \dfrac{P}{B}\sin\theta_{eq}(s) + \dfrac{R}{B}\sin\theta_{eq}(s) = 0, \qquad s \in \left(l_{eq}, \bar{l}\right) \\ \theta_{eq}\left(l_{eq}\right) = 0, \\ \left.\dfrac{\mathrm{d}\theta_{eq}(s)}{\mathrm{d}s}\right|_{s=\bar{l}} = 0, \\ \displaystyle\int_0^{\bar{l}} \sin\theta(s)\mathrm{d}s = 0, \\ \dfrac{\sqrt{P^2+R^2}}{B} = \dfrac{k}{B} l_{eq} + \dfrac{1}{2}\left[\theta'_{eq}(l_{eq})\right]^2. \end{cases} \tag{205}$$

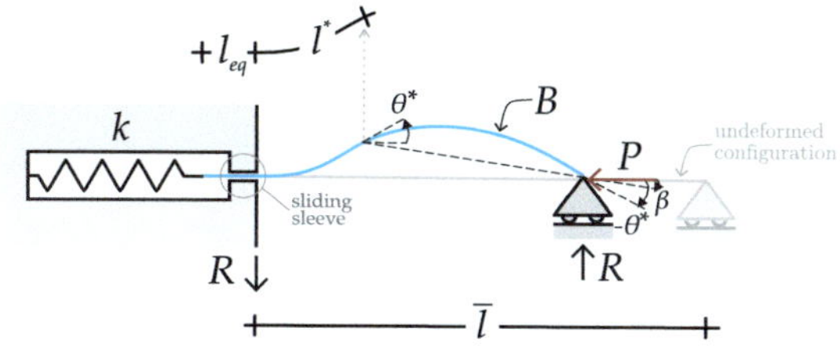

Figure 36. The penetrating blade with a simply supported end in a deformed configuration. An inflection point exists at $s = l^*$, where the rotation angle is denoted with $\theta(s = l^*) = \theta^*$. The angle β defines the inclination with respect to the horizontal direction of the resultant vector, sum of the the applied load P and of the vertical reaction of the support R.

Note that equation $(205)_5$ is the equilibrium equation in the sliding direction of the rod, where the 'Eshelby-like' force can be recognized. Introducing the dimensionless load $\gamma^2 = \sqrt{P^2 + R^2}/B$ and the angle $\psi(s) = \theta_{eq}(s) + \beta$, where β is the angle of inclination of the resultant vector sum of P and R (see Fig. 36), the differential problem (205) can be rewritten as

$$\begin{cases} \dfrac{\mathrm{d}^2\psi(s)}{\mathrm{d}s^2} + \gamma^2 \sin\psi(s) = 0, \qquad s \in \left(l_{eq}, \bar{l}\right) \\[2ex] \psi\left(l_{eq}\right) = \beta, \\[2ex] \left.\dfrac{\mathrm{d}\psi(s)}{\mathrm{d}s}\right|_{s=\bar{l}} = 0, \\[2ex] \displaystyle\int_0^{\bar{l}} \sin\left(\psi(s) - \beta\right) \mathrm{d}s = 0, \\[2ex] \dfrac{\sqrt{P^2 + R^2}}{B} = \dfrac{k}{B} l_{eq} + \dfrac{1}{2}\left[\psi'\left(l_{eq}\right)\right]^2. \end{cases} \tag{206}$$

Similarly to the antisymmetric buckling modes of a doubly clamped rod (Section 2.2), it is a standard expedient to operate the following change of variables

$$\eta = \sin\frac{\psi^*}{2}, \qquad \eta \sin\omega(s) = \sin\frac{\psi(s)}{2}, \tag{207}$$

where $\psi^* = \theta^* + \beta$. For the first bifurcation mode, integration of the differential problem (206) leads to the following relation between the dimensionless load parameter γ and the angle β

$$3\mathcal{K}\left(\eta\right)-\mathcal{K}\left(\eta,\omega_{\beta}\right)=\gamma\left[\bar{l}-\gamma^{2}\frac{B}{k}\left(1-2\eta^{2}\right)\right], \tag{208}$$

where $\omega_{\beta}=\arcsin\left(\eta^{-1}\sin\left(\frac{\beta}{2}\right)\right)$. Using the dimensionless parameters (193) it is possible to rewrite equation (208) in the following form

$$pq_{1}\left(1-p\left(1-2\eta^{2}\right)\right)^{2}=\left(1-2\sin^{2}\frac{\beta}{2}\right)\left[3\mathcal{K}(\eta)-\mathcal{K}\left(\eta,\omega_{\beta}\right)\right]^{2}. \tag{209}$$

The relation between the load P and the kinematic parameter θ^{*} can now be obtained by considering, in addition to eqn (209), the condition $(205)_4$, which provides a relation between θ^{*} and β, so that, similarly to equation (89), we obtain

$$\begin{aligned}0=&-2\eta\cos\omega_{\beta}\left(1-2\eta^{2}\sin^{2}\omega_{\beta}\right)\\&+2\eta\sin\omega_{\beta}\sqrt{1-\eta^{2}\sin^{2}\omega_{\beta}}\left\{3\left[2E(\eta)-K(\eta)\right]-2E(\omega_{\beta},\eta)+K(\omega_{\beta},\eta)\right\}.\end{aligned} \tag{210}$$

Equations (209) and (210) are highly non-linear and determine the non-trivial solution. The relation between β and θ^{*} (the former contained in ω_{β} and the latter in η) can be numerically obtained from equation (210) and used into equation (209) to obtain the relation between P and θ^{*}.

Finally, the coordinates $x_1(s)$ and $x_2(s)$ of the rod's axis in the deformed configuration for $s\in\left(l_{eq},\bar{l}\right)$ can be computed from the rotation field $\theta_{eq}(s)$ by integrating equation (11) as

$$
\begin{aligned}
x_1(s) = &+\sin\beta \left[-\frac{2\eta}{\gamma}\mathrm{cn}\big(\gamma\,(s - l_{eq}) + \mathcal{K}(\omega_\beta,\eta),\eta\big) + \frac{2\eta}{\gamma}\mathrm{cn}\big(\mathcal{K}(\omega_\beta,\eta),\eta\big)\right] \\
&\cos\beta \Big\{ -s + \frac{2}{\gamma}\left[E\left[\mathrm{am}\left(\gamma\,(s) + \mathcal{K}(\omega_\beta,\eta),\eta\right),\eta\right]\right. \\
&\left. - E\left[\mathrm{am}\left(\mathcal{K}(\omega_\beta,\eta),\eta\right),\eta\right]\right] \Big\},
\end{aligned}
$$

$$
\begin{aligned}
x_2(s) = &\cos\beta \left[-\frac{2\eta}{\gamma}\mathrm{cn}\big(\gamma\,(s) + \mathcal{K}(\omega_\beta,\eta),\eta\big) + \frac{2\eta}{\gamma}\mathrm{cn}\big(\mathcal{K}(\omega_\beta,\eta),\eta\big)\right] \\
&- \sin\beta \Big\{ -s + \frac{2}{\gamma}\left[E\left[\mathrm{am}\left(\gamma\,(s) + \mathcal{K}(\omega_\beta,\eta),\eta\right),\eta\right]\right. \\
&\left. - E\left[\mathrm{am}\left(\mathcal{K}(\omega_\beta,\eta),\eta\right),\eta\right]\right] \Big\}.
\end{aligned}
\tag{211}
$$

7 The elastica arm scale

The integration of the elastica and the concept of configurational forces so far developed are now employed in the design of an innovative weighing device. It is well-known that for millennia the equal and unequal arm balance scales have been used (for instance the classic Roman balance, see Fig. 37 left), and still are used (see the overview by Robens et al., 2014), to measure weight by exploiting the equilibrium of a rigid lever, so that a deformation of the arms would merely represent an undesired effect. On the other hand, the modern digital weighting systems, inspired by the principle of the spring balance (which was invented at the end of the 17th century by R. Hooke, Fig. 37 on the right), are based on the strain of an elastic element, so that equilibrium is always satisfied and a counterweight is not needed.

A new concept is now introduced of an 'elastica arm scale', based on the exploitation of nonlinear kinematics and configurational mechanics of elastic rods, following Bosi et al. (2014). The new scale has deformable arms, so that an inflected equilibrium configuration is employed to measure weight and the scale can work both *with or without a counterweight.* In a sense, this new type of balance is a combination of a rigid arm balance with a spring balance, because equilibrium and deformation are both simultaneously exploited. The prototype of the elastica arm scale is shown on the right of Fig. 38, as a realization of the scheme reported on the left of the same figure, where an elastic rod (of total of length $\bar{l} + l^*$ and inclined at

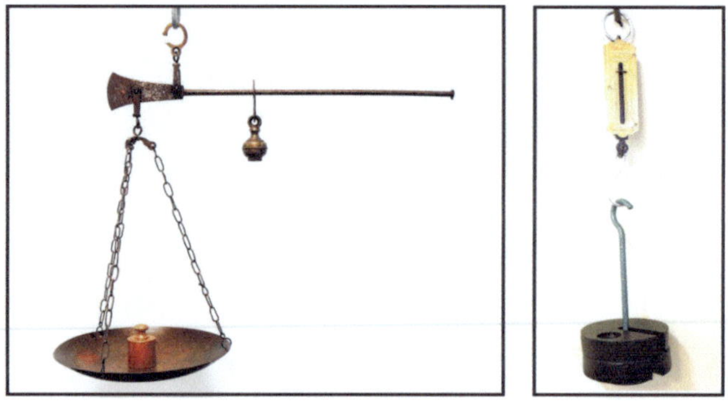

Figure 37. A steelyard based on rigid lever principle (left) and a spring balance based on deformation (right).

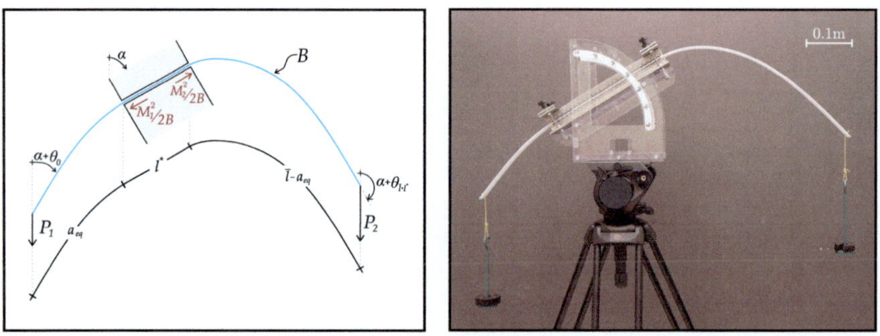

Figure 38. Scheme (left) and prototype (right) of the deformable arm scale. An elastic lamina of bending stiffness B is inserted into a sliding sleeve of length l^* inclined at an angle α with respect to the vertical direction. The length of the left part of the lamina is a_{eq}, while the length of the second is $\bar{l} - a_{eq}$.

an angle $\alpha \in [0, \pi/2]$ with respect to the vertical loads applied at its ends) is free to slide in a frictionless sleeve of length l^*. For given loads (P_1 and P_2), the scale admits an equilibrium configuration, possible by virtue of the flexural deformation of the arms.

The presence of 'Eshelby-like' forces (see section 5) at the ends of the sliding sleeve defines the nonlinear equilibrium equation in the sliding direction that can be written as

$$(P_1 + P_2)\cos\alpha + \underbrace{\frac{M_1^2 - M_2^2}{2B}}_{\text{Eshelby-like forces}} = 0, \tag{212}$$

or, equivalently, but with reference to the angles at the ends of the elastic rod, as the following 'geometrical condition' of equilibrium

$$P_1 \cos(\alpha + \theta_0) + P_2 \cos(\alpha + \theta_{\bar{l}+l^*}) = 0, \tag{213}$$

representing the balance of axial thrust of the deformable scale ($0 \leq \alpha + \theta_0 \leq \alpha$ and $\pi/2 \leq \alpha + \theta_{\bar{l}+l^*} \leq \pi$). The equilibrium equations of the two arms of the scale are

$$B\theta_{eq}^{''}(s) - P_j \sin\left[\theta_{eq}(s) - (-1)^j \alpha\right] = 0, \tag{214}$$

where $j = 1$ for the left arm ($s \in [0, a_{eq}]$) and $j = 2$ for the right one ($s \in \left[a_{eq} + l^*, \bar{l} + l^*\right]$). From integration of these two differential equations we can obtain the relations between the applied loads P_1 and P_2, the kinematical parameters θ_0 and $\theta_{\bar{l}+l^*}$ and the lengths of the two arm a_{eq} and $\bar{l} - a_{eq}$ as

$$a_{eq}\sqrt{\frac{P_1}{B}} = \mathcal{K}\left(\kappa_1\right) - \mathcal{K}\left(m_1, \kappa_1\right), \qquad \left(\bar{l} - a_{eq}\right)\sqrt{\frac{P_2}{B}} = \mathcal{K}\left(\kappa_2\right) - \mathcal{K}\left(m_2, \kappa_2\right), \tag{215}$$

where $\mathcal{K}\left(\kappa_j\right)$ and $\mathcal{K}\left(m_j, \kappa_j\right)$ are respectively the complete (39) and incomplete (83) elliptic integral of the first kind, and

$$\begin{aligned}
&\kappa_1 = \sin\frac{\theta_0 + \alpha + \pi}{2}, && m_1 = \arcsin\left[\frac{\sin\dfrac{\alpha + \pi}{2}}{\kappa_1}\right],\\
&\kappa_2 = \sin\frac{\theta_{\bar{l}+l^*} + \alpha}{2}, && m_2 = \arcsin\left[\frac{\sin\dfrac{\alpha}{2}}{\kappa_2}\right],\\
&\kappa_1 \sin\phi_1(s) = \sin\frac{\theta_{eq}(s) + \alpha + \pi}{2}, && \kappa_2 \sin\phi_2(s) = \sin\frac{\theta_{eq}(s) + \alpha}{2}.
\end{aligned} \tag{216}$$

Note that, when $\alpha + \theta_{\bar{l}+l^*} = \pi/2$ the equilibrium equation (213) implies $P_1 = 0$, so that *a counterweight is not needed.*

Furthermore, when the sliding sleeve is in the vertical direction, namely $\alpha = 0°$, the equilibrium is governed by the following *purely geometrical condition*, visibly satisfied in Fig. 39,

$$P_2 \cos\left(\theta_{\bar{l}+l^*}\right) + P_1 = 0, \tag{217}$$

where the two weights have to satisfied the condition

$$0 \leq \frac{P_1}{P_2} \leq 1. \tag{218}$$

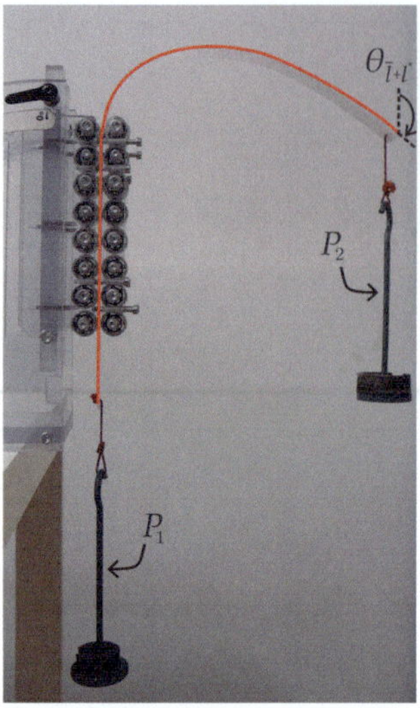

Figure 39. Example of use of the elastica arm scale when the sliding sleeve is in the vertical direction, $\alpha = 0$. The Eshelby-like force guarantees equilibrium of the elastic lamina subject to two loads $P_2 = 2$N and $P_1 = 3$N, so that the angle $\theta_{\bar{l}+l^*}$ is equal to $131.81°$.

7.1 Mode of use of the elastic scale

The following modes of use of the elastica arm scale can be envisaged.

- ◇ The easiest way to use the elastica arm scale is with reference to equation (213) and therefore measuring the two angles θ_0 and $\theta_{\bar{l}+l^*}$. In this way, assuming that P_1 and α are known, P_2 can be evaluated. Note that the knowledge of the bending stiffness B is not needed in this mode of use.
- ◇ Another mode of use of the elastica arm scale is through the measure of the length a_{eq}. Knowing P_1, B, and α, the weight P_2 can be determined in the following steps:
  - i) Equation $(215)_1$ gives θ_0;
  - ii) Equation (213) gives $\theta_{\bar{l}+l^*}$ as a function of the unknown P_2;
  - iii) Equation $(215)_2$ can be numerically solved for the unknown P_2.

Note that equations (215) define a_{eq} as a one-to-one function respectively of θ_0 (first equation) and of $\theta_{\bar{l}+l^*}$ (second equation), while equation (213) defines a unique relation between θ_0 and $\theta_{\bar{l}+l^*}$. Therefore, excluding all deformations of the elastica which would be unstable even for clamped ends,

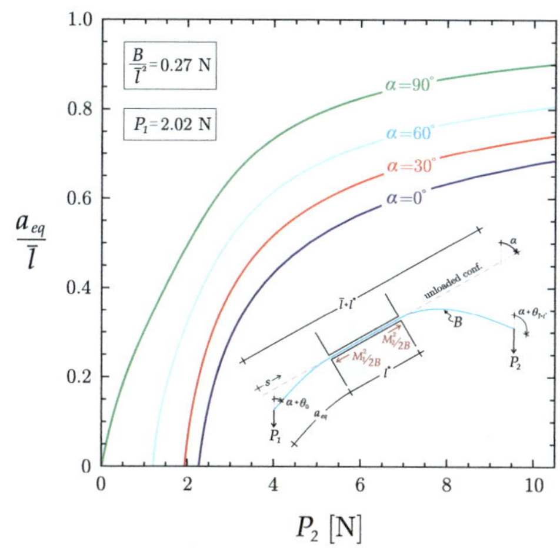

Figure 40. The elastica arm scale: equilibrium length a_{eq} versus weight P_2 for different inclinations $\alpha = \{0; 30; 60; 90\}^\circ$. The values of the counterweight P_1 and the parameter $B/\bar{l}$ are the same adopted in the experiments reported by Bosi et al. (2014).

when the equilibrium solution of equations (215) and (213) exists, it is also unique.

The graph reported in Fig. 40 can be used in the second of the above-listed modes of use of the elastica arm scale to obtain the value of P_2 from the measured length a_{eq}. The inclination of the scale α can be adjusted to obtain a desired range of variation for the measured weight P_2 or sensitivity $\mathcal{S}$ (see Section 7.2). In fact, when α increases from 0° (vertical configuration) to 90° (horizontal configuration), the range of measure for the weight tends to increase, up to the case when all possible values of P_2 can be covered, namely $\alpha = 90°$.

7.2 Sensitivity analysis and comparisons between different scales

The performance of the elastica arm scale can be appreciated through a comparison with the steelyard (Fig. 37, left), a simple device still used nowadays. The comparison is performed through the sensitivity parameter $\mathcal{S}$, when the properties of the scales (inclination angle α, bending stiffness B, length $\bar{l}$) are varied.

Steelyard The steelyard is based on the principle of the lever with two rigid arms, so that it operates exactly as the second mode of use of the elastica arm scale, presented in Section 7.1. In this scale the equilibrium equation is guaranteed by the moment balance at the fulcrum

$$P_1 a_{eq} = P_2 \left(\bar{l} - a_{eq} \right), \tag{219}$$

where P_1 is the counterweight associated with the left arm of length a_{eq}, whereas P_2 represents the weight (to be measured) linked with the right arm of length $\bar{l} - a_{eq}$ (see also the scheme in Fig. 41). Therefore, also for the unequal balance, once the values of the counterweight P_1 and the total length of the two rigid arms $\bar{l}$ are fixed, after measuring the length a_{eq}, the value of unknown load P_2 can be obtained from the linear equation (219).

Sensitivity analysis The concept of sensitivity associated to scales allows to compare the precisions in weighing (Robens et al. (2014)). According to DIN/ISO, sensitivity is defined as the response of a measuring instrument, which may be an angle or a length, divided by the corresponding change in the stimulus, that is in our case the weight to be evaluated. For the considered scale, the sensitivity $\mathcal{S}$ is defined as the ratio between the observed variation of a_{eq} and the corresponding variation of the measured weight P_2 for a fixed value of the counterweight P_1,

$$\mathcal{S} = \frac{\partial a_{eq}}{\partial P_2}. \tag{220}$$

For the steelyard, taking into account the equilibrium equation (219), the sensitivity can be evaluated as

$$\mathcal{S}_{\text{steelyard}} = \frac{P_1 \bar{l}}{\left(P_1 + P_2\right)^2}, \tag{221}$$

while the sensitivity $\mathcal{S}$ of the elastica arm scale can be evaluated only numerically due to the non-linearity of the equilibrium equations (215) and (213).

Comparison between the steelyard and the elastica arm scale A comparison between the unequal arm balance and the elastica arm scale is reported in Fig. 41, where the equilibrium length a_{eq} as a function of the unknown weight P_2 is shown on the left, while the sensitivity $\mathcal{S}$, representing the tangent to the curve on the left, is shown on the right. From Fig. 41 (left), it can be noted that, while the steelyard can measure every value of the weight P_2, for the elastica arm scale there is a minimum value for the weight P_2 that can be measured, except in the case when the device is in the horizontal position ($\alpha = 90°$). On the other hand, the sensitivity

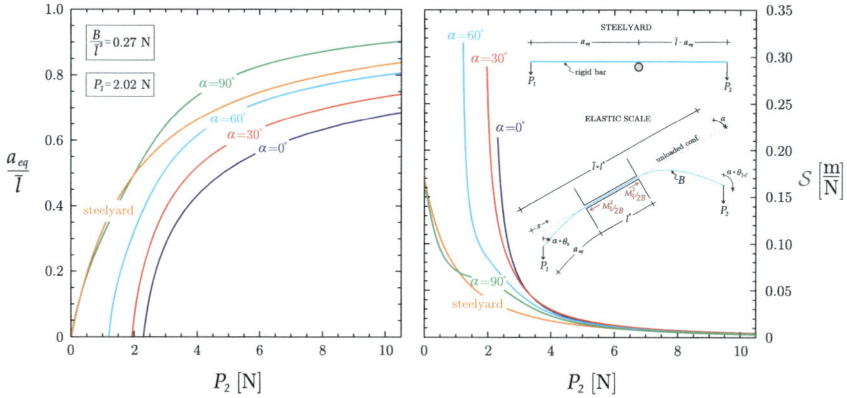

Figure 41. Comparison between the performances of a steelyard and of the elastica arm scale for different inclinations $\alpha = \{0°; 30°; 60°; 90°\}$: equilibrium length a_{eq} (left) and sensitivity $\mathcal{S}$ (right) are reported versus the weight to be measured P_2. The values of the counterweight P_1 and the parameter $B/\bar{l}^2$ have been assumed fixed and equal to the same values adopted in the experiments performed by Bosi et al. (2014).

analysis shows that the inclined elastica arm scale ($\alpha \neq 0$) can measure weights with a precision higher than that observed with the steelyard. In fact, once P_2 is fixed, the sensitivity $\mathcal{S}$ increases at increasing inclination towards the vertical configuration ($\alpha = 0°$), where possible effects due to friction are also considerably reduced.

Comparison between different elastica arm scales Finally, a comparison between elastica arm scales with two different values for $B/\bar{l}^2$ each with different inclinations α is shown in Fig. 42. The figure shows that a decrease (increase) in the parameter $B/\bar{l}^2$ leads to an increase (decrease) in both the range of measured weights P_2 and of sensitivity $\mathcal{S}$. Therefore, a more accurate device can be realized either by reducing the rod's bending stiffness (at fixed length $\bar{l}$) or by increasing the length $\bar{l}$ (at fixed bending stiffness B).

A proof-of-concept device showing how the elastica arm balance works was realized by Bosi et al. (2014), movies of the experiments can be found at http://ssmg.unitn.it/elasticscale.html.

7.3 A perspective view on configurational forces

Configurational or Eshelby-like forces emerge in a mechanical system when the possibility arises of a change in configuration with a consequent

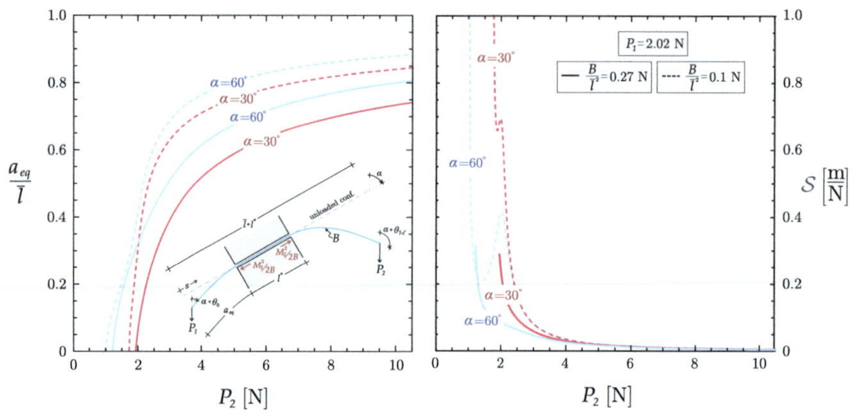

Figure 42. Comparison between two elastica arm scales differing in the parameter B/l^2 for two different inclinations $\alpha = \{30°; 60°\}$: equilibrium length a_{eq} (left) and sensitivity $\mathcal{S}$ (right) are reported versus the weight to be measured P_2.

release of elastic energy. These forces are therefore more widespread than the few simple structural examples shown above. For instance, configurational forces have been revealed under torsion (Bigoni et al., 2014c) and the same forces are responsible for snake locomotion; in fact the sliding sleeve used in the above structural systems can be viewed as a frictionless, narrow channel in which an elastic rod can move. Our results show that motion along this channel can be induced even when the applied forces are orthogonal to it, which is the essence of the locomotion strategy employed by a snake, which exploits lateral friction to generate a constraint (similar to the channel) and releases bending energy to generate a propulsive force, see Gray (1974), Gray and Lissmann (1950) and Gray (1953).

8 A concluding remark

The study of the Euler's planar elastica is useful from many points of view: it represents a nice introduction to the complex behaviour of nonlinear mechanical systems and provides an important tool in the design of flexible mechanisms such as in the emergent field of soft robotics. Indeed, we have used this tool to create elastic structures capable of displaying new mechanical behaviour, such as tensile instability, configurational forces, restabilization of the trivial path, and an innovative measuring device.

In a world where machines drive blind researchers along unknown and often false directions, our structures have been sketched on a piece of paper and solved usually by hand calculations, sometimes with the help of a numerical solver, before the development of experiments. The solutions have guided the design of prototypes capable of giving evidence to phenomena first discovered with the 'paper solution'. The evidence was usually so closely following predictions that we found our experiments loving our theory.

Acknowledgements Financial support of the FP7-PEOPLE-IDEAS-ERC-2013-ADG-340561-INSTABILITIES is gratefully acknowledged.

Bibliography

L.I. Balabukh, M.N. Vulfson, B.V. Mukoseev and Ya G. Panovko (1970). On work done by reaction forces of moving supports. *Research on Theory of Constructions*, Moscow. **18**, 190-200.

D. Bigoni and L. Deseri (2011). *Recent Progress in the Mechanics of Defects.* Springer.

D. Bigoni (2012). *Nonlinear Solid Mechanics. Bifurcation theory and material instability.* Cambridge University Press.

D. Bigoni, D. Misseroni, G. Noselli and D. Zaccaria (2012). Effects of the constraints curvature on structural instability. *Proc. Roy. Soc. A*, **468**, 2191-2209.

D. Bigoni, D. Misseroni, G. Noselli and D. Zaccaria (2014a) Surprising instabilities of simple elastic structures. In *Nonlinear Physical Systems - Spectral Analysis, Stability and Bifurcations, Kirillov, N. and Pelinovsky, D.E. Eds., Wiley*, 1-14; ISBN: 978-1-84821-420-0.

D. Bigoni, F. Bosi, F. Dal Corso and D. Misseroni (2014b). Instability of a penetrating blade. *J. Mech. Phys. Solids*, **64**, 411-425.

D. Bigoni, F. Dal Corso, D. Misseroni and F. Bosi (2014c). Torsional locomotion. *Proc. Roy. Soc. A*, **470** (2171), 20140599.

D. Bigoni, F. Dal Corso, F. Bosi and D. Misseroni (2015). Eshelby-like forces acting on elastic structures: theoretical and experimental proof. *Mech. Materials*, **80**, 368-374.

F. Bosi, D. Misseroni, F. Dal Corso and D. Bigoni (2014). An Elastica Arm Scale. *Proc. Roy. Soc. A*, **470** (2169), 20140232.

A. Broman (1970). *Introduction to Partial Differential Equations: From Fourier Series to Boundary-Value Problems.* Addison-Wesley, London.

P.F. Byrd and M.D. Friedman (1954). *Handbook of elliptic integrals for engineers and scientists.* Springer-Verlag, Berlin.

R. Courant and D. Hilbert (1962). *Methods of Mathematical Physics.* J. Wiley and Sons, New York.

C. Dascalu, G.A. Maugin and C. Stolz (2010). *Defect and Material Mechanics*. Springer.

G. Domokos (1994). Global description of elastic bars. *Z. Angew. Math. Mech.*, **74**, T289-T291.

J.D. Eshelby (1956). The continuum theory of lattice defects. *Solid State Phys.*, **3** (C), 79-144.

V.I. Feodosyev (1977). *Selected Problems and Questions in Strength of Materials*. MIR, Moscow.

A. Gajewski and R. Palej (1974). Stability and shape optimization of an elastically clamped bar under tension (in Polish). *Rozprawy Inzynierskie - Engineering Transactions*, **22**, 265-279.

Gray, J. (1946) The mechanism of locomotion in snakes. J. Exp. Biol. 23, 101-120.

Gray, J. and Lissmann H.W. (1950) The kinetics of locomotion of the grass-snake. J. Exp. Biol. 26, 354-367.

Gray, J. (1953) How animals move. Cambridge University Press.

M.E. Gurtin. (2000). *Configurational forces as basic concept of continuum physics*. Springer.

K.A. Hoffman, R.S. Manning and R.C. Paffenroth (2002). Calculation of the stability index in parameter-dependent calculus of variations problems: buckling of a twisted elastic strut. *SIAM J. Appl. Dyn. Syst.*, **1** (1), 115-145.

R. Kienzler and G. Herrmann (2000). *Mechanics in Material Space*. Springer.

V.M. Kuznetsov and S.V. Levyakov (1999). Secondary loss of stability of an euler rod. *J. Appl. Mech. Tech. Phy.*, **40** (6), 1161-1162.

V.M. Kuznetsov and S.V. Levyakov (2002). Complete solution of the stability problem for elastica of Euler's column. *Int. J. Non-Linear Mech.*, **37** (6), 1003-1009.

H. Lamb (1928). *Statics*. Cambridge University Press.

S.V. Levyakov and V.M. Kuznetsov (2010). Stability analysis of planar equilibrium configurations of elastic rods subjected to end loads. *Acta Mech.*, **211**, 73-87.

A.E.H. Love (1927). *A treatise on the mathematical theory of elasticity*. Cambridge University Press.

J.H. Maddocks (1984). Stability of nonlinear elastic rods. *Arch. Rat. Mech. Analysis*, **85** (4), 311-354.

C. Majidi (2007). Remarks on formulating an adhesion problem using Euler's elastica. *Mech. Res. Comm.*, **34** (1), 85-90.

C. Majidi, O.M. O'Reilly and J.A. Williams (2012). On the stability of a rod adhering to a rigid surface: Shear-induced stable adhesion and the instability of peeling. *J. Mech. Phys. Solids*, **60** (5), 827-843.

R.S. Manning, K.A. Rogers and J.H. Maddocks (1984). Isoperimetric conjugate points with application to the stability of DNA minicircles. *Proc. Roy. Soc. A*, **454** (1980), 3047-3074.

R.S. Manning (2009). Conjugate points revisited and Neumann-Neumann problems. *SIAM Review*, **51** (1), 193-212.

R.S. Manning (2014). A catalogue of stable equilibria of planar extensible or inextensible elastic rods for all possible dirichlet boundary conditions. *J. Elast.*, **115** (2), 105-130.

G.A. Maugin (1993). *Material Inhomogeneities in Elasticity*, Applied Mathematics and Mathematical Computation. Springer, London.

G.A. Maugin (2011). *Configurational forces: Thermodynamics, physics, mathematics and numerics.* Taylor and Francis Ltd, New York.

Y. Mikata (2007). Complete solution of elastica for a clamped-hinged beam, and its applications to a carbon nanotube. *Acta Mech.*, **190**, 133-150.

O.M. O'Reilly and D.M. Peters (2011). On Stability Analysis of Three Classical Buckling Problems for the Elastic Strut. *J. Elast.*, **105**, 117-136.

O.M. O'Reilly and D.M. Peters (2012). Nonlinear stability criteria for tree-like structures composed of branched elastic rods. *Proc. Roy. Soc. A*, **468**, 206-226.

M. Potier-Ferry (1987). Foundations of elastic postbuckling theory. *Buckling and Post-Bucklings*, Lecture Notes in Physics, Springer Berlin Heidelberg, **288**, 1-82.

E.L. Reiss (1969). Column buckling: An elementary example of bifurcation. *Bifurcation theory and nonlinear eigenvalue problems.* W.A. Benjamin Inc., New York, 1-16.

E. Robens, S.A.A. Jayaweera and S. Kiefer (2014). *Balances. Instruments, Manufacturers, History.* Springer Berlin Heidelberg.

J. Tarnai (1980). Destabilizing effect of additional restraint on elastic bar structures. *Int. J. Mech. Sci.*, **22** (6), 379-390.

N.M. Temme (1996). *Special functions.* John Wiley and Sons, New York.

S.P. Timoshenko and J.M. Gere (1961). *Theory of elastic stability.* McGraw-Hill, New York.

B. Van Brunt (2005). *The calculus of variations.* Springer.

M.A. Vaz and D.F.C. Silva (2003). Post-buckling analysis of slender elastic rods subjected to terminal forces. *Int. J. Nonlinear Mech.*, **38**, 483-492.

C.Y. Wang (1997). Post-buckling of a clamped-simply supported elastica. *Int. J. Nonlinear Mech.*, **32** (6), 1115-1122.

D. Zaccaria, D. Bigoni, G. Noselli and D. Misseroni (2011). Structures buckling under tensile dead load. *Proc. R. Soc.A*, **467**, 1686-1700.

H. Ziegler (1977). Principles of Structural Stability. *Birkhauser Verlag, Basel und Stuttgart.*

Dynamic curling of an Elastica: a nonlinear problem in elastodynamics solved by matched asymptotic expansions

B. Audoly [*], A. Callan-Jones [†] and P.-T. Brun [*‡]
[*] Sorbonne Universités, UPMC Univ Paris 06, CNRS, UMR 7190 Institut Jean Le Rond d'Alembert, F-75005 Paris, France
[†] Univ. Paris 7 Denis Diderot, CNRS, UMR 7057, Matière et Systèmes Complexes, F-75013 Paris, France
[‡] Department of Mathematics, Massachusetts Institute of Technology, Cambridge, Massachusetts 02139, USA

Abstract We consider the motion of an infinitely long, naturally curved, planar Elastica. The Elastica is flattened onto a rigid impenetrable substrate and held by its endpoints. When one of its endpoints is released, it is set off in a curling motion, which we seek to describe mathematically based on the non-linear equations of motions for planar elastic rods undergoing finite rotations. This problem is used to introduce the technique of matched asymptotic expansions. We derive a non-linear solution capturing the late dynamics, when a roll comprising many turns has formed: in this regime, the roll advances at an asymptotically constant velocity, whose selection we explain. This contribution presents an expanded version of the results published in Callan-Jones *et al.* (Phys. Rev. Lett. 2012).

1 Introduction

In this chapter, the dynamic curling of an elastic coil serves as an illustration of an classical method in mathematical physics, namely matched asymptotic expansions.

In biology, coiling can be observed in the tendrils of climbing vines such as *Bryonia dioica* in response to a mechanical stimulus (Engelberth, 2003): when stimulated mechanically, they respond by coiling on the timescale of minutes, which is rapid for plants. Unlike the elastic coil we study here, this coiling phenomenon is limited by viscosity.

D. Bigoni (Ed.), *Extremely Deformable Structures*,
CISM International Centre for Mechanical Sciences
DOI 10.1007/ 978-3-7091-1877-1_3 © CISM Udine 2015

We will analyze the curling for an elastic coil using the non-linear equations for a dynamic Elastica. A linear version of these equations is classically used to analyze dynamic buckling (Golubovic et al., 1998; Gladden et al., 2005; Schindler and Kolsky, 1983; Audoly and Neukirch, 2005) as well as the stability of equilibrium solutions (Caflisch and Maddocks, 1984). The dynamics of an Elastica has also been characterized by means of amplitude equations (Goriely and Tabor, 1996, 2000), which is a type of a weakly non-linear expansion. For problems involving stationary solutions in a rotating frame, such as the coiling of elastic ropes, the dynamical problem can be reformulated as an equilibrium problem with inertial forces (Mahadevan and Keller, 1996; Habibi et al., 2007). However, problems of elasticity that are genuinely dynamical (*i.e.* are not stationary in *any* frame) and non-linear (*i.e.* cannot be approached by perturbation methods) cannot be solved analytically in general. The curling of an elastic coil is a problem of this kind, and we derive below an analytical solution method which fully characterizes its long-time dynamics.

2 Model

2.1 Experiments

We carried out experiments using a steel strip, as described by Callan-Jones et al. (2012). The strip is reference CF0250043 (stainless steel type 301) from Ressort Spec, and has length $L = 635$ mm, thickness $a = 0.13$ mm, width $b = 9.5$ mm, and radius of natural curvature $\kappa_0^{-1} = 9.3$ mm. It makes approximately $L/(2\,\pi\,\kappa_0^{-1}) \approx 11$ turns in its undeformed (coiled) configuration. The material properties of the strip are mass per unit length $\rho = 9.732 \times 10^{-3}$ kg/m, Young's modulus $E = 193$ GPa, Poisson's ratio $\nu = .25$, and bending modulus $B = E\,a^3\,b/(12(1-\nu^2)) = 0.358 \times 10^{-3}$ N.m^2. In the following, the center-line position of the strip is parametrized by the arc-length variable s.

In the experiments, the spring was laid flat on an approximately 2 meter-long Norcan bar, secured at the end $s = L$ and released at the other end, $s = 0$, and its motion was imaged using a Photron fast camera at 7000 frames per second. Curling of the spring occurs in the xy-plane, as shown in figure 1a.

2.2 Scaling analysis

Using the geometrical and physical constants of the spring, one can construct a typical velocity $\sqrt{\frac{B\,\kappa_0^2}{\rho}} \sim 20$ m/s. This yields the correct order of magnitude: the experimentally measured velocity is $v = 12.5$ m/s. The goal

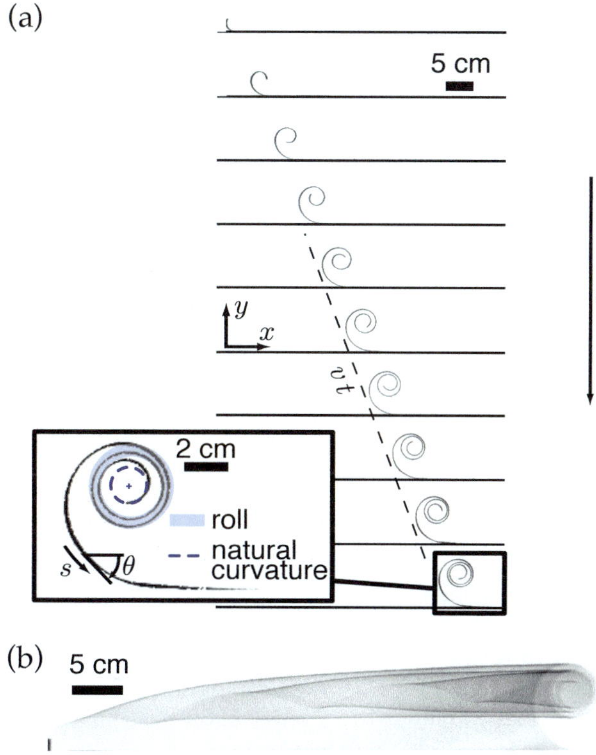

Figure 1. Experiments with a steel strip, reprinted from Callan-Jones et al. (2012). (a) Snapshots with a time interval of 2.85 ms. The dashed curve drawn across images represents a propagation of the front at constant velocity $v = 12.5$ m/s. The inset shows the structure of the curled region in the final snapshot, with a roll of constant curvature (light blue overlay) connected to a boundary layer where the natural curvature is recovered (dashed circle). (b) Long-exposure photograph until time $t = 30.85$ ms.

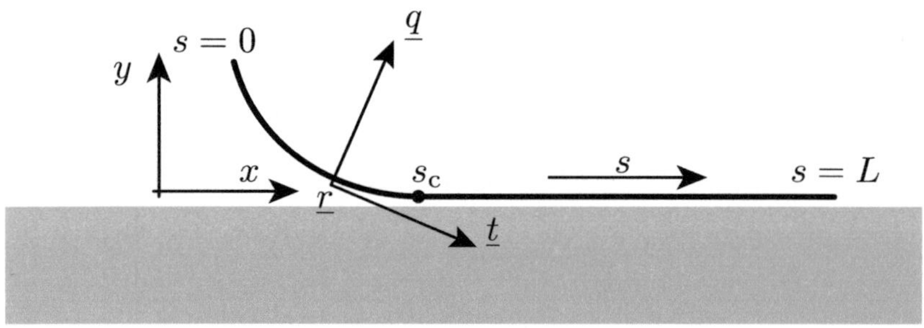

Figure 2. Geometry of a 2D Elastica

of the rest of the chapter is to construct an explicit solution and propose a quantitative prediction for the coiling velocity.

2.3 Equations of motion for a non-linear planar Elastica

With the aim to describe the shapes observed in the experiments, and in particular to explain the selection of the velocity v, we represent the coil as a naturally curved planar Elastica, *i.e.* a one-dimensional slender elastic structure moving in a two-dimensional world.

We use a Cartesian frame, with the axis x parallel to the edge of the rigid plane, and $y > 0$ corresponding to the half-plane in which the Elastica can move, see figure 2. Let $\underline{r}(s,t)$ denote the centerline position as a function of arclength s (which we use as Lagrangian coordinate) and time t. We use an inextensible Elastica model, and the centerline $\underline{r}(s,t)$ is subject to the inextensibility condition

$$|\underline{t}(r,s)| = 1 \tag{1}$$

where $\underline{t}$ is the tangent

$$\underline{t}(r,s) = \underline{r}'(s,t). \tag{2}$$

Primes denote derivation with respect to arclength s.

Denote the unit normal by $\underline{q}(s,t)$,

$$\underline{q}(s,t) = \underline{e}_z \times \underline{t}(s,t), \tag{3}$$

where $\underline{e}_z\times$ denotes the rotation of $+\pi/2$ in the plane (x,y).

The signed curvature strain $\kappa(s,t)$ is defined by the Frénet equation,

$$\underline{t}'(s,t) = \kappa(s,t)\, \underline{q}(s,t). \tag{4}$$

The component on the other vector $\underline{t}$ in the right-hand is zero, as $\underline{t}' \cdot \underline{t} = \frac{1}{2}\frac{\mathrm{d}\underline{t}^2}{\mathrm{d}s} = 0$ by the inextensibility condition.

The linear constitutive law reads

$$m(s,t) = B\,(\kappa(s,t) - \kappa_0), \tag{5}$$

where B is the bending modulus and κ_0 the natural curvature.

Neglecting rotational inertia, we write the balance of moment as

$$m'(s,t) + \underline{q}(s,t) \cdot \underline{n}(s,t) = 0. \tag{6}$$

This equation can be obtained by projecting the Kirchhoff equation $\underline{m}' + \underline{t} \times \underline{n} = \underline{0}$ along the perpendicular direction $\underline{e}_z$. Here, $\underline{n}(s,t)$ is the internal force, representing the resultant of the internal stress across a cross-section s. In our inextensible, unshearable rod model, the internal force $\underline{n}(s,t)$ is a Lagrange multiplier and has to be determined by solving the equations of equilibrium.

The balance of linear momentum is written as

$$\underline{n}'(s,t) + \underline{p}(s,t) = \rho\,\underline{\ddot{r}}(s,t), \tag{7}$$

where $\underline{p}(s,t)$ is the density of contact force coming from the substrate per unit length, and ρ the lineic mass.

Near the free end $s = 0$, we have the equilibrium conditions

$$\underline{n}(0,t) = \underline{0}, \qquad m(0,t) = 0. \tag{8}$$

2.4 Equations of motion in local basis

Define the normal force as the tangent component of the internal force, $N = \underline{n} \cdot \underline{t}$: positive values of N correspond to a tensile force, and negative values to a compressive force. Then, equation (6) implies

$$\underline{n}(s,t) = N(s,t)\,\underline{t}(s,t) - m'(s,t)\,\underline{q}(s,t). \tag{9}$$

The other component $(-m')$ is called the shear force.

We can then rewrite equation (7) as

$$(N' + \kappa\, m')\,\underline{t} + (-m'' + \kappa\, N)\,\underline{q} + \underline{p}(s,t) = \rho\,\underline{\ddot{r}}(s,t) \tag{10}$$

Projection onto the local basis $(\underline{t}, \underline{q})$ yields the balance of linear and angular momentum:

$$N' + \kappa\, m' + \underline{p} \cdot \underline{t} = \rho\,\underline{\ddot{r}} \cdot \underline{t} \tag{11a}$$

$$-m'' + \kappa\, N + \underline{p} \cdot \underline{q} = \rho\,\underline{\ddot{r}} \cdot \underline{q} \tag{11b}$$

A complete set of equations is obtained by combining this with the constitutive law (5), the free end boundary conditions (8) and the various geometric relations.

2.5 Decomposition of acceleration in the local basis

Let introduce a decomposition in the local frame $(\underline{t}, \underline{q})$ of the velocity measured in a travelling frame moving at constant velocity $v\,\underline{e}_x$, which will be the velocity of the coil,

$$\dot{\underline{r}} - v\,\underline{e}_x = v_{\mathrm{t}}\,\underline{t} + v_{\mathrm{q}}\,\underline{q}, \tag{12}$$

where a dot denotes differentiation with respect to time t.

This decomposition is well adapted to the analysis of a rolling motion: we will show that v_{q} is then much smaller than v_{t}. By differentiating with respect to time, we find the acceleration in the form

$$\ddot{\underline{r}} = \gamma_{\mathrm{t}}\,\underline{t} + \gamma_{\mathrm{q}}\,\underline{q} \tag{13}$$

where

$$\gamma_{\mathrm{t}} = -\omega\,v_{\mathrm{q}} + \dot{v}_{\mathrm{t}}, \qquad \gamma_{\mathrm{q}} = \omega\,v_{\mathrm{t}} + \dot{v}_{\mathrm{q}} \tag{14}$$

In this equation we have introduced the angular velocity $\omega = \dot{\theta}$ of the material frame defined by

$$\dot{\underline{t}} = \omega\,\underline{q}, \qquad \dot{\underline{q}} = -\omega\,\underline{t}, \tag{15}$$

where θ is the angle of the tangent $\underline{t} = \cos\theta\,\underline{e}_x + \sin\theta\,\underline{e}_y$. The angular velocity satisfies the compatibility condition

$$\omega' = \dot{\theta}' = \dot{\kappa}. \tag{16}$$

2.6 Compact form of the equations of motion

We focus on the free region, where $\underline{p} = 0$. The equation of motion (11b) and the constitutive law (5) yield the normal force as

$$N = \rho\,\frac{\gamma_{\mathrm{q}}}{\kappa} + B\,\frac{\kappa''}{\kappa} \tag{17}$$

Plugging then into equation (11a), we find the equation of motion in the form

$$B\,\frac{\mathrm{d}}{\mathrm{d}s}\left(\frac{\kappa''}{\kappa} + \frac{1}{2}\kappa^2\right) = \rho\left(\gamma_{\mathrm{t}} - \frac{\mathrm{d}}{\mathrm{d}s}\left(\frac{\gamma_{\mathrm{q}}}{\kappa}\right)\right), \tag{18}$$

where the left-hand side involves elastic bending forces, and the right-hand side involves inertia.

2.7 Contact with a rigid impenetrable plane

In the presence of a planar obstacle, there is one additional unknown, the position of the moving boundary $s_c(t)$, see figure 2. The position of the moving boundary is set by the equation

$$\kappa(s_c^-(t), t) = 0, \tag{19}$$

where $s_c^-(t)$ denotes a values of s slightly less than s_c. The equation above expresses the fact that the adhesion energy with the substrate is zero.

The force applied by the substrate is a pointwise force, which we represent using the Dirac distribution δ_D

$$\underline{p}(s,t) = p_c(t)\,\delta_D(s, s_c(t))\,\underline{e}_y. \tag{20}$$

Here, the reaction force is normal to the plane as we assume a frictionless contact.

2.8 Dimensionless form

Using the radius of curvature $L^* = 1/\kappa_0$ as the length scale and the parameter $T^* = \sqrt{\frac{\rho}{B}}\,\kappa_0^{-2}$ as a time scale, we introduce rescaled units such that $\kappa_0 = 1$, $B = 1$, $\rho = 1$. Our numerical simulations were set up with these rescaled parameter values. Note that the dynamics of curling is a parameter-free problem when expressed in these units. In the following, we build solutions in the limit of a late time, $t \gg T^*$.

Note that the short time asymptotics has been analyzed by Callan-Jones et al. (2012) based on a self-similar solution of the linearized equations of motion, see figure 3.

3 Asymptotic solution at long times

We seek an asymptotic solution of the above set of equations in the limit $t \gg T^*$, and for an infinitely long rod, $L = \infty$.

3.1 Simulation results

We set up simulations of the curling problem using the Discrete Elastic Rod method (Bergou et al., 2008). The impenetrability constraint $y > 0$ is treated by intercalating a projection step after each dynamic step: in the dynamic step, the inequality constraint is ignored; in the projection step, the configuration is projected onto the manifold of admissible configurations, which in the present case simply involves repositioning vertices entering into the substrate ($y < 0$) back on its boundary ($y = 0$).

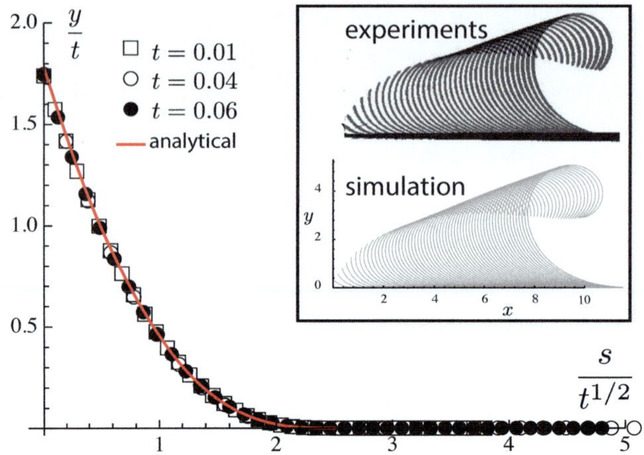

Figure 3. Short-time behavior, for $t \ll 1$ in rescaled units: comparison of simulation (symbols) with self-similar solution (curve). Reprinted from Callan-Jones et al. (2012)

The results of the simulation are summarized in figure 4. Exponents characterizing the long-time dynamics are identified in the simulations. They will be recovered later from the solution obtained by matched asymptotic expansions.

3.2 Absence of simple front solutions

We start by seeking front-like solutions to our problem. We assume that the moving boundary has an asymptotic velocity v, *i.e.* $s_\mathrm{c} \sim v\,t$, and introduce an arclength coordinate $\tilde{u} = v\,t - s$ in the frame following the front (at any time, the moving boundary is at $\tilde{u} = 0$).

We seek a front solution of the form

$$\underline{r}(s,t) = v\,t\,\underline{e}_x + \underline{R}(\tilde{u}),$$

where $\underline{R}$ describes the stationary front shape in the frame moving with the coil. This yields $\dot{\underline{r}} - v\,\underline{e}_x = v\,\underline{R}' = -v\,\underline{t}$. Identifying with equation (12), we find $v_\mathrm{t} = -v$ and $v_\mathrm{q} = 0$, hence $\gamma_\mathrm{t} = 0$ and $\gamma_\mathrm{q} = -\omega\,v$ by equation (14). Besides, $\underline{t} = \underline{r}' = \underline{R}'\,\frac{\mathrm{d}\tilde{u}}{\mathrm{d}s} = -\underline{R}'$ yields $\kappa\,\underline{q} = \underline{t}' = +\underline{R}''$ and $\omega\,\underline{q} = \dot{\underline{t}} = -\frac{\mathrm{d}\tilde{u}}{\mathrm{d}t}\,\underline{R}'' = -v\,\underline{R}''$; taking the ratio, we find $\frac{\omega}{\kappa} = -v$ and so $\frac{\gamma_\mathrm{q}}{\kappa} = -\frac{\omega\,v}{\kappa} = v^2$ is constant and the inertial term $\mathrm{d}(\gamma_\mathrm{q}/\kappa)/\mathrm{d}s = 0$ in equation (18). In fact, the entire right-hand side of (18) cancels, since we have also shown $\gamma_\mathrm{t} = 0$. This right-hand side contains all the inertial terms, and we are left with the

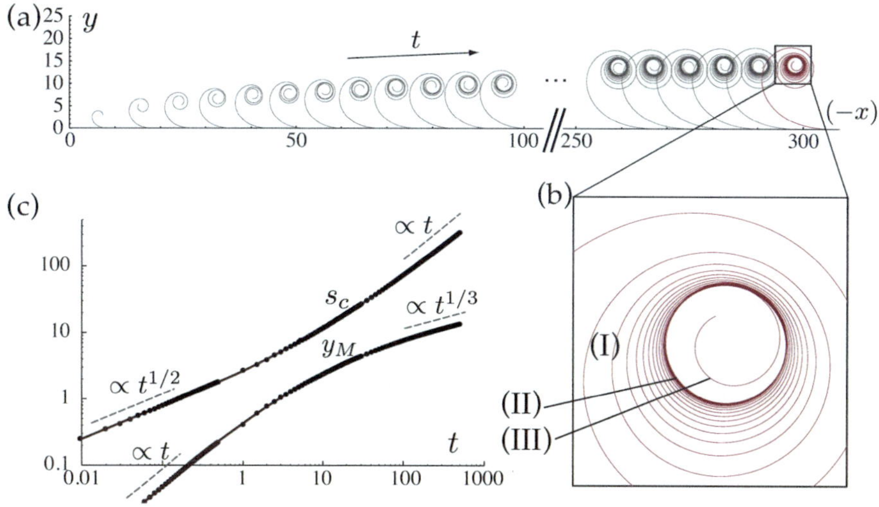

Figure 4. Simulation results, reprinted from Callan-Jones et al. (2012). (a) Simulation snapshots. (b) Internal structure of the coil at a late time. (c) Log-log plot of the position s_c of the moving boundary and of the highest elevation y_M as a function of time. The asymptotic regimes corresponding to both early time $t \ll T^*$ and late time $t \gg T^*$ are shown, along with the corresponding exponents.

elastic terms in the left-hand side, which are insufficient by themselves to select a velocity v. Therefore, the simple front construction fails.

This failure can be interpreted from a balance of vertical momentum on the curl. The simple front solution assumes that the center of mass has a constant elevation, hence that the vertical momentum of the curl is zero. This is in contradiction with the presence of the pointwise contact force $p_c(t)$ in equation (20), which for a front solution is a constant force pushing in the vertical direction.

3.3 Scaling analysis

In the following, we use a more general trial form for $\underline{r}(s,t)$ than in the equation above. Instead of assuming that the size of the curl is independent of time, we allow it to grow like t^β for large time, with some exponent β to be determined later. As a result, the center of mass of the curl moves constant upwards and the consistency with the balance of vertical momentum is restored. As a matter of fact, one can observe that the curl is very slowly but steadily raising in the numerical simulations, see figure 4a.

A scaling argument based on the balance of vertical momentum suggests that $\beta = 1/3$, see Callan-Jones et al. (2012). This will be confirmed later. This exponent is indeed compatible with the simulation results, see figure 4c. For a reason that will become clear later, however, it is preferable to leave the exponent β unspecified for the moment.

We will continue to assume that the moving boundary moves with an asymptotically constant velocity v, as this follows from an energy argument given by Callan-Jones et al. (2012).

3.4 Self-similar solution

Reduced variables We consider a self-similar variable u, defined by

$$u = \frac{v\,t - s}{t^\beta}. \tag{21}$$

This is the arc-length measured from the moving boundary, and rescaled with the expected typical size t^β of the curl.

Its derivatives with respect to arclength s and time t read, respectively,

$$u'(s,t) = -\frac{1}{t^\beta}, \qquad \dot{u}(s,t) = \frac{v}{t^\beta} - \frac{\beta\,u}{t} = \frac{v\,f(w)}{t^\beta}, \tag{22}$$

where

$$w(s,t) = \frac{v\,t - s}{v\,t} = \frac{u}{v\,t^{1-\beta}}, \tag{23}$$

and

$$f(w) = 1 - \beta\,w. \tag{24}$$

Compare (23) with the previous of u in (21): we have introduced a second rescaling for the arc-length in the definition of w above, based this time on the curvilinear length $\sim v\,t$ of the curl.

The derivatives of w will be needed as well:

$$w'(s,t) = -\frac{1}{v\,t}, \qquad \dot{w}(s,t) = \frac{1-w}{t}. \tag{25}$$

In the following we consider the limit $t \gg T^*$. With $s = \mathcal{O}(v\,t)$ in the curl, we have the estimates $u = \mathcal{O}(t^{1-\beta})$ and $w = \mathcal{O}(1)$.

Self-similar Ansatz We seek a solution in a more general form than a simple front (§3.2): we postulate the following self-similar form in a frame moving with the curl,

$$\underline{r}(s,t) = v\,t\,\underline{e}_x + t^\beta\,\underline{R}(u) \tag{26}$$

where u is the self-similar variable. Note that the simple front solution appears as the particular case $\beta = 0$.

We require $\underline{R}(u) = -u\,\underline{e}_x$ for $u \leq 0$, warranting that the rod is unperturbed in the region of contact beyond the moving boundary, *i.e.* $\underline{r}(s,t) = s\,\underline{e}_x$ if $s \geq v\,t$.

Computing the tangent $\underline{t} = \underline{r}'$, we have

$$\underline{t}(s,t) = t^\beta \frac{\mathrm{d}\underline{R}}{\mathrm{d}u}\,u' = -\underline{T}(u(s,t)), \tag{27}$$

where

$$\underline{T}(u) = \underline{R}'(u), \tag{28}$$

and a prime on a rescaled function such as $\underline{R}(u)$ denotes a derivative *with respect to its argument* u and not with respect to s. The inextensibility condition (1) yields

$$|\underline{T}(u)| = 1. \tag{29}$$

Then, one can define the rescaled curvature $K(u)$ by

$$\underline{T}'(u) = K(u)\,\underline{Q}(u) \tag{30}$$

where $\underline{Q}(u) = \underline{e}_z \times \underline{T}(u) = -\underline{q}(u(s,t))$.

A consequence of equation (27) is that the tangent orientation is a function of u only:

$$\theta(s,t) = \Theta(u) + \pi, \tag{31}$$

where Θ is the angle of the tangent $\underline{T} = \cos\Theta\,\underline{e}_x + \sin\Theta\,\underline{e}_y$. In view of equation (30), $K(u) = \Theta'(u)$. We can then compute

$$\omega(s,t) = \dot{\theta}(s,t) = \Theta'\,\dot{u} = \frac{v\,f(w)}{t^\beta}\,K(u), \tag{32}$$

and

$$\kappa(s,t) = \theta'(s,t) = \Theta'\,u' = -\frac{K(u)}{t^\beta}. \tag{33}$$

These equations imply that the following quantity, which can be interpreted as the rolling velocity, is a function of only w at late times ($t \gg T^*$):

$$-\frac{\omega(s,t)}{\kappa(s,t)} = v\,f(w). \tag{34}$$

This prediction is confirmed by the collapse of the simulation data in figure 5 onto a straight line in the plane $\left(-\frac{\omega}{\kappa}, w\right)$; recall that $f(w)$ has been defined as an affine function of w in equation (24). Note that, taken separately,

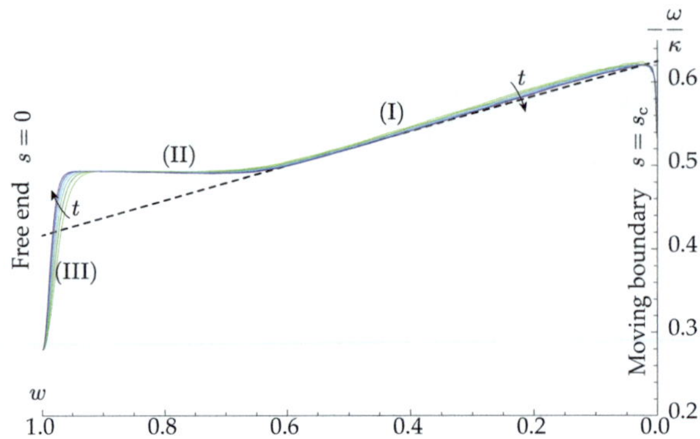

Figure 5. Test of the collapse predicted by equation (34) using numerical solutions at late time. Dashed line is expected value, $v\,f(w)$, with an adjusted value of v and $\beta = 1/3$. The collapse only takes place on about two thirds of the interval, which points to the fact that the actual solution has a more complex structure than we implicitly assumed (see below).

the functions $\omega(s,t)$ and $\kappa(s,t)$ are highly oscillatory as they depend on the 'fast' variable u: the collapse observed in figure 5 is a first confirmation of the relevance of the self-similar Ansatz in equation (26).

However, this collapse, which was expected to take place in the entire region where the rod has lifted off from the substrate, $0 < w < 1$, actually takes place in a *fraction* of this interval, corresponding roughly to $0 < w \lesssim 2/3$. We'll come back to this later.

We return to the equation of motion (18) and start by expressing its left-hand side in terms of the rescaled quantities $K(u)$. Using equation (33), we find $\kappa' = u' \frac{\mathrm{d}}{\mathrm{d}u}\left(-\frac{K(u)}{t^\beta}\right) = -\frac{K'(u)}{t^{2\beta}}$. Similarly, $\kappa'' = -\frac{K''(u)}{t^{3\beta}}$. Continuing a similar algebra, we end up rewriting the compact equation of motion (18) as

$$-\frac{B}{t^{3\beta}}\frac{\mathrm{d}}{\mathrm{d}u}\left(\frac{K''(u)}{K(u)} + \frac{1}{2}K^2(u)\right) = \rho\left(\gamma_\mathrm{t} - \frac{\mathrm{d}}{\mathrm{d}s}\left(\frac{\gamma_\mathrm{q}}{\kappa}\right)\right). \tag{35}$$

Again, the left-hand side arises from elasticity and the right-hand side from inertia.

We now proceed to insert the self-similar Ansatz into the right-hand side of the equation above. Differentiating (26) now with respect to time, we compute the velocity in the lab frame

$$\dot{\underline{r}}(s,t) = v\,\underline{e}_x + \frac{\beta}{t^{1-\beta}}\,\underline{R}(u) + v\,f(w)\,\underline{T}(u). \tag{36}$$

Note that we expect $\underline{R}(u)$ to be bounded, and so the term proportional to $\underline{R}$ is subdominant (negligible) at long times if $\beta > 0$. We must keep it, however, as it will contribute to the acceleration at the dominant order.

Differentiating one more time with respect to t, we calculate the acceleration as

$$\ddot{\underline{r}}(s,t) = -\frac{\beta\,(1-\beta)}{t^{2-\beta}}\,\underline{R}(u) + \frac{\beta}{t^{1-\beta}}\,\frac{v\,f(w)}{t^{\beta}}\,\underline{T}(u)\cdots$$
$$+\,v\,\frac{1-w}{t}\,f'(w)\,\underline{T}(u) + \frac{v^2\,f^2(w)}{t^{\beta}}\,K(u)\,\underline{Q}(u). \quad (37)$$

Reading off the projections from equation (13), we have

$$\gamma_{\mathrm{t}} = -\ddot{\underline{r}}\cdot\underline{T} = -\frac{v}{t}\,((1-w)\,f'(w) + \beta\,f(w)) + \frac{\beta\,(1-\beta)}{t^{2-\beta}}\,\underline{R}(u)\cdot\underline{T}(u) \quad (38)$$

and

$$\gamma_{\mathrm{q}} = -\ddot{\underline{r}}\cdot\underline{Q} = \kappa(s,t)\,\left(v^2\,f^2(w) - \frac{\beta\,(1-\beta)}{t^{2(1-\beta)}}\,\frac{\underline{R}(u)\cdot\underline{Q}(u)}{K(u)}\right) \quad (39)$$

We would like to compute the right-hand side of the equation of motion (35), and need the quantity $\gamma_{\mathrm{t}} - \frac{\mathrm{d}}{\mathrm{d}s}\left(\frac{\gamma_{\mathrm{q}}}{\kappa}\right)$. The dominant contribution is of order $1/t$, and comes from the terms that depend on w but not on u in the expressions of γ_{t} and γ_{q} above. The terms depending on u yield contributions that of relative magnitude $\sim |\underline{R}|/(t^{1-\beta}\,|K|)$. This number goes to zero assuming that $\underline{R}$ and K are bounded. Neglecting these other terms, we find

$$\rho\,\left(\gamma_{\mathrm{t}} - \frac{\mathrm{d}}{\mathrm{d}s}\left(\frac{\gamma_{\mathrm{q}}}{\kappa}\right)\right) = -\frac{\rho\,v}{t}\,((1-w)\,f'(w) + \beta\,f(w) - 2\,f(w)\,f'(w)). \quad (40)$$

Simplifying the right-hand side using the explicit form of f in (24) and inserting into equation (35), we have

$$\frac{B}{t^{3\beta}}\,\frac{\mathrm{d}}{\mathrm{d}u}\left(\frac{K''(u)}{K(u)} + \frac{1}{2}\,K^2(u)\right) = \frac{\rho\,v\,\beta}{t}\,(2 + (1-3\,\beta)\,w). \quad (41)$$

This equation characterizes the self-similar curvature profile $K(u)$ in terms of u or w—recall that u and w are related by equation (23)—, and sets the front velocity v, as we show later. For this equation to have a solution $K(u)$, the time parameter t must factor out. This yields two possibilities: either $\beta = 1/3$ (in which case the $1/t$ dependence on both sides of the equation can be eliminated), or $\beta = 0$ (in which case the left-hand side is independent of time while the right-hand side is zero because of the factor β in the numerator). Note that the value $\beta = 1/3$ has been anticipated earlier based on a scaling argument, while the value $\beta = 0$ corresponds to the simple front solution which we tried in §3.2.

For either one of the possible values of β, the self-similar equation of motion (41) reads

$$B\,\frac{\mathrm{d}}{\mathrm{d}u}\left(\frac{K''(u)}{K(u)} + \frac{1}{2}\,K^2(u)\right) = \begin{cases} \frac{2\,\rho\,v}{3} & \text{for } \beta = \frac{1}{3} \\ 0 & \text{for } \beta = 0 \end{cases}. \tag{42}$$

Integrating with respect to $u = v\,t^{1-\beta}\,w$, and returning to the original variable $\kappa(s,t)$, we find

$$\frac{\kappa''(s,t)}{\kappa(s,t)} + \frac{1}{2}\,\kappa^2(s,t) = \begin{cases} \frac{\rho}{B}\,\frac{2\,v^2\,w}{3} + \left[\frac{\mathrm{Const}}{t^{2/3}}\right] & \text{for } \beta = \frac{1}{3} \\ \mathrm{Const}' & \text{for } \beta = 0 \end{cases}. \tag{43}$$

The constant of integration in square brackets can be neglected at large times and will be dropped from now on.

In figure 6, we plot the quantity appearing in the left-hand side of equation (43) using the numerical solution. Even though the raw functions $\kappa(s,t)$ are highly oscillatory, see figure 6a, the numerical curves for $\frac{\kappa''(s,t)}{\kappa(s,t)} + \frac{1}{2}\,\kappa^2(s,t)$ are smooth and collapse onto a master curve, see figure 6b, as predicted by the the equation above. This master curve is piecewise linear, each linear piece corresponding to a possible value of β in the right-hand side.

Therefore, both possible exponents $\beta = 1/3$ and $\beta = 0$ are realized in the numerical solutions, in two subintervals of the domain $0 < w < w_r$ and $w_r < w < 1$ respectively. These subintervals are the regions labelled (I) and (II) in the figures 4–6. The other region labelled (III) is a boundary layer near the free end, where the self-similar Ansatz does not apply; it will be discussed later on.

So far, we have predicted the shape of the master curve corresponding to the dashed line in figure 6b and to the right-hand side of equation (6)b, but not the position w_r of the boundary between the regions (I) and (II), nor the constant curvature κ_r in region (II), nor the value of the velocity v.

In the remaining sections, we show that these remaining unknowns can be selected by matching the self-similar solutions with the boundary conditions.

4 Matching regions self-similar region (I) and roll (II)

In region (I), $0 < w < w_r$, equation (43) reads

$$\frac{\kappa''(s,t)}{\kappa(s,t)} + \frac{1}{2}\,\kappa^2(s,t) = \frac{\rho}{B}\,\frac{2\,v^2\,w}{3}.$$

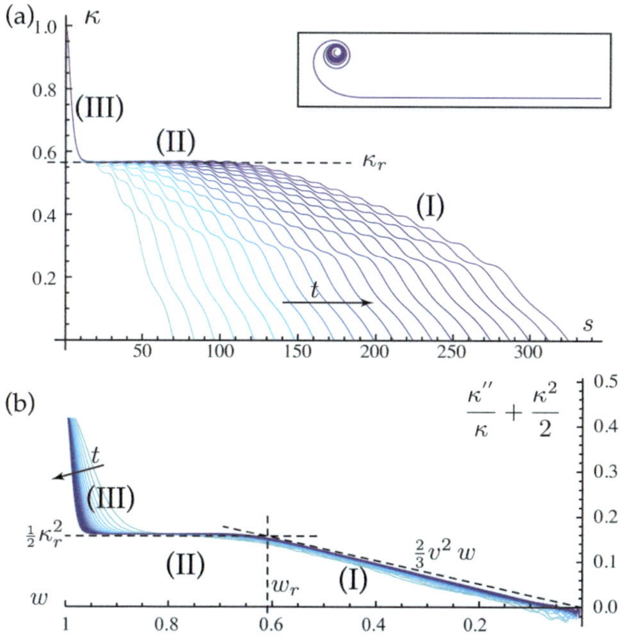

Figure 6. Curvature from the numerical simulation, reprinted from Callan-Jones et al. (2012). (a) Curvature profiles $\kappa(s, t)$ as a function of arclength s at different fixed times t; the support of the curvature $\kappa(s, t)$ is the time-dependent domain $(0, s_c(t))$ because of the moving boundary. (b) Collapse of the same data onto a master, piecewise linear curve as predicted by equation (43). The region labelled (I,II,III) correspond to those shown in figure 4b.

The derivative in the left-hand side introduces fast oscillations, with a typical length-scale $\Delta u \sim 1$. At long time, the right-hand side varies on a much slower length-scale $\Delta w \sim 1$. We shall admit without proof that the fast oscillations of $\kappa(s, t)$ are limited to a neighborhood of $s = s_c$ (*i.e.* of $w = 0$) whose size is much smaller than the size of the region $0 < w < w_r$ which has lifted off the substrate. The consequence is that these oscillations are damped out at the boundary $w = w_r$ between regions (I) and (II) at sufficiently long times. Therefore, the oscillatory term is negligible there, and we find

$$\frac{1}{2}\kappa_r^2 = \frac{\rho}{B}\frac{2\, v^2\, w_r}{3}, \tag{44}$$

where κ_r is also, by continuity, the curvature in region (II), see figure 6b.

Similarly, one can calculate the angular velocity ω_r at the endpoint $w = w_r$ of region (I) using equation (32). The result is

$$\omega_r = -\kappa_r\, v\left(1 - \frac{w_r}{3}\right) \tag{45}$$

By continuity, this is also the (constant) angular velocity ω in the entire region (II), as the latter undergoes a rigid-body motion ($\beta = 0$).

So far, we have found two equations for the four remaining unkowns $(v, \kappa_r, \omega_r, w_r)$, see equations (44–45). Two more equations will be found in the following section by analyzing the boundary layer near the free end ($s = 0$, $w = 1$).

5 Matching roll (II) and the free end (III)

In regions (II) and (III), for $w > w_r$, the motion is a rigid-body tumbling motion with angular velocity ω_r. The self-similar Ansatz is not applicable in region (III), and we revert to the equations of §2.3. We work in the rotating frame where the shape is stationary, *i.e.* is at equilibrium in the presence of inertial forces: this is the frame moving at velocity $v\,\underline{e}_x$ with its origin at the center of the roll. In this frame, the equilibrium equations read

$$m(s) = B\,(\kappa(s) - \kappa_0) \tag{46a}$$
$$m'(s) = -\underline{q}(s)\cdot\underline{n}(s) \tag{46b}$$
$$\underline{n}'(s) = -\rho\,\omega_r^2\,\underline{r}(s) \tag{46c}$$
$$\kappa(s) = \theta'(s) \tag{46d}$$
$$\underline{t}(s) = \{\cos\theta(s), \sin\theta(s)\} \tag{46e}$$
$$\underline{q}(s) = \{-\sin\theta(s), \cos\theta(s)\} \tag{46f}$$
$$\underline{r}'(s) = \underline{t}(s) \tag{46g}$$

The right-hand side of equation (46c) is an inertial force (we assume that the origin of the coordinates $\underline{r} = \underline{0}$ is the center of the roll).

We seek a solution of the above list of equations with the stressfree boundary conditions $m(0) = 0$ and $\underline{n}(0) = \underline{0}$, which match with a roll of constant curvature, $\kappa(s) \to \kappa_r$ for $s \to \infty$. This defines an eigenvalue problem, where ω_r and κ_r are eigenvalues.

To ease the solution of this problem, one can note the existence of an invariant

$$I_1(s) = m(s) + (\underline{r}(s)\times\underline{n}(s))\cdot\underline{e}_z. \tag{47}$$

Its conservation follows from the fact that the inertia term in the right-hand side of equation (46c) is a central force: $(\underline{r}\times\underline{n}')\cdot\underline{e}_z = 0$.

At the free end $s = 0$, the stress-free boundary conditions read $m(0) = 0$ and $\underline{n}(0) = \underline{0}$. Therefore, the value of the invariant is $I_1(0) = 0$. Combining this result with equations (46b) and (47) taken for $s \to \infty$, one first finds $B\,(\kappa_r - \kappa_0) = -\kappa_r^{-1}\,N_\infty$. The tension in region (II) tends to $N_\infty = \rho\,\omega_r^2\,\kappa_r^{-2}$, according to equation (46c). One thus obtains ω_r in terms of κ_r, leaving only one independent eigenvalue:

$$\omega_r = -\sqrt{\frac{B}{\rho}\,\kappa_r^3\,(\kappa_0 - \kappa_r)}\,. \tag{48}$$

Note that a second invariant follows from the fact that $\underline{n}' \cdot \underline{t}$ is an exact derivative by equation (46c),

$$I_2(s) = \frac{\rho}{2}\,\omega_r^2\,\underline{r}^2(s) + \frac{B}{2}\,\kappa^2(s) + \underline{t}(s) \cdot \underline{n}(s). \tag{49}$$

This invariant I_2 is less useful, however, as its value cannot be found directly from the boundary or asymptotic conditions.

Solving equations (46) by a shooting method with κ_r as a single shooting parameter, we find a unique solution, corresponding to the eigenvalues

$$\kappa_r = 0.5642 \times \kappa_0, \tag{50}$$

and

$$\omega_r = (-0.2798) \times \sqrt{\frac{B}{\rho}}\,\kappa_0^2 \tag{51}$$

These are the last two equations we need to solve the problem: knowing the values of κ_r and ω_r, one can solve equations (44) and (45) for the last two unknowns, v and w_r. The result is

$$w_r = 0.6142 \tag{52}$$

and

$$v = \sqrt{\frac{B}{\rho}}\,\kappa_0 \times 0.6235. \tag{53}$$

6 Conclusion

We have found a unique solution describing the asymptotic behavior at long time. Overall, the size of the curl grows slowly, like $t^{1/3}$, which explains the asymptotic behavior for y_M in figure 4c. However, a fixed fraction

$(1 - w_r)$ of the curl (around a third of it, region II) undergoes a rigid-body tumbling motion with a constant curvature set by the balance of bending force and centrifugal force. The structure of the solution turned out to be more complex that we had anticipated: the solution of the full problem, and in particular the selection of the velocity v, requires identifying three regions, solving in these regions using different approximations [self-similar Ansatz in (I), tumbling motion with is stationary in a moving frame in (II), and a boundary layer near the free end in (III)], and finally matching them. The final prediction for the velocity in equation (53) yields $v = 12.8$ m/s, and agrees well with the experimentally measured value $v = 12.5$ m/s.

Bibliography

B. Audoly and S. Neukirch. Fragmentation of rods by cascading cracks: Why spaghetti does not break in half. *Physical Review Letters*, 95(9): 095505, 2005.

M. Bergou, M. Wardetzky, S. Robinson, B. Audoly, and E. Grinspun. Discrete elastic rods. *ACM Transactions on Graphics*, 27(3):63:1–63:12, 2008.

R.E. Caflisch and J.H. Maddocks. Nonlinear dynamical theory of the Elastica. *Proc. Roy. Soc. Edinburgh A*, 99:1–23, 1984.

A. C. Callan-Jones, P.-T. Brun, and B. Audoly. Self-similar curling of a naturally curved Elastica. *Physical Review Letters*, 108(17):174302, 2012.

J. Engelberth. Mechanosensing and signaltransduction in tendrils. *Advances in Space Research*, 32(8):1611–1619, 2003.

J. R. Gladden, N. Z. Handzy, A. Belmonte, and E. Villermaux. Dynamic buckling and fragmentation in brittle rods. *Physical Review Letters*, 94 (3):035503, 2005.

L. Golubovic, D. Moldovan, and A. Peredera. Dynamics of the euler buckling instability. *Physical Review Letters*, 81(16):3387, 1998.

A. Goriely and M. Tabor. New amplitude equations for thin elastic rods. *Phys. Rev. Lett.*, 77(17):3537–3540, Oct 1996.

Alain Goriely and Michael Tabor. The nonlinear dynamics of filaments. *Nonlinear Dynamics*, 21(1):101–133, 2000.

M. Habibi, N. M. Ribe, and D. Bonn. Coiling of elastic ropes. *Physical Review Letters*, 99(15):154302, 2007.

L. Mahadevan and Joseph B. Keller. Coiling of flexible ropes. *Proceedings of the Royal Society A: Mathematical, Physical and Engineering Sciences*, 452, 1996.

H. J. Schindler and H. Kolsky. Multiple fractures produced by the bending of brittle beams. *Journal of the Mechanics and Physics of Solids*, 31(5): 427–437, 1983.

Stability of periodic porous structures

Katia Bertoldi[‡]

[‡] School of Engineering and Applied Sciences, Harvard University, Cambridge, Massachusetts 02138, USA

Abstract In this chapter we focus on the mechanics of two-dimensional periodic elastomeric cellular structures and present numerical techniques for investigating their finite deformations. We then use them to show that in an elastic matrix with a periodic array of pores instabilities with wavelengths that are of the order of the size of the microstructure can be triggered. Interestingly, these instabilities can be utilized to design a novel class of responsive materials. Possible applications include materials with unusual properties such as negative Poisson's ratio, phononic and photonic switches and colorful and reconfigurable displays.

1 Introduction

Periodic and porous structures may significantly change their architecture in response to the applied deformation. When subjected to excessive deformation, they may eventually become unstable and beyond the instability threshold, rapid and dramatic changes of the structural geometry occur. Traditionally, mechanical instabilities have been viewed as failure modes and a number of numerical methods, including the finite-difference method (Forman and Hutchinson, 1970; Renton, 1973; Wah, 1965; Wah and Calcote, 1970), the finite elements method (Noor et al., 1978; Papka and Kyriakides, 1999) and the Bloch-wave method (Anderson, 1981; Anderson and Williams, 1986; Geymonat et al., 1993; Schraad and Triantafyllidis, 1997; Hutchinson and Fleck, 2006; Bertoldi et al., 2008) have been developed and used to determine the conditions of the onset of bifurcation.

Interestingly, a new trend is emerging in which instabilities are harnessed to enable new functionalities. In fact, for elastic materials the geometric reorganization occurring at instability is both reversible and repeatable and it occurs over a narrow range of the applied load, providing opportunities for the design of tunable/adaptive structures (i.e. structures that can change their properties in response to variations in their environment) with applications in sensors, microfluidics, bioengineering, robotics, acoustics and

D. Bigoni (Ed.), *Extremely Deformable Structures*,
CISM International Centre for Mechanical Sciences
DOI 10.1007/ 978-3-7091-1877-1_4 © CISM Udine 2015

photonic (Horozov et al., 2006; Chan et al., 2008; Jang et al., 2009; Yang et al., 2010; Kim et al., 2012; Shim et al., 2012).

In particular, instabilities in periodic porous structures comprising square and triangular arrays of circular holes have been found to lead to the transformation of the pores into ordered arrays of high-aspect ratio (almost closed) ellipses (Mullin et al., 2007; Zhang et al., 2008; Singamaneni et al., 2009) and have been demonstrated to be instrumental for the design of phononic switches (Jang et al., 2009; Wang and Bertoldi, 2012), color displays (Li et al., 2012) and materials with unusual properties such as large negative Poisson's ratio (Overvelde et al., 2012; Bertoldi et al., 2010).

Here, we focus on a square array of pores embedded in an elastomeric matrix and investigate the effect of the pore shape on the response of the structure. First, the pore shapes, material model and loading conditions used in this study are presented (Sections 2, 3 and 4). Then, in Section 5 the numerical analyses that are used to investigate the stability of both finite-sized and infinite periodic and porous structures are introduced. Finally, in Section 6 we present the non-linear analysis employed to capture the post-buckling behavior of the structures.

2 Geometry

In this study, we consider a square array of pores in an elastomeric matrix and focus on pores with four-fold symmetry. Taking the circular pore shape as a starting point, we make use of Fourier series expansion to alter their contour according to

$$x_1 = r(\theta)\cos\theta, \qquad x_2 = r(\theta)\sin\theta, \qquad (1)$$
$$\text{with } r(\theta) = r_0\left[1 + c_1\cos(4\theta) + c_2\cos(8\theta)\right],$$

where $0 \leq \theta \leq 2\pi$ and three parameters have been introduced to control the pore size (r_0) and shape (c_1 and c_2). While $c_1 = c_2 = 0$ in Eqn. (1) provides a description of a circle of radius r_0, by changing c_1 and c_2 a variety of shapes can be obtained (Overvelde et al., 2012; Overvelde and Bertoldi, 2014). Since we focus on pores arranged on a square array, r_0 is related to the initial porosity of the structure ψ_0 through

$$r_0 = \frac{L_0\sqrt{2\psi_0}}{\sqrt{\pi(2 + c_1^2 + c_2^2)}}, \qquad (2)$$

with L_0 denoting the center-to-center distance between neighboring pores in the undeformed configuration.

While the results of an extensive numerical study on the effect of c_1 and c_2 on the non-linear material response have been recently reported

(Overvelde and Bertoldi, 2014), here we focus on two shapes showing qualitative and quantitative remarkably different behaviors. In particular, we investigate the mechanical behavior of structures with initial porosity $\psi_0 = 0.5$ and compare the response of a system with circular holes defined by $(c_1, c_2) = (0, 0)$ (see Fig. 1) to that of a structure with star-shaped pores defined by $(c_1, c_2) = (-0.21, 0.28)$ (see Fig. 2).

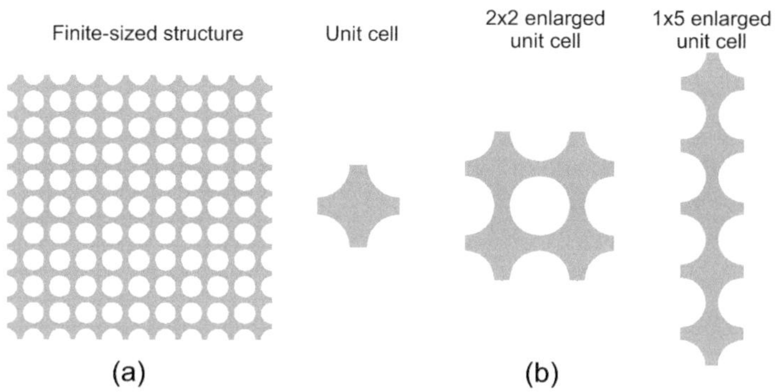

Figure 1. (a) Schematic of a finite-sized periodic structure with initial porosity $\psi_0 = 0.5$ and circular pores (i.e. $c_1 = c_2 = 0.0$). (b) Schematic of the unit cell and $2\times$ 2 and $1\times$ 5 enlarged unit cells.

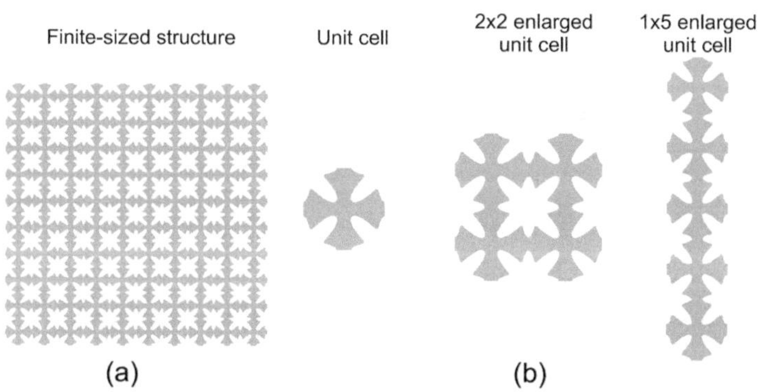

Figure 2. (a) Schematic of a finite-sized periodic structure with initial porosity $\psi_0 = 0.5$ and star-shaped pores defined by $(c_1, c_2) = (-0.21, 0.28)$. (b) Schematic of the unit cell and $2\times$ 2 and $1\times$ 5 enlarged unit cells.

3 Material

We consider structures made of an hyperelastic material. Let

$$\mathbf{F} = \frac{\partial \mathbf{x}}{\partial \mathbf{X}}, \tag{3}$$

be the deformation gradient, mapping a material point from the reference position $\mathbf{X}$ to its current location $\mathbf{x}$ and J be its determinant, $J = \det\mathbf{F}$. For an isotropic hyperelastic material the strain energy density W can be expressed as a function of the invariants of the right Cauchy-Green tensor $\mathbf{C} = \mathbf{F}^T\mathbf{F}$ (or, alternatively, also the left Cauchy-Green tensor $\mathbf{B} = \mathbf{F}\mathbf{F}^T$). In particular, the behavior of nearly incompressible materials is effectively described by splitting the deformation locally into volume-changing ($J^{1/3}\mathbf{I}$) and distortional ($\overline{\mathbf{F}}$) components as

$$\mathbf{F} = (J^{1/3}\mathbf{I})\overline{\mathbf{F}}, \tag{4}$$

where $\mathbf{I}$ denotes the identity matrix.

Here, stress-strain behavior of the material is modeled using a Neo-Hookean model, modified to include compressibility (with a high bulk modulus):

$$W = \frac{\mu_0}{2}(\overline{I}_1 - 3) + \frac{K_0}{2}(J-1)^2, \tag{5}$$

where μ_0 and K_0 are the initial shear and bulk moduli and $\overline{I}_1 = \mathrm{tr}(\overline{\mathbf{F}}^T\overline{\mathbf{F}})$. The nominal (first Piola-Kirchoff) stress is then given by

$$\mathbf{S} = \frac{\partial W}{\partial \mathbf{F}} = \left[\mu_0 \mathrm{dev}\overline{\mathbf{B}} + K_0 J(J-1)\right]\mathbf{F}^{-T}, \tag{6}$$

where $\overline{\mathbf{B}} = \overline{\mathbf{F}}\,\overline{\mathbf{F}}^T$ and dev is the deviatoric operator. Note that in this study the material is modeled as nearly incompressible, characterized by $K_0/\mu_0 \approx 2500$.

4 Loading conditions

In this study two different loading conditions are considered: (*i*) uniaxial compression and (*ii*) equibiaxial compression. Here, we describe the boundary conditions used to simulate such loadings in both the finite-sized and infinite periodic structures.

Finite-sized periodic structures To simulate uniaxial loading conditions for the finite-sized periodic specimen the bottom edge is fixed in vertical direction (the bottom left corner is also fixed in horizontal direction to prevent a rigid body motion), whereas the top edge is uniformly compressed in vertical direction.

Similarly, to simulate equibiaxial loading conditions the bottom and left edge are fixed in vertical and horizontal direction, respectively, whereas the top and right edges are uniformly compressed in the vertical and horizontal direction.

Infinite periodic structures The response of the infinite periodic structure is investigated using a unit cell (see Figs. 1-b and 2-b). To subject the unit cell to a macroscopic deformation gradient $\overline{\mathbf{F}}$ periodic boundary conditions are imposed on all cell boundaries (Danielsson et al., 2002; Bertoldi and Boyce, 2008)

$$u_\alpha^{A_i} - u_\alpha^{B_i} = (\overline{F}_{\alpha\beta} - \delta_{\alpha\beta})(X_\beta^{A_i} - X_\beta^{B_i}), \quad i = 1, 2, ..., N \tag{7}$$

where $\delta_{\alpha\beta}$ is the Kronecker delta, $u_\alpha^{A_i}$ and $u_\alpha^{B_i}$ ($\alpha = 1, 2$) are displacements of points periodically located on the boundary of the unit cell. Moreover, N denotes the number of pairs of nodes periodically located on the boundary of the unit cell. For the square lattice considered in this study $N = 2$ when the unit cell is considered, while $N = 4$ and $N = 6$ for the enlarged 2×2 and 1×5 unit cells, respectively. Note that the components of $\overline{\mathbf{F}}$ can be conveniently prescribed within the finite element framework using a set of virtual nodes. The corresponding macroscopic first Piola-Kirchhoff stress $\overline{\mathbf{P}}$ is then obtained through virtual work considerations (Danielsson et al., 2002; Bertoldi and Boyce, 2008).

In this study the following macroscopic loading conditions are considered:

- Uniaxial compression in the x_2 direction, so that the macroscopic deformation gradient $\overline{\mathbf{F}}$ is given by

$$\overline{\mathbf{F}} = \lambda_{11}\ \hat{\mathbf{e}}_1 \otimes \hat{\mathbf{e}}_1 + (1+\varepsilon)\ \hat{\mathbf{e}}_2 \otimes \hat{\mathbf{e}}_2 + \hat{\mathbf{e}}_3 \otimes \hat{\mathbf{e}}_3, \tag{8}$$

  where ε denotes the applied strain in x_2 direction, and λ_{11} is determined from $\sigma_{11} = 0$.

- Equibiaxial compression, so that the macroscopic deformation gradient $\overline{\mathbf{F}}$ is given by

$$\overline{\mathbf{F}} = (1+\varepsilon)\ \hat{\mathbf{e}}_1 \otimes \hat{\mathbf{e}}_1 + (1+\varepsilon)\ \hat{\mathbf{e}}_2 \otimes \hat{\mathbf{e}}_2 + \hat{\mathbf{e}}_3 \otimes \hat{\mathbf{e}}_3, \tag{9}$$

  where ε denotes the applied strain.

5 Stability analysis

5.1 Stability of finite-sized specimens

The stability of periodic finite-sized specimens is examined using eigenvalue analyses. A linear perturbation procedure is used and is accomplished within the commercial finite element code ABAQUS/Standard using the `*BUCKLE` module.

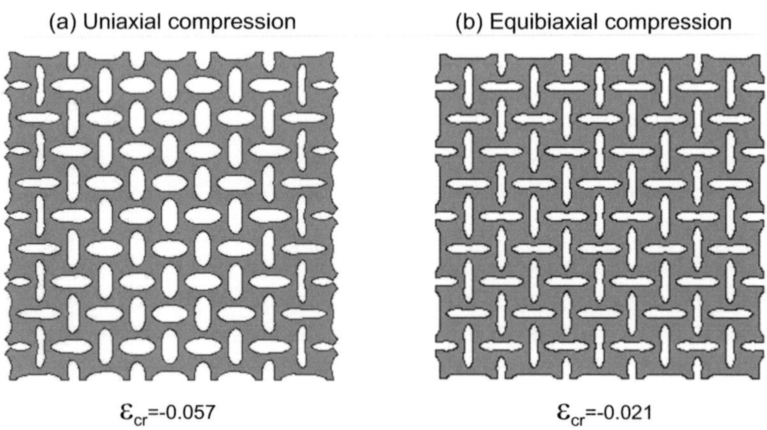

Figure 3. Critical buckling modes and corresponding strains for a finite-sized periodic porous structure comprising an array of circular pores under (a) uniaxial and (b) equibiaxial compression. The structure comprises an array of 10×10 unit cells.

In Fig. 3 we focus on the structure with circular pores and report the critical eigenmode and corresponding critical strain obtained under uniaxial (Fig. 3-a) and equibiaxial (Fig. 3-b) compression. Interestingly, the first eigenmode of the finite-sized specimen for both loading cases is characterized by an array of alternating mutually orthogonal ellipses and is homogeneous throughout the specimen. By contrast, for the structure with star-shaped holes loaded under uniaxial compression we find a buckling mode with a wavelength equal to the size of the sample (see Fig. 4-a), reminiscent of the twinning observed in austenite to martensite phase transformations in shape memory alloys. Such long wavelength mode is not compatible with the boundary conditions experienced by the sample during equibiaxial compression, so that the critical mode for that loading case is characterized by a wavelength equal to two unit cells in both vertical and horizontal direction and leads to the formation of a checkerboard pattern (see Fig. 4-b).

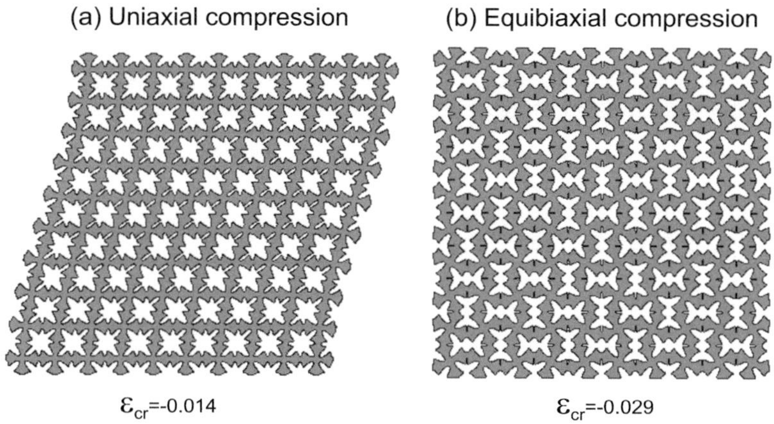

Figure 4. Critical buckling modes and corresponding strains for a finite-sized periodic porous structure comprising an array of pores defined by $(c_1, c_2) = (-0.21, 0.28)$ under (a) uniaxial and (b) equibiaxial compression. The structure comprises an array of 10×10 unit cells.

5.2 Stability of infinite periodic specimens

Upon application of deformation, an infinite periodic structure can suddenly change its periodicity due to mechanical instability. Such instability could be either microscopic (*i.e.*, with wavelength that is of the order of the size of the microstructure) or macroscopic (*i.e.*, with much larger wavelength than the size of the microstructure) (Bertoldi and Boyce, 2008; Triantafyllidis and Maker, 1985; Geymonat et al., 1993).

Microscopic instability Microscopic (local) buckling modes are characterized by wavelengths that are of the order of the size of the microstructure and may alter the initial periodicity of the solid. The simplest, but computationally expensive path for investigating them is to construct enlarged unit cells of various size and to use a linear perturbation procedure to calculate their critical strains and corresponding modes. The critical strain of the infinite periodic structure is then defined as the minimum of the critical strains on all possible enlarged unit cells (Bertoldi et al., 2008).

Interestingly, the computational cost of such analysis can be significantly reduced considering a single unit cell and applying Bloch-periodic boundary conditions (Kittel, 2005)

$$\mathbf{u}(\mathbf{x}+\mathbf{r}) = \mathbf{u}(\mathbf{x})\mathrm{e}^{i\mathbf{k}\cdot\mathbf{r}}, \tag{10}$$

where $\mathbf{k}$ denotes the wave vector and

$$\mathbf{r} = r_{a1}\mathbf{a}_1 + r_{a2}\mathbf{a}_2, \tag{11}$$

r_{a1} and r_{a2} being arbitrary integers and $\mathbf{a}_1$ and $\mathbf{a}_2$ denoting the lattice vectors spanning the single unit cell. In fact, the response of an enlarged cell comprising $m_1 \times m_2$ unit cells can be investigated using Eq. (10) with (Triantafyllidis et al., 2006; Bertoldi and Boyce, 2008)

$$\begin{aligned} \mathbf{r} &= m_1\mathbf{a}_1 + m_2\mathbf{a}_2 \ , \quad \text{and} \\ \mathbf{k} &= \frac{1}{m_1}\mathbf{b}_1 + \frac{1}{m_2}\mathbf{b}_2 \ , \end{aligned} \tag{12}$$

$\mathbf{b}_1$ and $\mathbf{b}_2$ being the reciprocal primitive vectors defined as

$$\mathbf{b}_1 = 2\pi \frac{\mathbf{a}_2 \times \hat{\mathbf{e}}_3}{||\mathbf{a}_1 \times \mathbf{a}_2||} \ , \quad \mathbf{b}_2 = 2\pi \frac{\hat{\mathbf{e}}_3 \times \mathbf{a}_1}{||\mathbf{a}_1 \times \mathbf{a}_2||}, \tag{13}$$

where $\hat{\mathbf{e}}_3 = (\mathbf{a}_1 \times \mathbf{a}_2) / ||\mathbf{a}_1 \times \mathbf{a}_2||$.

For this particular choice of $\mathbf{r}$ and $\mathbf{k}$, it is easy to see that (10) reduces to:

$$\mathbf{u}(\mathbf{x} + \mathbf{r}) = \mathbf{u}(\mathbf{x}) \ , \tag{14}$$

indicating that a single primitive unit cell can be used to investigate the response of an enlarged unit cell spanned by the lattice vectors $m_1\mathbf{a}_1$ and $m_2\mathbf{a}_2$, when boundary conditions specified by Eqns. (10) and (12) are applied.

Therefore, the stability of the enlarged unit cell comprising $m_1 \times m_2$ unit cells can be investigated within the finite element framework by detecting when the tangent stiffness matrix of the corresponding single unit cell subjected to Bloch-type boundary conditions defined by Eqns. (10) and (12) becomes singular along the loading path (*i.e.*, $\det [K] = [0]$). Alternatively, it can be also investigated by detecting the applied load at which the smallest eigenfrequency associated to a non-trivial eigenmode of the single unit cell subjected to Bloch-type boundary conditions is zero (Bathe, 1996).

Finally, the onset of instability for the infinite periodic structure is defined as the minimum critical strain on all the considered enlarged unit cells defined by $m_1\mathbf{a}_1$ and $m_2\mathbf{a}_2$. Here, we investigated the stability of 25 enlarged primitive units by choosing $m_1 = 1, \ldots, 5$ and $m_2 = 1, \ldots, 5$ in Eqn. (12). To detect the onset of instability for each enlarged unit cell, we perform eigenfrequency analysis along the loading path at increasing values of applied deformation and detect the smallest load for which an eigenfrequency associated with a non-trivial eigenmode becomes zero[1]. The critical strain of the infinite periodic structure is then defined as the minimum of such loads on all 25 considered enlarged unit cells.

To work with the complex-valued relations of the Bloch-periodic conditions (Eqn. (10)) in a commercial software such as ABAQUS/Standard, we split all fields into real and imaginary parts. In this way, the equilibrium equations are divided into two sets of uncoupled equations for the real and imaginary parts (Aberg and Gudmundson, 1997; Bertoldi and Boyce, 2008). Thus, the problem is solved using two identical finite-element meshes for the unit cell, one for the real part and one for the imaginary part, and coupling them by Bloch-periodic displacement boundary conditions:

$$\begin{aligned}\text{Real}(\mathbf{u}^{B_i}) &= \text{Real}(\mathbf{u}^{A_i})\cos[\mathbf{k}\cdot\mathbf{r}_{A_iB_i}] - \text{Imag}(\mathbf{u}^{A_i})\sin[\mathbf{k}\cdot\mathbf{r}_{A_iB_i}] \\ \text{Imag}(\mathbf{u}^{B_i}) &= \text{Real}(\mathbf{u}^{A_i})\sin[\mathbf{k}\cdot\mathbf{r}_{A_iB_i}] + \text{Imag}(\mathbf{u}^{A_i})\cos[\mathbf{k}\cdot\mathbf{r}_{A_iB_i}] \quad (15)\end{aligned}$$

where $\mathbf{r}_{A_iB_i} = \mathbf{x}^{B_i} - \mathbf{x}^{A_i}$ denotes the distance in the current/deformed configuration between the two nodes A_i and B_i periodically located on the boundary. Note that Eqns. (15) can be easily implemented within ABAQUS/Standard using a user-subroutine MPC.

The results obtained for a square lattice of circular pores subjected to uniaxial and equibiaxial compression are shown in Figs. 5 and 6, respectively. In the undeformed configuration (i.e. $\varepsilon = 0$) all eigenfrequencies ω associated with the considered wave vectors $\mathbf{k}$ are positive. However, as ε increases, the eigen-frequencies associated with each wave vector $\mathbf{k}$ gradually decrease and eventually become negative. The critical strain parameter ε_{cr} associated with each wave vector can be easily extracted from the plot, since it corresponds to the intersection point between each curve and the horizontal line $\omega = 0$.

Similar trends are observed for the case of uniaxial and equibiaxial loading. The lowest critical strain for the 25 considered enlarged unit cells is $\varepsilon_{cr}^{uni} = -0.073$ and $\varepsilon_{cr}^{equi} = -0.025$ and is associated with $(m_1, m_2) = (2, 2)$, yielding the mode shown in Fig. 5-b. We note that the critical mode for the infinite structure is characterized by the same checkerboard pattern found investigating the stability of the corresponding finite-sized structure. Hence, these results indicate that the instability event does not rely on a finite size lattice and, hence, will also occur in very large arrays.

In Figs. 7 and 8 we report the results obtained for an infinite square lattice of star-shaped pores subjected to uniaxial and equibiaxial compres-

[1]Note that this procedure is significantly simpler than calculating det $[K] = [0]$, since it does not require to export the stiffness matrix calculated by ABAQUS and to evaluate its determinant using a numerical package

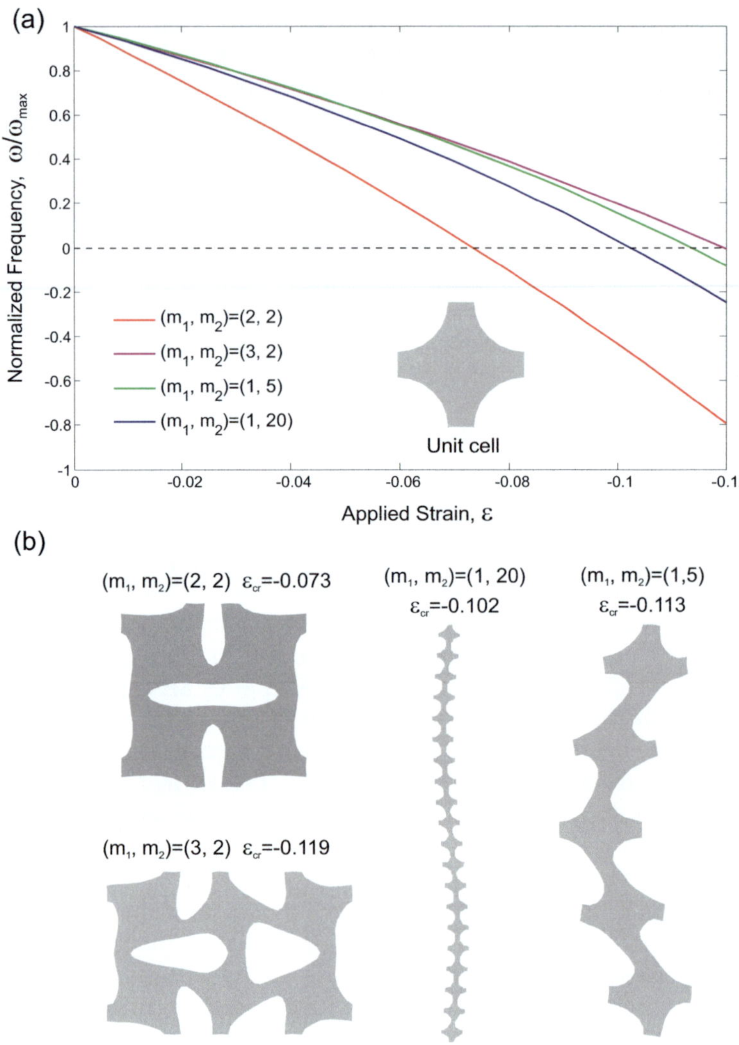

Figure 5. Top: Evolution of the frequency parameter as a function of the applied strain for an infinite square array of circular pores under uniaxial loading conditions. Bottom: Mode shapes associated to $(m_1, m_2) = (2, 2)$, (c) $(m_1, m_2) = (3, 2)$, $(m_1, m_2) = (1, 5)$ and $(m_1, m_2) = (1, 20)$ obtained at their corresponding critical load.

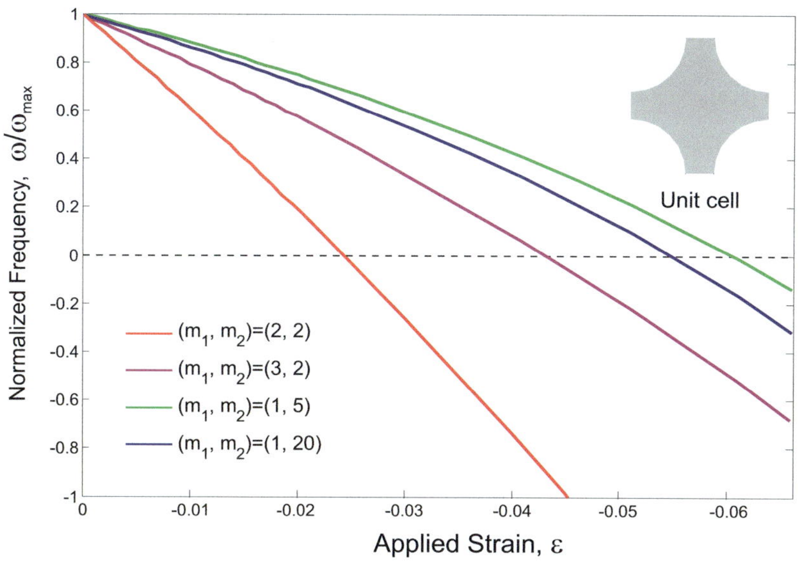

Figure 6. Evolution of the frequency parameter as a function of the applied strain for an infinite square array of circular pores under equibiaxial loading conditions. The corresponding critical modes are very similar to those found under uniaxial compression and shown in Fig. 5

sion, respectively. The lowest critical strains for the 25 considered enlarged unit cells are $\varepsilon_{cr}^{uni} = -0.020$ and $\varepsilon_{cr}^{equi} = -0.015$ and are associated with $(m_1, m_2) = (1, 5)$. Moreover, slightly lower values of critical strain are found when the number of unit cells in vertical direction is increased (as shown for the case $(m_1,m_2)=(1, 20)$ in Figs. 7 and 8). The eigenmodes associated to the periodicities $(m_1, m_2) = (1, 5)$ and $(m_1, m_2) = (1, 20)$ are shown in Fig. 7 and are reminiscent of the long wavelength mode found when study the stability under uniaxial compression of the corresponding finite-sized structure (see Fig. 4-a).

Focusing on such long wavelength modes, we first note that they are compatible with the boundary conditions experienced by the finite-sized specimen during uniaxial compression, so that we expect they to be observed during the loading of real (finite-sized) samples. By contrast, the boundary conditions applied during an equibiaxial compression test to finite-sized specimens are not compatible with such modes, so that we do not expect them to be triggered during equibiaxial loading.

To identify the critical mode for the infinite periodic structure with

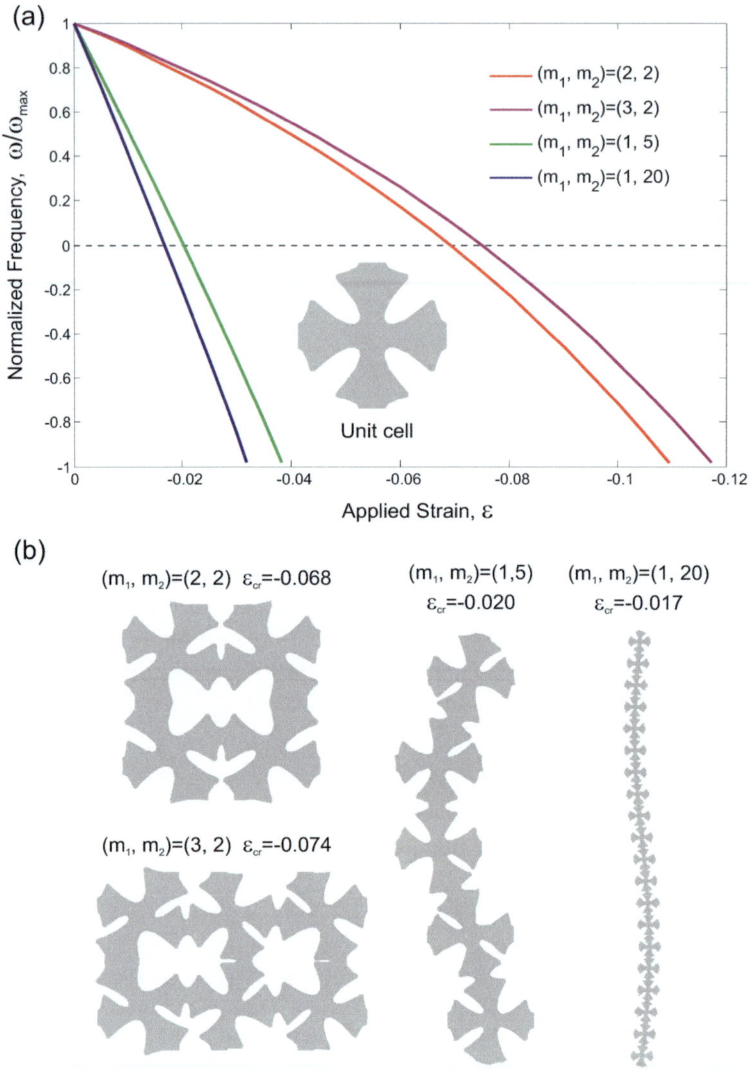

Figure 7. Top: Evolution of the frequency parameter as a function of the applied strain for an infinite square array of star-shaped pores under uniaxial loading conditions. Bottom: Mode shapes associated to $(m_1, m_2) = (2, 2)$, (c) $(m_1, m_2) = (3, 2)$, $(m_1, m_2) = (1, 5)$ and $(m_1, m_2) = (1, 20)$ obtained at their corresponding critical load.

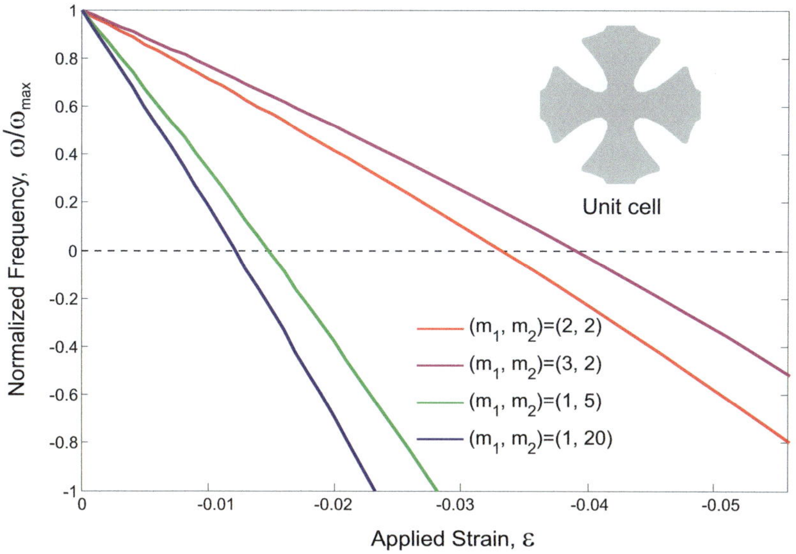

Figure 8. Evolution of the frequency parameter as a function of the applied strain for an infinite square array of star-shaped pores under equibiaxial loading conditions. The corresponding critical modes are very similar to those found under uniaxial compression and shown in Fig. 7

star-shaped pores compatible with the boundary conditions experienced by finite-sized samples under equibiaxial compression, we then fix the displacement of the central node of the unit cell. In this case we find that the critical strain $\varepsilon_{cr}^{equi} = -0.033$ is that associated to $(m_1, m_2) = (2, 2)$, resulting in a pattern similar to that found for the corresponding finite-sized sample (see Fig. 7-b).

Macroscopic instability As shown above for the case of the structure with star-shaped holes, macroscopic (or long wavelength) instabilities can be detected considering $k_i \to 0$ in Eqn. (12) when performing the microscopic instability analysis. Alternatively, it has been rigorously shown that macroscopic instabilities can also be detected from the loss of strong ellipticity of the overall response of the material (Geymonat et al., 1993). Specifically, for the cellular structure considered in this study macroscopic instability may occur whenever the condition (Marsden and Hughes, 1983)

$$(\mathbf{m} \otimes \mathbf{N}) : \overline{\mathbb{L}} : (\mathbf{m} \otimes \mathbf{N}) > 0 \,, \qquad \overline{\mathbb{L}}_{iJkL}\, m_i\, N_J\, m_k N_L > 0 \tag{16}$$

is first violated along the loading path, $\mathbf{m}$ and $\mathbf{N}$ denoting unit vectors defined in the current and the initial configurations, respectively. Note that the homogenized mixed elasticity tensor $\overline{\mathbb{L}}$ relates the macroscopic deformation gradient increment $\dot{\overline{\mathbf{F}}}$ to the macroscopic first Piola-Kirchhoff stress increment $\dot{\overline{\mathbf{P}}}$ as:

$$\dot{\overline{\mathbf{P}}} = \overline{\mathbb{L}} : \dot{\overline{\mathbf{F}}} \,, \qquad \dot{\overline{P}}_{iJ} = \overline{\mathbb{L}}_{iJkL} \dot{\overline{F}}_{kL} \,. \tag{17}$$

In this study, 2-D finite element simulations on a single primitive unit cell with spatially periodic boundary conditions (Eqns. (7)) are performed to monitor the loss of ellipticity of the homogenized tangent modulus $\overline{\mathbb{L}}$. After determining the principal solution, the components of $\overline{\mathbb{L}}$ are identified by subjecting the unit cells to four independent linear perturbations of the macroscopic deformation gradient $\dot{\overline{\mathbf{F}}}$, calculating the corresponding averaged stress components $\dot{\overline{\mathbf{P}}}$ and comparing to Eq. (17). Then, the loss of ellipticity condition is examined by checking the positive definite condition (16) with $\mathbf{m}$ and $\mathbf{N}$ separately explored at every $\pi/360$ radian increment.

For the infinite periodic structure with circular holes considered in this study we find that the critical strain associated with macroscopic instability is $\varepsilon_{cr}^{uni} = -0.11$ and $\varepsilon_{cr}^{equi} = -0.11$ for uniaxial and equibiaxial loading conditions, respectively. Note that this values of critical strain are in good agreement to those predicted by the microscopic instability analysis for $(m_1,m_2)=(1,20)$.

Focusing on the infinite periodic structure with star-shaped holes, we find that the critical strain associated with macroscopic instability is $\varepsilon_{cr}^{uni} = -0.017$ and $\varepsilon_{cr}^{equi} = -0.013$ for uniaxial and equibiaxial loading conditions, respectively, showing also in this case a nice agreement with the results obtained from the microscopic instability analysis. Again, we note that the long wavelength mode is compatible with the boundary conditions applied to the finite-sized sample when loaded uniaxially, so that we expect it to be triggered along the loading path. By contrast, given the incompatibility of such mode with the boundary conditions applied to the finite-sized sample for equibiaxial compression, we do not expect it to emerge when the samples are deformed equibiaxially.

6 Post-buckling analysis

Load-displacement analyses for both the finite-sized and infinite periodic structures are then performed with ABAQUS/Standard to capture the post-transformation behavior. After determining the pattern transformation (the lowest eigenmode) from the eigen analysis, an imperfection in the form of the critical eigenmode is introduced into the mesh. [2]

Note that to study the postbuckling behavior of the infinite period structure, we use enlarged unit cells with size dictated by the new periodicity introduced by buckling. In particular, for the structure with circular pores we use an enlarged unit cell with $(m_1,m_2) = (2, 2)$ for both uniaxial and equibiaxial compression. Differently, for the structure with star-shaped pores we consider an enlarged unit cell with $(m_1,m_2) = (1, 5)$.

The macroscopic stress-strain relationships for the structure with a square array of circular holes subjected to uniaxial and equibiaxial compression are shown in Figs. 9-a and -b, respectively. The finite-sized and infinite model results are found to be in very good agreement, indicating that the response of the system is not significantly affected by boundary effects. For both loading cases, the lattice exhibits an initial linear elastic behavior with a sudden departure from linearity to a plateau stress. This behavior is the result of a sudden instability which alters the initial architecture of the lattice (see insets in Fig. 9). As predicted by the stability analysis, when the lattice is compressed an homogeneous pattern transformation is induced by buckling. Here, we emphasize the uniformity and robustness of the transformation where the change occurs essentially uniformly throughout the structure. In other words, the instability does not localize deformation in a row or diagonal band, but instead results in a homogeneous pattern transformation throughout the structure. The transformed structure is then accentuated with continuing deformation and the entire process is reversible and repeatable.

Interestingly, these elastic instabilities can be harnessed to create novel effects within materials with periodic microstructure. For example, the microscopic instability triggered under uniaxial compression in the structure with circular pores can be exploited to design materials with macroscopic negative Poisson's ratio (Overvelde et al., 2012; Bertoldi et al., 2010). The evolution of the macroscopic strain in transverse direction, ε_{11}, and macroscopic Poisson's ratio, $\overline{\nu}$, as a function on the applied strain are presented in Fig. 10. Interestingly, noticeable lateral contraction induced by buckling is observed, eventually leading to negative values of Poisson' s ratio.

Finally, in Figs. 11-a and -b we report the macroscopic stress-strain relationships for the structure with a square array of star-shaped holes subjected to uniaxial and equibiaxial compression. Similarly to the case of the structure with circular pores, the structure exhibit an initial linear elastic behavior with a sudden departure from linearity to a plateau stress,

[2]Note that we have also considered imperfections in the form of linear combination of the first three modes. However, we found these to lead to the same post-transformation behavior.

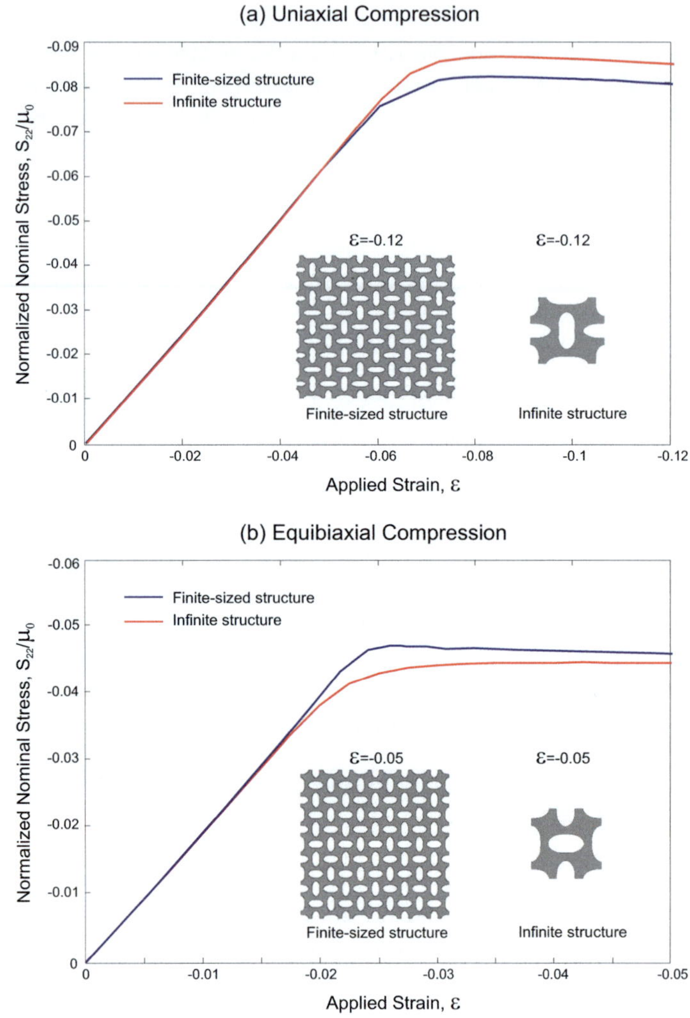

Figure 9. Nominal stress vs. nominal strain curves for a finite-sized and infinite structure with a square array of circular holes under (a) uniaxial and (b) equibiaxial compression.

which is induced by a mechanical instability. However, in this case we find a significant discrepancy between the response of the infinite and finite-sized structure. In fact, under uniaxial compression the finite-sized periodic

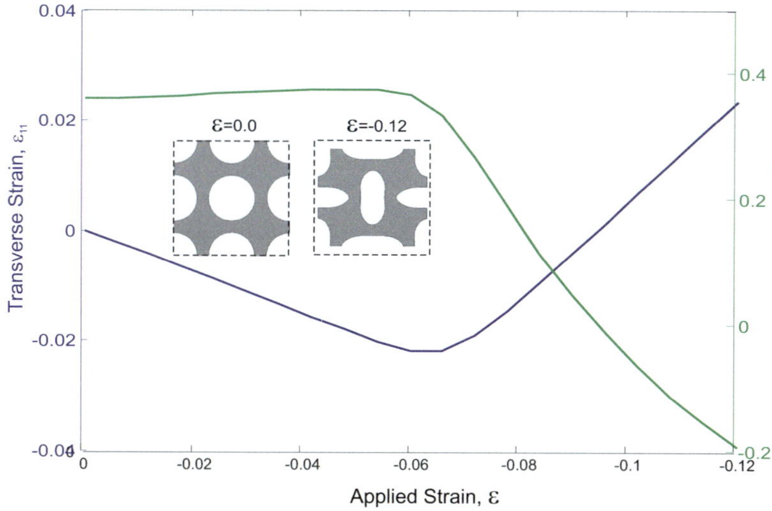

Figure 10. Evolution of the macroscopic strain in transverse direction and macroscopic Poisson's ratio as a function on the applied strain for an infinite structure with a square array of circular holes.

solid exhibits an earlier departure from linearity than the infinite model. This discrepancy is dictated by the choice of the enlarged unit cell used to represent the infinite periodic system. As shown in Section 5, a long wavelength instability is triggered when the system is compressed and the critical strain decreases as the number of unit cells in vertical direction increases, approaching $\varepsilon_{cr} = -0.017$ for very long systems. Therefore, the strain at the onset of instability is lower for the 10×10 finite-sized model than for the infinite model whose response is here represented by a 1×5 enlarged unit cell.

Interestingly, an opposite trend is found when the structure with star-shape holes is compressed equibiaxially (see Fig. 9-b), with the infinite periodic solid exhibiting an earlier departure from linearity. This discrepancy is due the different postbuckling patterns predicted by the two models. In fact in the infinite model we observe the formation of a pattern with a long wavelength, while in the finite-sized solid a checkerboard pattern is formed. As found in Section 5, the critical strain associated to the long wavelength mode is the lowest, but such mode is not compatible with the boundary conditions experienced by a sample during equibiaxial compression, preventing its formation in finite-sized models.

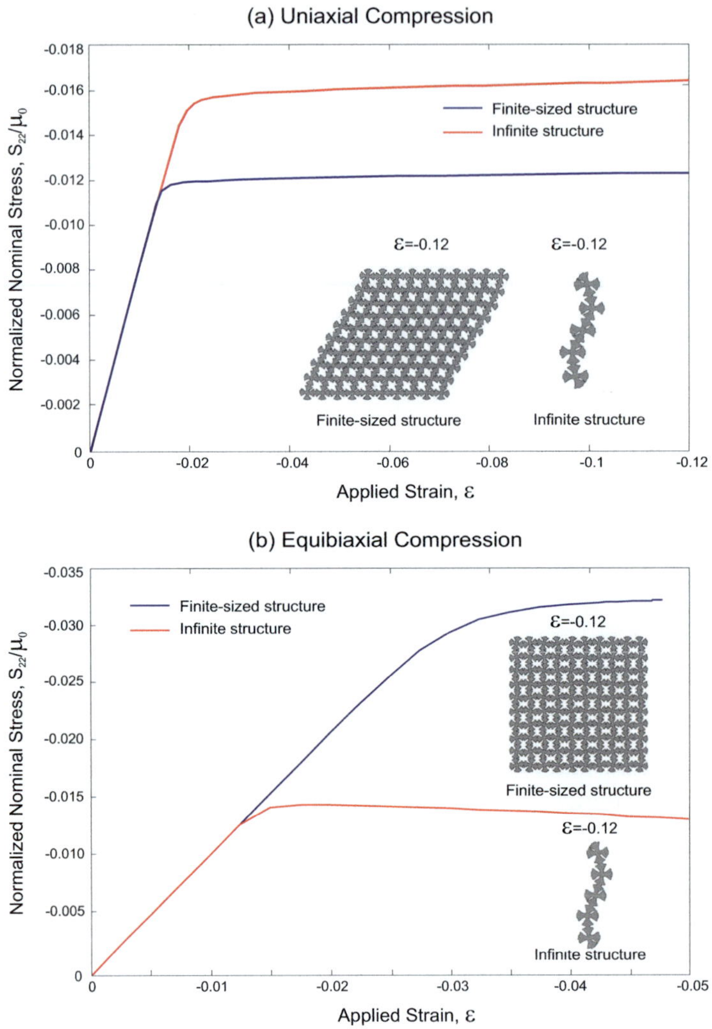

Figure 11. Nominal stress vs. nominal strain curves for a finite-sized and infinite structure with a square array of star-shaped holes under (a) uniaxial and (b) equibiaxial compression.

7 Conclusions

To summarize, we presented robust numerical tools to study the stability and non-linear response of porous periodic solids and used them to investigate the response under compression of a structure with a square array of pores. In particular, we numerically studied the effect of pore shape on the non-linear response of a square array of pores in an elastomeric matrix. Our results show that the pore shape strongly affects structural stability, not only changing the critical strain, but most importantly the wavelength of the critical mode. Interestingly, it has been recently shown that porous solids for which microscopic instabilities are critical open avenues for the design of phononic switches (Jang et al., 2009; Wang and Bertoldi, 2012), color displays (Li et al., 2012) and materials with unusual properties such as large negative Poisson's ratio (Overvelde et al., 2012; Bertoldi et al., 2010).

Bibliography

M. Aberg and P. Gudmundson. The usage of standard finite element codes for computation of dispersion relations in materials with periodic microstructure. *Journal of the Acoustical Society of America*, 102(4):2007–2013, 1997.

Melvin S Anderson. Buckling of periodic lattice structures. *AIAA journal*, 19(6):782–788, 1981.

MS Anderson and FW Williams. Natural vibration and buckling of general periodic lattice structures. *AIAA journal*, 24(1):163–169, 1986.

Klaus-Jürgen Bathe. *Finite element procedures*. Prentice Hall, Englewood Cliffs, N.J., 1996.

K. Bertoldi and M. C. Boyce. Wave propagation and instabilities in monolithic and periodically structured elastomeric materials undergoing large deformations. *Physical Review B*, 78(18), 2008.

K. Bertoldi, M. C. Boyce, S. Deschanel, S. M. Prange, and T. Mullin. Mechanics of deformation-triggered pattern transformations and superelastic behavior in periodic elastomeric structures. *Journal of the Mechanics and Physics of Solids*, 56(8):2642–2668, 2008.

K. Bertoldi, P. M. Reis, S. Willshaw, and T. Mullin. Negative poisson's ratio behavior induced by an elastic instability. *Advanced Materials*, 22 (3):361–+, 2010.

Edwin P Chan, Erica J Smith, Ryan C Hayward, and Alfred J Crosby. Surface wrinkles for smart adhesion. *Advanced Materials*, 20(4):711–716, 2008.

M. Danielsson, D. M. Parks, and M. C. Boyce. Three-dimensional micromechanical modeling of voided polymeric materials. *Journal of the Mechanics and Physics of Solids*, 50(2):351–379, 2002.

Steven E Forman and John W Hutchinson. Buckling of reticulated shell structures. *International Journal of Solids and Structures*, 6(7):909–932, 1970.

G. Geymonat, S. Muller, and N. Triantafyllidis. Homogenization of nonlinearly elastic-materials, microscopic bifurcation and macroscopic loss of rank-one convexity. *Archive for Rational Mechanics and Analysis*, 122 (3):231–290, 1993.

Tommy S Horozov, Bernard P Binks, Robert Aveyard, and John H Clint. Effect of particle hydrophobicity on the formation and collapse of fumed silica particle monolayers at the oil–water interface. *Colloids and Surfaces A: Physicochemical and Engineering Aspects*, 282:377–386, 2006.

RG Hutchinson and NA Fleck. The structural performance of the periodic truss. *Journal of the Mechanics and Physics of Solids*, 54(4), 2006.

Ji-Hyun Jang, Cheong Yang Koh, Katia Bertoldi, Mary C Boyce, and Edwin L Thomas. Combining pattern instability and shape-memory hysteresis for phononic switching. *Nano letters*, 9(5):2113–2119, 2009.

Jungwook Kim, James A Hanna, Myunghwan Byun, Christian D Santangelo, and Ryan C Hayward. Designing responsive buckled surfaces by halftone gel lithography. *Science*, 335(6073):1201–1205, 2012.

Charles Kittel. *Introduction to solid state physics.* Wiley, Hoboken, NJ, 8th edition, 2005.

Jie Li, Jongmin Shim, Justin Deng, Johannes TB Overvelde, Xuelian Zhu, Katia Bertoldi, and Shu Yang. Switching periodic membranes via pattern transformation and shape memory effect. *Soft Matter*, 8(40):10322–10328, 2012.

Jerrold E. Marsden and Thomas J. R. Hughes. *Mathematical foundations of elasticity.* Prentice-Hall civil engineering and engineering mechanics series. Prentice-Hall, Englewood Cliffs, N.J., 1983.

Tom Mullin, S Deschanel, K Bertoldi, and MC Boyce. Pattern transformation triggered by deformation. *Physical review letters*, 99(8):084301, 2007.

Ahmed K Noor, Melvin S Anderson, and William H Greene. Continuum models for beam-and platelike lattice structures. *AIAA Journal*, 16(12): 1219–1228, 1978.

JTB Overvelde and K Bertoldi. Relating pore shape to the non-linear response of periodic elastomeric structures. *Journal of the Mechanics and Physics of Solids.*, 64:351–366, 2014.

JTB Overvelde, S Shan, and K Bertoldi. Compaction through buckling in 2d periodic, soft and porous structures: Effect of pore shape. *Advanced Materials*, 24(17):2337–2342, 2012.

SD Papka and S Kyriakides. In-plane biaxial crushing of honeycombs: Part ii: Analysis. *International Journal of Solids and Structures*, 36(29):4397–4423, 1999.

JD Renton. Buckling of long, regular trusses. *International Journal of Solids and Structures*, 9(12):1489–1500, 1973.

M Schraad and N Triantafyllidis. Scale effects in media with periodic and nearly periodic microstructuresii. failure mechanisms. *the: Journal*, 64: 762–771, 1997.

J. Shim, C. Perdigou, E. R. Chen, K. Bertoldi, and P. M. Reis. Buckling-induced encapsulation of structured elastic shells under pressure. *Proceedings of the National Academy of Sciences of the United States of America*, 109(16):5978–5983, 2012.

Srikanth Singamaneni, Katia Bertoldi, Sehoon Chang, Ji-Hyun Jang, Seth L Young, Edwin L Thomas, Mary C Boyce, and Vladimir V Tsukruk. Bifurcated mechanical behavior of deformed periodic porous solids. *Advanced Functional Materials*, 19(9):1426–1436, 2009.

N. Triantafyllidis and B. N. Maker. On the comparison between microscopic and macroscopic instability mechanisms in a class of fiber-reinforced composites. *Journal of Applied Mechanics-Transactions of the ASME*, 52(4):794–800, 1985.

N Triantafyllidis, MD Nestorovic, and MW Schraad. Failure surfaces for finitely strained two-phase periodic solids under general in-plane loading. *Journal of Applied Mechanics*, 73:505–515, 2006.

Thein Wah. The buckling of gridworks. *Journal of the Mechanics and Physics of Solids*, 13(1):1–16, 1965.

Thein Wah and Lee R Calcote. *Structural analysis by finite difference calculus*. Van Nostrand Reinhold New York, 1970.

Lifeng Wang and Katia Bertoldi. Mechanically tunable phononic band gaps in three-dimensional periodic elastomeric structures. *International Journal of Solids and Structures*, 49(19):2881–2885, 2012.

Shu Yang, Krishnacharya Khare, and Pei-Chun Lin. Harnessing surface wrinkle patterns in soft matter. *Advanced Functional Materials*, 20(16): 2550–2564, 2010.

Ying Zhang, Elisabetta A Matsumoto, Anna Peter, Pei-Chun Lin, Randall D Kamien, and Shu Yang. One-step nanoscale assembly of complex structures via harnessing of an elastic instability. *Nano letters*, 8(4): 1192–1196, 2008.

Folding and Deployment of Thin Shell Structures

Sergio Pellegrino
Graduate Aerospace Laboratories, California Institute of Technology
Pasadena, CA 91125, USA

Abstract Thin shells made of high modulus material are widely used as lightweight deployable space structures. The focus of this chapter is the most basic deployable thin shell structure, namely a straight, transversely curved strip known as a *tape spring*. Following a review of the materials used for the construction of deployable thin shell structures, including constitutive models and failure criteria developed specifically for this type of structures, this chapter provides an introduction to the mechanics of tape springs and tape spring hinges. Finite element techniques to model deployable structures containing tape springs are presented and the ability of these models to accurately simulate experimentally observed behavior is demonstrated. These tools can be used to design structures able to achieve specific behaviors. As an example, the design of a two-hinge boom that can be wrapped around a small spacecraft without any damage, and can dynamically deploy and smoothly latch into the deployed configuration is presented.

1 Introduction and Background

Thin shells are widely used in deployable space structures. Open section cylindrical shells have been used on spacecraft for deployable booms for many years and, more recently, a wide range of singly- and doubly-curved this shells have been developed for deployable reflector antennas and radar structures. Some examples are shown in Figs 1, 2.

This chapter presents a detailed study of the mechanics, design, realization and testing of a particular type of deployable thin-shell structures, known as *tape springs*.

Tape springs are straight, thin-walled, elastic strips with curved cross-section. The simplest and most common form, shown in Fig. 3, is a cylindrical shell of uniform thickness t and transverse radius of curvature R, subtending an angle α.

D. Bigoni (Ed.), *Extremely Deformable Structures*,
CISM International Centre for Mechanical Sciences
DOI 10.1007/ 978-3-7091-1877-1_5 © CISM Udine 2015

Figure 1: Two spring back reflector antennas on MSAT-2 spacecraft. Courtesy: Communications Research Centre (CRC), Industry Canada.

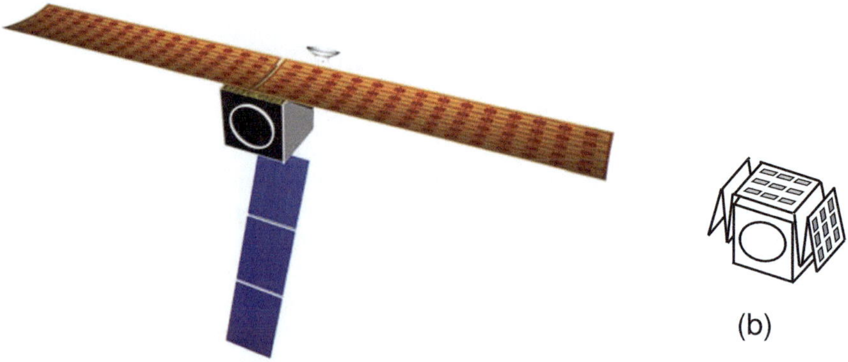

(a) (b)

Figure 2: Folding Large Antenna Tape Spring concept (Soykasap et al., 2008) (a) deployed and (b) folded.

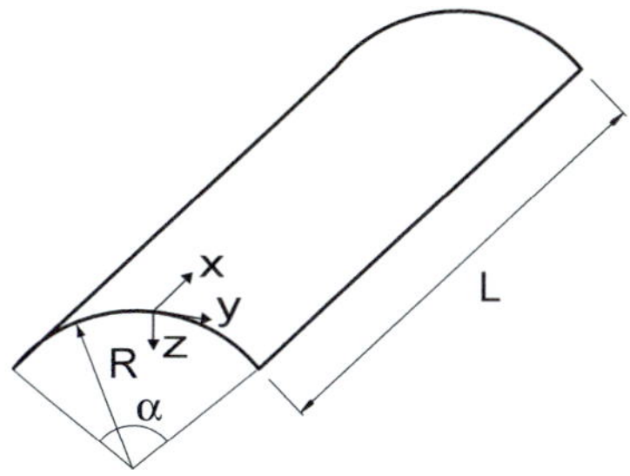

Figure 3: Tape spring geometry.

The most common application of tape springs is in tape measures, also known as carpenter tapes, which were invented in the late 1920's. Tape measures can be wound in a small case, and are strained elastically in this configuration, but they become essentially strain-free when they are extended. Typical geometric parameters of a tape measure are $R \approx 20$ mm and $\alpha \approx 50$ deg, i.e. 0.87 rad. The values of α will be given in degrees, for clarity, but this value is converted to radians for the calculation of the arc-length of the tape spring cross section.

Figure 4: Early tape measure with cup-shaped case and 38 mm wide FatMax tape measure.

A more specialized application of the same structural form is found in a type of deployable boom known as the STEM (Rimrott, 1966). In its simplest configuration this boom has $\alpha > 360$ deg, to form a slit circular tube with an overlap region. Variants of this configuration include tape springs with $\alpha \approx 360$ deg and with interlocking tabs along the straight edges, that form closed-section tubes with higher torsional stiffness, and also booms formed by two or more STEM's nested inside one another, as in the first set of solar arrays for the Hubble Space Telescope shown in Fig. 5. STEM's have been used as low-gain antennas, to position scientific instruments away from spacecraft, to deploy solar arrays, etc. (Pellegrino,

Figure 5: Four Bi-STEM booms were used in the Hubble Telescope first set of solar arrays (image courtesy of NASA).

1995).

Both the tape measure and the STEM store elastic strain energy during folding and, in principle, they would both freely deploy into the straight, unstrained configuration when all constraints are released. However, free deployment of tape springs is uncommon: in the tape measure friction between the tape and the case that contains it dissipates much of the stored strain energy, while in the STEM deployment out of the housing mechanism is usually controlled by an electric motor. The same motor is also used to retract the STEM into its housing. A type of self-deploying emergency beacon consists of a STEM coiled inside a cylindrical box; when the lid of the box is opened, the STEM deploys from the center of the coil, forming a complex transition shape which gradually develops into the unstressed, extended tubular shape as shown in Fig. 6. Rimrott (1966, 1980) has shown that this deployment mode is stable and that the extension speed and angular velocity of the tip are approximately constant.

2 Materials, Constitutive Models and Failure Criteria

A general requirement common to all of the structures described in this chapter is that they are made from high modulus, high yield or failure stress materials that can be manufactured to the required shape and with a small thickness. Both metals and fiber-reinforced composites have been used and several examples are presented in this section.

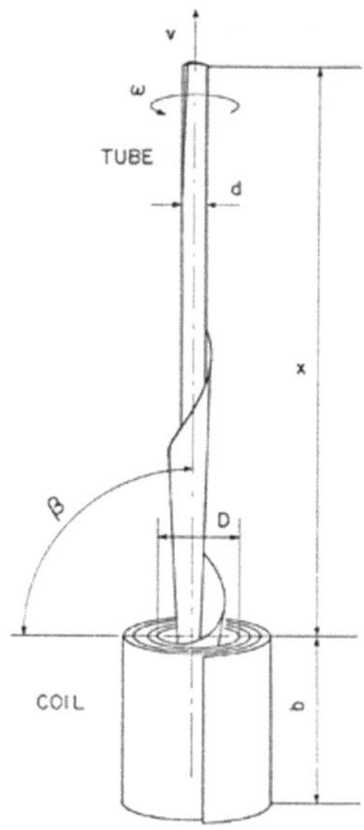

Figure 6: "Jack-in-the-box" STEM demployment (image courtesy of Rimrott (1980)).

2.1 Materials

Tape springs were initially made of metal, such as spring steel, stainless steel or beryllium-copper (CuBe). For example, CuBe 1565 alloy increases its yield stress from 200 MPa to 1100 MPa through age-hardening for 3 hours at 310 deg. However the elastic modulus of this material is 131 GPa and its density 8200 kg/m^3, corresponding to a relatively low specific stiffness of $E/\rho = 15.8 \times 10^6$ m^2/s^2.

Since the 1990's tape springs have increasingly been made from carbon-fiber reinforced plastics, which have much lower density that metals, have larger failure strains and also can be manufactured in a variety of different shapes. Composite tape springs have been made by co-curing unidirectional laminas of thin-ply composites, which are available as pre-impregnated sheets of parallel fibers and uncured resin. Thin-plies are made by spreading individual tows of carbon fiber and available in thicknesses down to 15 μm. A six-ply laminate made from 30 μm thick prepregs is shown in Fig. 7; multi-ply laminates with overall thickness on the order of 50 μm thickness can be made.

An alternative approach is ultra-thin laminates made from carbon-fiber

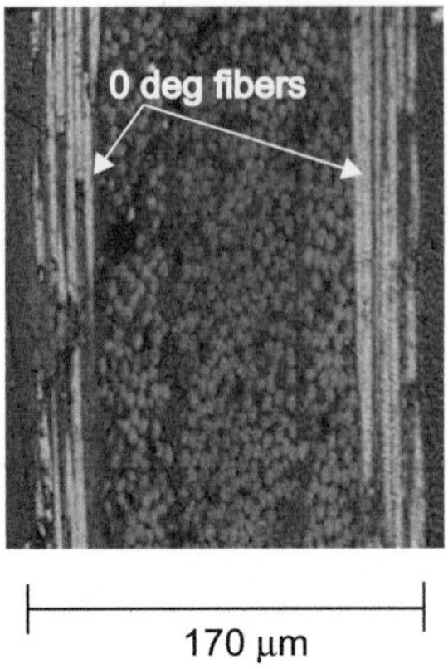

Figure 7: Micrograph of 6-ply laminate $[0/+60,-60]_s$.

fabric, but they tend to be thicker. For example, Fig. 8 shows a 0.2 mm thick ±45 two-ply plain weave laminate of T300-1k carbon fibre fabric and HexPly 913 resin. The fiber and resin properties are given in Table 1.

Table 1: Fiber and resin properties (Hexcel (2007); Toray (2015))

Properties	T300 fiber	HexPly 913 resin
Longitudinal stiffness, E_1 (N/mm^2)	233,000	3,390
Transverse stiffness, E_2 (N/mm^2)	23,100	3,390
Shear stiffness, G_{12} (N/mm^2)	8,963	1,210
Poisson′s ratio, ν_{12}	0.2	0.41
Density, ρ (kg/m^3)	1,760	1,230
Areal weight of fabric/film, W (g/m^2)	98	30

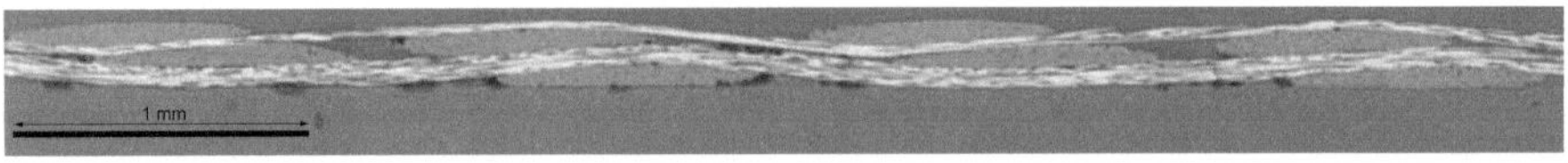

Figure 8: Micrograph of T300-1k/913 two-ply plain weave laminate.

2.2 Constitutive Models

Thin shells can be modeled as linear-elastic Kirchhoff thin plates. Kinematically, they are described by the deformation of their mid-surface and by the assumption that lines perpendicular to the mid-surface remain straight and perpendicular to the deformed mid-surface.

In the following analysis several coordinate systems are used. For isotropic materials, a single system x', y', z based on the principal directions of curvature of the shell structure and the $x'y'$ plane coinciding with the mid-surface of the structure. For orthotropic materials, two additional coordinate systems are used, a first material-based system with the 1-axis parallel to the direction of the fibers and the 2-axis and 3-axis perpendicular to the fibers, and a second material-based system with the x-axis and the y-axis parallel to the directions of the tows in the woven fabric. The xy plane coincides with the mid-surface of the laminate.

The constitutive model for a Kirchhoff plate takes the form of a 6×6 ABD matrix that relates the three in-plane stress resultants, $N_{x'}, N_{y'}, N_{x'y'}$, and the three out-of-plane stress resultants, $M_{x'}, M_{y'}, M_{x'y'}$, to the mid-surface strain components, $\epsilon_{x'}, \epsilon_{y'}, \gamma_{x'y'}$, and the mid-surface curvatures, $\kappa_{x'}, \kappa_{y'}, \kappa_{x'y'}$:

$$\left\{\begin{array}{c} N_{x'} \\ N_{y'} \\ N_{x'y'} \\ -- \\ M_{x'} \\ M_{y'} \\ M_{x'y'} \end{array}\right\} = \left(\begin{array}{ccccccc} A_{11} & A_{12} & A_{16} & | & B_{11} & B_{12} & B_{16} \\ A_{21} & A_{22} & A_{26} & | & B_{21} & B_{22} & B_{26} \\ A_{61} & A_{62} & A_{66} & | & B_{61} & B_{62} & B_{66} \\ -- & -- & -- & -- & -- & -- & -- \\ B_{11} & B_{21} & B_{61} & | & D_{11} & D_{12} & D_{16} \\ B_{12} & B_{22} & B_{62} & | & D_{21} & D_{22} & D_{26} \\ B_{16} & B_{26} & B_{66} & | & D_{61} & D_{62} & D_{66} \end{array}\right) \left\{\begin{array}{c} \epsilon_{x'} \\ \epsilon_{y'} \\ \gamma_{x'y'} \\ -- \\ \kappa_{x'} \\ \kappa_{y'} \\ \kappa_{x'y'} \end{array}\right\} \quad (1)$$

Denoting each 3×3 sub-matrix and the corresponding kinematic and static variables with a single letter, Eq. 1 is written in the compact form:

$$\begin{bmatrix} \mathbf{N} \\ \mathbf{M} \end{bmatrix} = \begin{bmatrix} \mathbf{A} & \mathbf{B} \\ \mathbf{B} & \mathbf{D} \end{bmatrix} \begin{bmatrix} \epsilon \\ \kappa \end{bmatrix} \quad (2)$$

It should be noted that the *engineering shear strain* and *twice the surface twist* have been used in the definition of these deformation variables.

For an isotropic plate of modulus E, Poisson's ratio ν, and thickness t the three matrices have the expressions (Calladine, 1983):

$$\mathbf{A} = \frac{Et}{1-\nu^2} \begin{bmatrix} 1 & \nu & 0 \\ \nu & 1 & 0 \\ 0 & 0 & \frac{1-\nu}{2} \end{bmatrix}; \mathbf{B} = 0; \mathbf{D} = \frac{Et^3}{12(1-\nu^2)} \begin{bmatrix} 1 & \nu & 0 \\ \nu & 1 & 0 \\ 0 & 0 & \frac{1-\nu}{2} \end{bmatrix} \quad (3)$$

For a laminated plate with n orthotropic laminas, see Fig. 10, with longitudinal modulus E_1, transverse modulus E_2, Poisson's ratios ν_{12}, ν_{21},

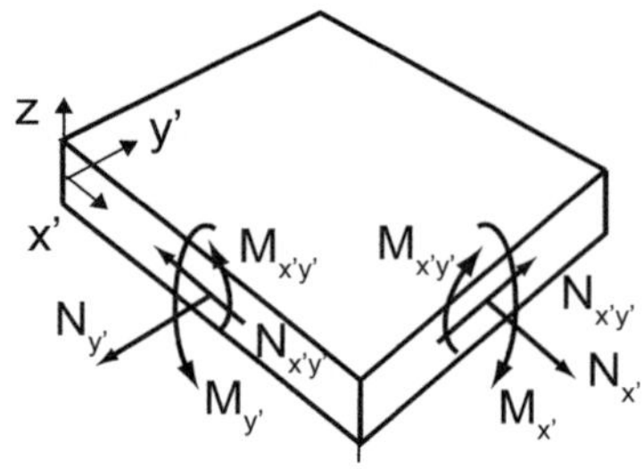

Figure 9: Stress resultants for Kirchhoff plate model in the structure coordinate system.

and shear modulus G_{12} (the subscript k denotes the layer number), the three matrices have the expressions (Daniel and Ishai, 2006):

$$\mathbf{A} = \sum_{k=1}^{n} \overline{\mathbf{Q}}_k (z_k - z_{k-1}) \tag{4}$$

$$\mathbf{B} = \frac{1}{2}\sum_{k=1}^{n} \overline{\mathbf{Q}}_k (z_k^2 - z_{k-1}^2) \tag{5}$$

$$\mathbf{D} = \frac{1}{3}\sum_{k=1}^{n} \overline{\mathbf{Q}}_k (z_k^3 - z_{k-1}^3) \tag{6}$$

where the lamina stiffness in the x', y', z coordinate system is given by

$$\overline{\mathbf{Q}}_k = \mathbf{T}_k \mathbf{Q}_k \mathbf{T}_k^T \tag{7}$$

and is obtained from the lamina stiffness in the 1,2,3 coordinate system

$$\mathbf{Q}_k = \begin{bmatrix} \frac{E_1}{1-\nu_{12}\nu_{21}} & \frac{\nu_{21}E_1}{1-\nu_{12}\nu_{21}} & 0 \\ \frac{\nu_{12}E_2}{1-\nu_{12}\nu_{21}} & \frac{E_2}{1-\nu_{12}\nu_{21}} & 0 \\ 0 & 0 & G_{12} \end{bmatrix}_k \tag{8}$$

and the transformation matrix

$$\mathbf{T}_k = \begin{bmatrix} \cos^2\theta & \sin^2\theta & -2\sin\theta\cos\theta \\ \sin^2\theta & \cos^2\theta & 2\sin\theta\cos\theta \\ \sin\theta\cos\theta & -\sin\theta\cos\theta & \cos^2\theta - \sin^2\theta \end{bmatrix}_k \tag{9}$$

where the angle θ for lamina k is the angle from the x'-axis to the fiber direction (1-axis). For a two-ply plain-weave laminate the ABD matrix

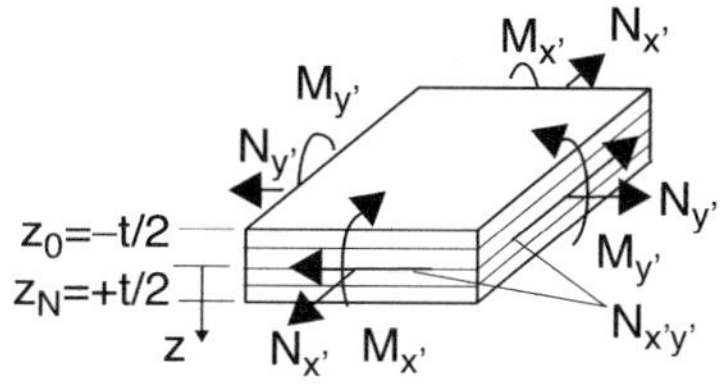

Figure 10: Multi-ply laminate.

can be obtained from a finite element analysis of a representative unit cell, shown in Fig. 11. Note that the two plies are in phase and the surfaces of each tow are defined with respect to the unit-cell-based reference system by

$$z = a\left(\pm\, 2 \pm \sqrt[4]{\left|\sin\left(\frac{2\pi x}{\Delta L}\right)\right|} \pm \sqrt[4]{\left|\sin\left(\frac{2\pi y}{\Delta L}\right)\right|}\right) \tag{10}$$

where the sign of the first term is + for the upper ply and − for the lower ply; the signs of the second and third terms are respectively $(+,+)$ for the top surface of the ply, $(+,-)$ or $(-,+)$ for the interface between the two tows, and $(-,-)$ for the lower surface of the ply.

This representation of the tow surfaces ensures a perfect geometric match between tows in the same ply, however tows in different plies touch only at four points and hence there is a gap that has been filled with triangular prisms modeling additional neat resin.

The particular model shown in Figure 11 consists of 1520 and 3200, 6-node triangular prisms elements, respectively for each tow and for the additional resin. The boundary nodes were attached by means of rigid vertical beams to dummy nodes lying in the mid-plane. Each dummy node was related to the corresponding dummy node opposite by means of a constraint equation to enforce periodic boundary conditions. These equations prescribe the relative displacements and rotations of the dummy nodes in terms of the mid-plane strains and out-of-plane curvatures

$$\varepsilon_x = \frac{\partial u}{\partial x} \tag{11a}$$

$$\varepsilon_y = \frac{\partial v}{\partial y} \tag{11b}$$

$$\gamma_{xy} = \frac{\partial u}{\partial y} + \frac{\partial v}{\partial x} \tag{11c}$$

$$\kappa_x = -\frac{\partial^2 w}{\partial x^2} \tag{11d}$$

$$\kappa_y = -\frac{\partial^2 w}{\partial y^2} \tag{11e}$$

$$\kappa_{xy} = -2\frac{\partial^2 w}{\partial x \partial y} \tag{11f}$$

of a homogenized Kirchhoff plate.

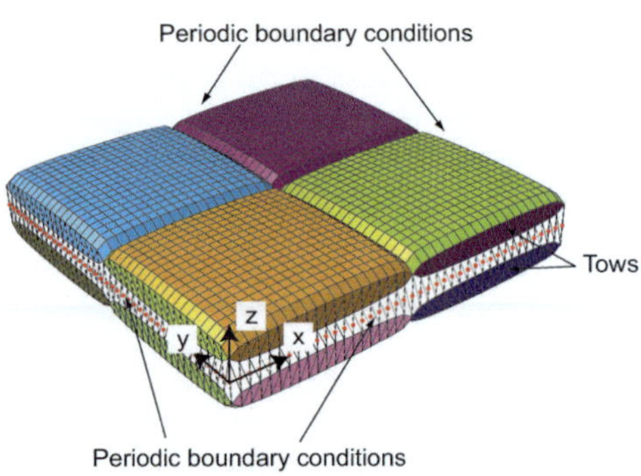

Figure 11: Finite element model for unit cell of two-ply laminate (with 3X magnification in z-direction).

The constraint equations are as follows

$$\Delta u^x = \varepsilon_x \, \Delta L \tag{12a}$$
$$\Delta v^x = \frac{1}{2} \, \gamma_{xy} \, \Delta L \tag{12b}$$
$$\Delta u^y = \frac{1}{2} \, \gamma_{xy} \, \Delta L \tag{12c}$$
$$\Delta v^y = \varepsilon_y \, \Delta L \tag{12d}$$
$$\Delta w^x = -\frac{1}{2} \, \kappa_{xy} \, y \, \Delta L \tag{12e}$$
$$\Delta w^y = -\frac{1}{2} \, \kappa_{xy} \, x \, \Delta L \tag{12f}$$
$$\Delta \theta_x^x = -\frac{1}{2} \, \kappa_{xy} \, \Delta L \tag{12g}$$
$$\Delta \theta_y^x = \kappa_x \, \Delta L \tag{12h}$$
$$\Delta \theta_x^y = -\kappa_y \, \Delta L \tag{12i}$$
$$\Delta \theta_y^y = \frac{1}{2} \, \kappa_{xy} \, \Delta L \tag{12j}$$
$$\Delta \theta_x^z = 0 \tag{12k}$$
$$\Delta \theta_y^z = 0 \tag{12l}$$

where the subscripts denote the deformation directions and the superscripts denote the direction of a pair of boundary nodes, with equal x or y coordinates.

The entries of the ABD matrix for the two-ply laminate can then be computed with Virtual Work (Karkainen and Sankar, 2006; Kueh and Pellegrino, 2007) after carrying out six separate ABAQUS/Standard analyses, each corresponding to a unit amplitude of the six deformation variables $\varepsilon_x, \ldots, \kappa_{xy}$.

For example, entry 1,1 is obtained by writing the equation of virtual work for the case $\varepsilon_x = 1$, with the forces/moments computed for this deformation mode. Hence the equation reads

$$N_{xx} \varepsilon_x \Delta L^2 = \sum_{b.n.} (F_x u + F_y v + F_z w + M_x \theta_x + M_y \theta_y + M_z \theta_z) \tag{13}$$

where the summation is extended to all boundary nodes (b.n.). Then, substituting $\varepsilon_x = 1$ and comparing with Eq. 1 gives

$$A_{11} = \frac{\sum_{b.n.} (F_x u + F_y v + F_z w + M_x \theta_x + M_y \theta_y + M_z \theta_z)}{\Delta L^2} \tag{14}$$

2.3 Failure Criteria

For metallic tape-springs a standard failure criterion, such as the Tresca yield criterion can be used. This was used by Rimrott (1966) for the calculation of the packaging limits of STEM booms.

For laminates of unidirectional plies the Tsai-Wu criterion can be used (Daniel and Ishai, 2006). For a unidirectional lamina, under plane stress and subject to normal stress components σ_1 in the fiber direction, σ_2 in the direction perpendicular to the fibers, and in-plane shear σ_{12}, the Tsai-Wu criterion has the expression:

$$f_1\sigma_1 + f_2\sigma_2 + f_{11}\sigma_1^2 + f_{22}\sigma_2^2 + f_{66}\sigma_{12}^2 + 2f_{12}\sigma_1\sigma_2 = 1 \tag{15}$$

where the coefficients have the following expressions:

$$f_1 = \frac{1}{F_{1t}} - \frac{1}{F_{1c}} \tag{16a}$$

$$f_2 = \frac{1}{F_{2t}} - \frac{1}{F_{2c}} \tag{16b}$$

$$f_{11} = \frac{1}{F_{1t}F_{1c}} \tag{16c}$$

$$f_{22} = \frac{1}{F_{2t}F_{2c}} \tag{16d}$$

$$f_{66} = \frac{1}{F_{12}^2} \tag{16e}$$

$$f_{12} = -\frac{1}{2}\sqrt{\frac{1}{F_{1t}F_{1c}F_{2t}F_{2c}}} \tag{16f}$$

and the characteristic strengths that appear in Eq. 16 are defined as:

- F_{1t}, F_{1c} tensile, compressive failure stress of lamina in the fiber direction;
- F_{2t}, F_{2c} tensile, compressive failure stress of lamina in the direction perpendicular to the fibers;
- F_{12} failure shear stress of lamina.

For ultra-thin laminates made of woven fabric a failure criterion has been developed by Mallikarachchi and Pellegrino (2013). The criterion is defined in terms of the in-plane stress resultants N_x, N_y, N_{xy} and the out-of-plane stress resultants M_x, M_y, M_{xy} *for the laminate*. Note that, because we are dealing with a woven fabric, the resultants are with respect to the directions of the weave.

The failure locus is defined by five characteristic strength values that can be obtained from tensile, compressive, shear, bending and twisting failure tests. For a symmetric two-ply plain weave laminate made from Toray T300-1k fibers and Hexply 913 epoxy resin, these strengths were measured and their values are given in the second column of Table 2.

Table 2: Characteristic strengths of two-ply plain weave T300-1k/Hexply 913 laminate (Mallikarachchi and Pellegrino, 2013)

Strength	Value
Tensile, $F_{1t} = F_{2t}$ (N/mm)	133.60
Compressive, $F_{1c} = F_{2c}$ (N/mm)	57.36
Shear, F_3 (N/mm)	16.91
Bending, $F_4 = F_5$ (Nmm/mm)	4.54 8
Twisting, F_6 (Nmm/mm)	1.10

The failure locus is defined by three failure indices, FI-1, FI-2 and FI-3:

$$\text{FI-1} = f_1(N_x + N_y) + f_{11}(N_x^2 + N_y^2) + f_{12}N_xN_y + f_{33}N_{xy}^2 \quad < 1 \quad (17a)$$

$$\text{FI-2} = f_{44} \times \max(M_x^2, M_y^2) + f_{66}M_{xy}^2 \quad < 1 \quad (17b)$$

$$\text{FI-3} = \max\left(\frac{N_x}{F_x}, \frac{N_y}{F_y}\right) + \frac{\max(|M_x|, |M_y|)}{F_4} \quad < 1 \quad (17c)$$

Failure occurs when at least one of these indices becomes equal to or greater than one.

FI-1 deals with the effects of in-plane forces in the laminate, and corresponds to the well known Tsai-Wu failure criterion for a lamina but formulated in stress-resultant space. FI-2 considers failure due to bending and twisting moments, whereas FI-3 considers the coupling between in-plane loading and bending of the laminate.

It will be seen in Section 7 that in ultrathin booms with tape spring hinges, the dominant effects are bending and coupled in-plane and bending effects, and hence FI-2 and FI-3 are the most dominant failure indices, with FI-3 usually being critical.

The coefficients in Eq. 17 have the following expressions:

$$f_1 = \frac{1}{F_{1t}} - \frac{1}{F_{1c}} \tag{18a}$$

$$f_{11} = \frac{1}{F_{1t}F_{1c}} \tag{18b}$$

$$f_{33} = \frac{1}{F_3^2} \tag{18c}$$

$$f_{44} = \frac{1}{F_4^2} \tag{18d}$$

$$f_{66} = \frac{1}{F_6^2} \tag{18e}$$

$$f_{12} = -\frac{f_{11}}{2} \tag{18f}$$

where

- F_{1t}, F_{1c} are tensile, compressive strengths per unit width of the laminate in the x-direction;
- F_{2t}, F_{2c} are tensile, compressive strengths per unit width of the laminate in the y-direction;
- F_3 is the shear strength of the laminate, N/mm;
- $F_4 = F_5, F_6$ are bending and twisting strengths of the laminate;

and F_x, F_y are the axial failure strengths, given by

$$F_x = \frac{-(f_1 + f_{12}N_y) \pm \sqrt{(f_1 + f_{12}N_y)^2 - 4f_{11}(f_1 N_y + f_{11}N_y^2 + f_{33}N_{xy}^2 - 1)}}{2f_{11}} \tag{19a}$$

$$F_y = \frac{-(f_1 + f_{12}N_x) \pm \sqrt{(f_1 + f_{12}N_x)^2 - 4f_{11}(f_1 N_x + f_{11}N_x^2 + f_{33}N_{xy}^2 - 1)}}{2f_{11}} \tag{19b}$$

More details on this failure locus, including a description of the test procedure to measure the strength parameters are available in Mallikarachchi and Pellegrino (2013). This paper also shows that the stress resultants obtained form a finite element analysis that uses a constant ABD stiffness matrix, as derived above, neglects the gradual reduction in the bending stiffness of the laminate, caused by changes in the geometry of the tows. The stiffness reduction is on the order of 20% when the laminate is close to failure. Because the simulations overestimate the bending and twisting moments near failure, a simple correction is introduced for the bending and twisting strengths, by multiplying them by a factor of 1.25.

3 Mechanics of Tape Springs

A tape spring can be bent either in *equal sense*, or in *opposite sense*, see Fig. 12. For equal sense bending the tape spring has longitudinal and transverse curvatures in the same direction; in this case the edges of the tape spring are subject to compressive stresses and the applied bending moment is defined to be positive. Conversely, for opposite-sense bending the tape spring has opposite longitudinal and transverse curvatures; in this case the edges of the tape spring are subject to tensile stresses and the applied bending moment is defined to be negative.

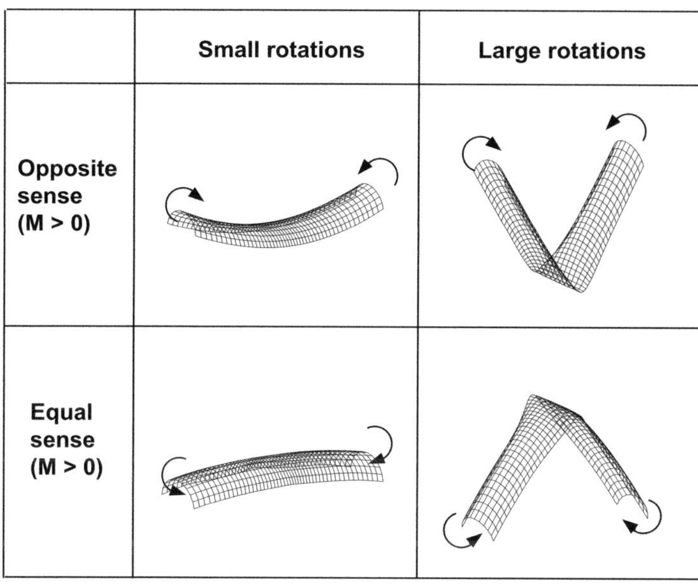

Figure 12: Two ways of bending/folding a tape spring.

The details of the moment-rotation relationship for a tape spring have been studied with the large-rotation test apparatus shown in Fig. 13. 200 mm long tape springs were potted with epoxy resin into acrylic blocks. The left-hand-side block was connected to a gear box fixed to the base of the apparatus. The right-hand-side block was connected to an identical gear box, but mounted on ball bearings so that it can move towards the first gear box when the large-deflection bending of the tape causes the distance between the end blocks to decrease. The adaptors that connect each block to a gear box consist of a U-shaped element supported by a short, thin-walled circular shaft with strain gauges mounted on its surface; these strain gauges measure the torque in the shaft, which is equal to the bending moment applied to the end of the tape spring. The elastic compliance of the apparatus is small in comparison with that of a shallow tape spring, but for springs subtending angles greater than $\alpha = 180$ deg the compliance of the apparatus needs to be accounted for. The input shafts of the gear boxes are turned manually, through small knobs and the angle turned through by each block is displayed on a dial.

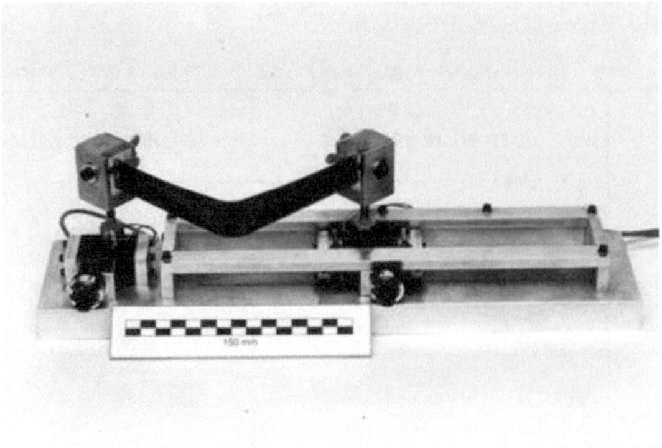

Figure 13: Apparatus for measuring the moment-rotation relationship of a tape spring.

Figure 14 shows the measured behavior of a 0.1 mm thick, 200 mm long tape spring, with a transverse radius of curvature of 13.3 mm and subtending an angle of 106 deg. The test measurements are overlaid with a finite-element simulation described in detail in Seffen and Pellegrino (1998).

When an initially straight tape spring is subject to gradually increasing equal and opposite end rotations, initially it takes a uniform longitudinally curved shape. Its moment-rotation relationship is linear for sufficiently small rotations. If the tape spring is subject to opposite-sense bending, as the end rotations are increased the tape spring suddenly snaps and forms an elastic fold that is approximately straight in the transverse direction and has approximately uniform longitudinal curvature, Fig. 14. Then, if the rotations are further increased, the arc-length of the fold increases while its curvature remains constant.

If, however, the tape spring is subject to equal-sense bending, it deforms by gradually twisting over two adjacent, but separate regions whose lengths grow until the two folds merge into a single, localized fold, Fig. 14. Once this single fold has formed, further increasing the end rotation results —again— only in an increase of the arc-length of the fold region.

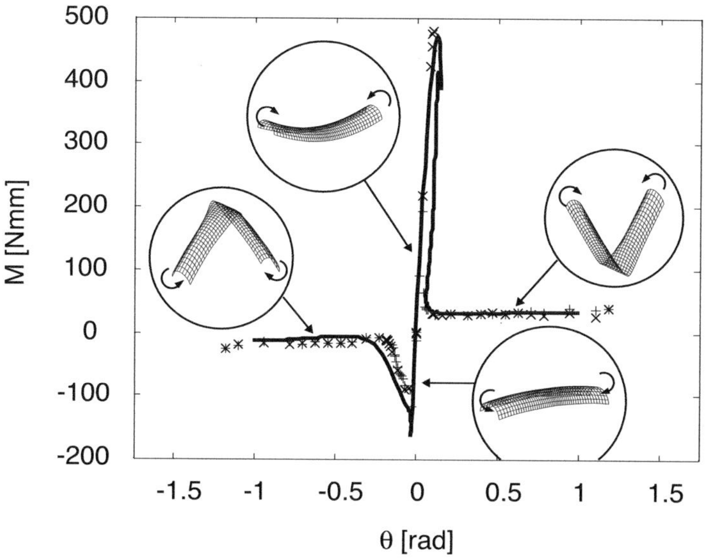

Figure 14: Moment-rotation relationship for a CuBe tape spring with $R =$ 13.3 mm, $\alpha = 106$ deg, and $t = 0.1$ mm.

Important characteristics of this behavior are that: (i) the elastic folds for both senses of bending have a longitudinal radius of curvature that is independent of the end rotations imposed on the ends of the tape spring; (ii) the peak stresses and strains in a folded tape spring are largely independent of the end rotations imposed on the ends; (iii) upon unfolding a tape spring snaps back into the straight configuration, this is most noticeable for a tape spring that has been bent in the opposite sense. This last characteristic makes tape springs ideal as self-latching components in deployable structures.

A schematic depiction of a tape spring folded in equal sense is provided in Fig. 15. The longitudinal fold radius is denoted by r. Note the x', y', z' coordinate system, with the x'-axis along the longitudinal direction of the tape spring.

The process of formation and growth of a localized elastic fold in a tape spring is a type of *propagating instability*. Examples of propagating insta-

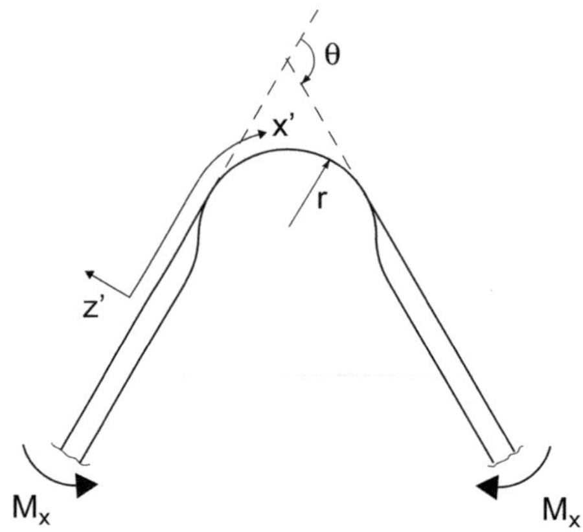

Figure 15: Tape spring with central fold.

bilities include the growth and propagation of bulges during the inflation of rubber balloons and the bucking of subsea pipelines.

Seffen and Pellegrino (1998) have shown that the tape spring instability can be analyzed in therms of the bending moment M applied at the two ends of the tape and the corresponding relative rotation between the two ends, θ. Note that for uniform bending $\theta = \kappa_x L$ and hence, apart from a scaling factor, the required function $M(\theta)$ is the same as $M(\kappa_x)$.

There are two characteristic values of the moment, the *maximum moment* M^{max} and the *steady-state moment* M^* at which the fold grows in length under a constant moment. Both of these values can be obtained from the expression of $M(\theta)$ for *uniform longitudinal bending* of the same tape spring.

An analytical solution to this problem was found by Wuest (1954) together with other solutions for "cross-curved leaf springs." His solution was laterincluded in a Russian monograph, later translated into English, by Panovko and Gubanova (1965). Wuest made the hypothesis that a relatively shallow spring, with α not much bigger than 60 deg, loaded by end moments M which impose a uniform curvature $\kappa_x = 1/r$ can be analyzed as a slightly distorted axi-symmetric cylindrical shell, as shown in Fig. 16. In this figure note that the axis of symmetry corresponds to the longitudinal direction of the tape spring. Also note that the x- and y-axes correspond to the longitudinal and transverse directions of the tape spring, respectively, while w is the out-of-plane deflection measured from the uniformly curved, axisymmetric configuration. Wuest's analysis was extended to orthotropic thin shells by Yee et al. (2004).

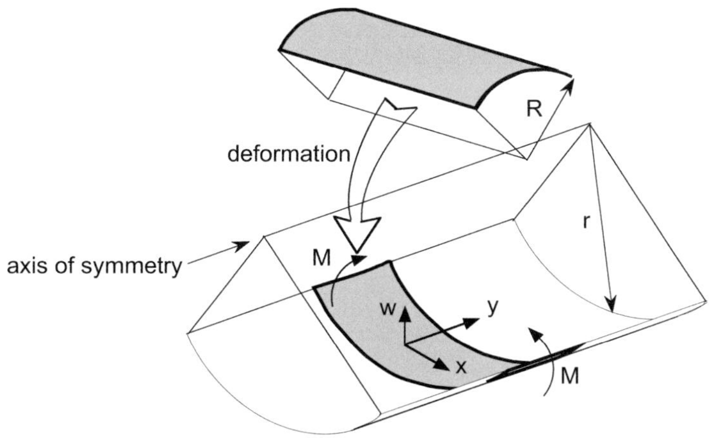

Figure 16: Definition of coordinate system for axisymmetric shell analyzed by Wuest.

The bending moments per unit length in an orthotropic thin shell are related to the changes of transverse and longitudinal curvature by Eq. 1 and, assuming $\mathbf{B} = 0$ and since the twisting curvatures are zero in both the initial and final configurations, shown in Fig. 16, $\kappa_{xy} = 0$. Hence only the bending moments are non-zero:

$$M_x = D_{11}\kappa_x + D_{12}\kappa_y \tag{20}$$

$$M_y = D_{12}\kappa_x + D_{22}\kappa_y \tag{21}$$

The curvature changes in the deformed configuration of the tape are

$$\kappa_x = \frac{1}{r} \tag{22}$$

$$\kappa_y = \frac{1}{R} - \frac{d^2w}{dy^2} \tag{23}$$

where w is the out-of-plane deflection of the deformed tape, in the z-direction. Substituting Eqs. 22 and 23 into Eqs. 20 and 21

$$M_x = D_{11}\frac{1}{r} + D_{12}\left(\frac{1}{R} - \frac{d^2w}{dy^2}\right) \tag{24}$$

$$M_y = D_{12}\frac{1}{r} + D_{22}\left(\frac{1}{R} - \frac{d^2w}{dy^2}\right) \tag{25}$$

An analysis of the deflected shape w, followed by the integration of the moment and longitudinal in-plane stress resultant (Wuest, 1954; Yee et al., 2004) leads to the following expression for the end moment, $M(1/r)$:

$$\begin{aligned} M(1/r) &= \int_{-s/2}^{s/2} (M_x + N_x w)dy \\ &= sD_{11}\left[\frac{1}{r} + \frac{\beta}{R} - \beta\left(\frac{1}{R} + \frac{\beta}{r}\right)F_1 \right. \\ &\quad \left. + \left(1 + \frac{\beta R}{r}\right)^2 \frac{r}{R^2} F_2\right] \end{aligned} \tag{26}$$

where $s = 2R\sin(\alpha/2)$, $\beta = D_{12}/D_{22}$, and

$$F_1 = \frac{2}{\lambda}\frac{\cosh\lambda - \cos\lambda}{\sinh\lambda + \sin\lambda} \tag{27}$$

$$F_2 = \frac{F_1}{4} - \frac{\sin\lambda\sin\lambda}{(\sinh\lambda + \sin\lambda)^2} \tag{28}$$

$$\lambda = \frac{ns}{r} \tag{29}$$

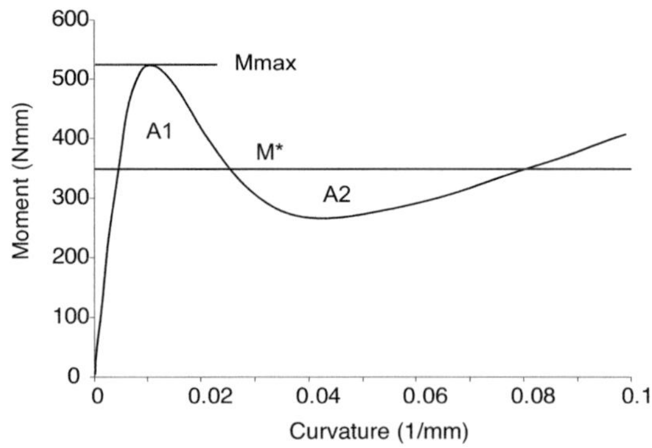

Figure 17: Moment vs. curvature plot showing Maxwell construction.

Figure 17 shows such a plot for a tape spring (0.42 mm thick, 2-ply plain weave T300 carbon fiber reinforced plastic, subtending an angle $\alpha = 80$ deg). Note that the function $M(\kappa_x)$ has a characteristic up-down-up relationship with a local maximum M_{max}.

The maximum moment occurs when a localized fold begins to form, and this value coincides with the maximum moment for purely cylindrical deformation of the shell. The steady-state propagation moment has a value such that the two shaded areas in Fig. 17, A_1 and A_2, are equal: this equal-area line is known as Maxwell's construction.

The above approach has provided a relatively simple method of estimating the maximum moment, M_+^{max}, by maximizing M with respect to r. It has also provided some insight into why the moment remains constant when the tape-spring is folded and the rotation of the elastic hinge is varied.

A further set of simple analytical predictions and insights into the behavior of an elastic fold in a tape spring can be obtained by observing that the fold region is approximately cylindrical, i.e. the transverse radius of curvature in the fold region is zero, and from the following analysis —proposed by Calladine (1988)— based on this observation.

Consider tape springs based on symmetric laminates; these laminates have $\mathbf{B} = 0$. and hence Eq. 1 can be decoupled into in-plane and out-of-plane effects.

During folding the tape spring undergoes changes of principal curvature (note that there is no twist along the longitudinal and transverse directions)

$$(\Delta\kappa_x, \Delta\kappa_y) = \left(\mp\frac{1}{r}, \frac{1}{R}\right) \tag{30}$$

where the positive and negative signs in the first term correspond to equal and opposite sense bending, respectively. The corresponding bending moments can then be obtained from the curvature changes

$$M_x = D_{11}\kappa_x + D_{12}\kappa_y \tag{31}$$

$$M_y = D_{12}\kappa_x + D_{22}\kappa_y \tag{32}$$

The total bending strain energy over the area of the fold region, equal to $Rr\theta\alpha$, is

$$U = \frac{R\theta\alpha}{2}\left[\begin{array}{cc}\Delta\kappa_x & \Delta\kappa_y\end{array}\right]\left[\begin{array}{c}M_x \\ M_y\end{array}\right] \tag{33}$$

Substituting Eqs. 30 to 32 into Eq. 33, and minimizing the total bending energy with respect to r yields the following expression for the radius of the fold region (Schulgasser, 1992)

$$r = \sqrt{\frac{D_{11}}{D_{22}}}R \tag{34}$$

The steady-state moments for equal and opposite sense bending of a tape spring are thus obtained by substituting Eqs. 34 and 30 into Eq. 31, and multiplying by the arc length of the tape, $R\theta$

$$M_{+}^{*} = (\sqrt{D_{11}D_{22}} + D_{12})\alpha \tag{35}$$

$$M_{-}^{*} = (-\sqrt{D_{11}D_{22}} + D_{12})\alpha \tag{36}$$

If, in addition to being a symmetric laminate, the tape spring has equal bending stiffnesses $D_{11} = D_{22}$, which of course is the case for isotropic tape springs, then, Eq. 34 yields

$$r = R. \tag{37}$$

The moment-rotation behavior of a tape spring under opposite sense bending and *before the deformation localizes at a fold* coincides with the analytical expression in Eq. 29. Both the initial slope of the moment-curvature plot (initial stiffness of tape spring) and the value of the maximum moment depend on both the material properties and the geometry of the tape spring.

Figure 18 shows the effect of varying the subtended angle for a particular laminate (two-ply of plain weave carbon-fiber fabric $[0°, 90°]_2$ made of T300-3K/913, while maintaining the arc length of the cross-section constant and equal to $R\alpha = 14$ mm; the thickness of this laminate is 0.42 mm). The characteristic tape spring behavior for opposite-sense bending, involving a sudden snap accompanied by the formation of an elastic fold, requires the moment-curvature relationship to have an up-down-up shape in order for the deformation to localize and then propagate, see Seffen and Pellegrino (1998) for a more in-depth explanation. Note that when the subtended angle is less than 40 deg for this particular laminate, the tape spring behavior approximates a flat strip.

4 Tape Spring Hinges

The material presented in this section was originally developed in collaboration with C. Green, S.D. Guest and A.M. Watt and was published in Pellegrino et al. (2000).

Two or more tape springs can be combined to obtain self deploying hinges with high deployed stiffness and able to carry relatively large moments (Luhmann et al., 1989; Szyszkowski et al., 1997). In order to fold these tapes a large moment is required to initially buckle the tape, after buckling a much lower moment is required to continue folding the tape. During deployment the tapes provide a small but constant restoring moment and "lock" into the straight position. Different ways of arranging the tape springs are shown in Fig. 19.

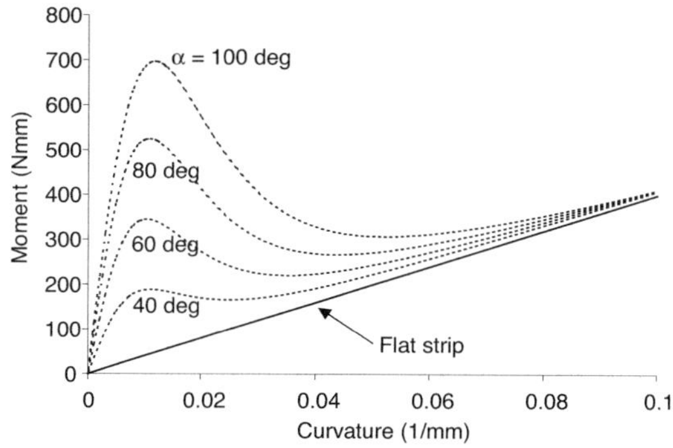

Figure 18: Moment vs. curvature relationships for tape springs with different subtended angles but equal cross-sectional arc length.

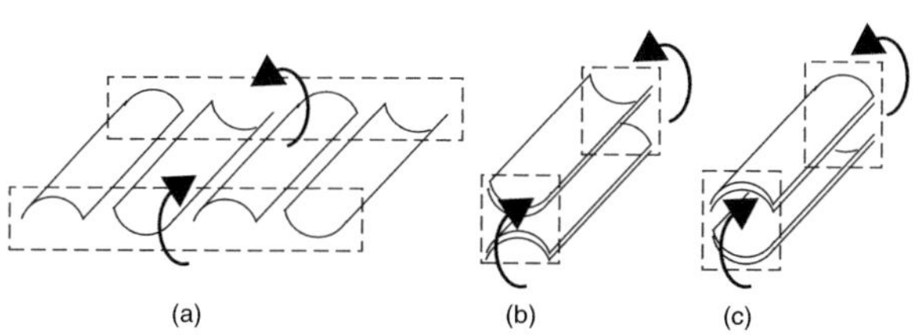

Figure 19: Different arrangements of four tape springs in a self-locking hinge.

Hinges based on any of these configurations will have the basic characteristics described above and in certain respects they perform in the same way. For example the deployment torque applied by these hinges will be approximately equal to the sum of the steady-state folding moments. However, the hinge in Fig. 19(c) is superior for the following reasons.

- By connecting two parallel tape springs at a small separation distance, the tape springs begin to act predominantly in tension and compression, instead of bending, and hence the moment required to snap the hinge increase.
- The stiffness of the "locked" hinge is much higher, and can also be varied by varying the separation of the tape springs, as well as their dimensions. Unlike the arrangement in Fig. 19(b) the edges of the top tape springs are not subject to high compression, and hence are less prone to buckle locally.
- A kick-off restoring moment can be obtained at the beginning of deployment, due to the interaction between the two tape springs.

Some of the results presented in Section 3 can be extrapolated to hinge consisting of two or more tape springs. The moment-rotation relationship for a hinge consisting of tape springs, as shown in Fig. 19, has the general shape of Fig. 20. There are several important points to note.

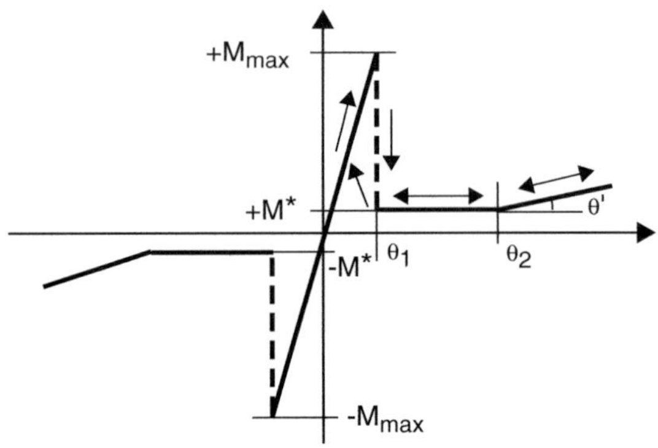

Figure 20: Typical moment-rotation relationship of tape-spring hinge.

- The $M - \theta$ curve is symmetric about the origin, because the behavior of the hinge does not change when the sign of the moment is reversed; this is because the arrangement of the tape springs is symmetric with respect to the centre plane.
- For small θ's the relationship is approximately linear, provided that $|M| < M_{max}$.
- As the rotation angle is further increased one of the tape springs snaps while M suddenly decreases, and an elastic "fold" starts forming in the middle of the hinge.
- As θ is further increased, M remains approximately constant, $M \approx M^*$. At a value between one and two orders of magnitude less than M_{max}.
- For even larger θ's M slowly increases, as the folds in different tape springs interact with each other and/or the folds reach the end of the springs.

Next, it will be shown how to estimate M_{max} and M^*.

Consider a hinge consisting of two tape springs, as shown in Figure 21. The main geometric parameters are the length L measured between the ends of the attachment blocks, and the separation h of the tape springs. In addition, of course, the cross-sectional and material properties of the tape springs. Here E, ν have the usual meaning, t is the thickness of a tape spring, and α is the angle subtended by its cross section.

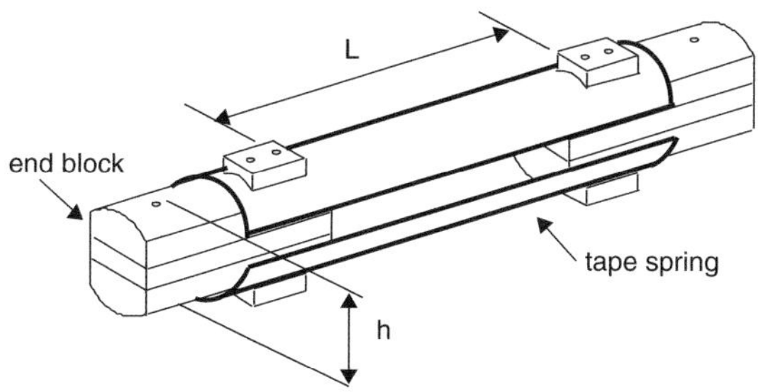

Figure 21: Tape-spring hinge geometry.

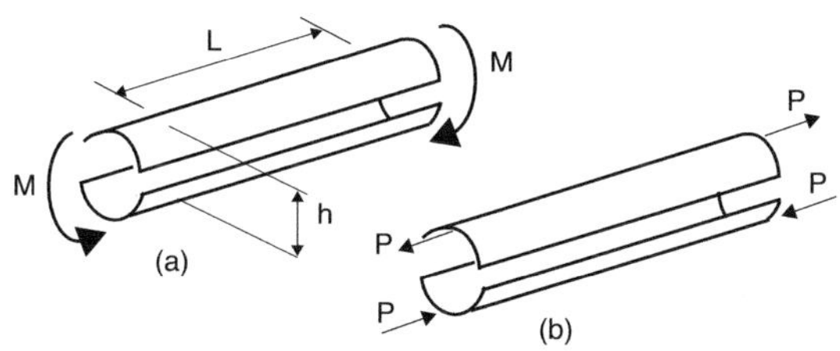

Figure 22: Loading mode tape-springs in deployed configuration of tape-spring hinge.

M_{max} can be estimated by multiplying the buckling load P_{cr} of the compression tape spring by the separation h, hence

$$M_{max} = P_{cr}h \tag{38}$$

where P_{cr} will be estimated later in this section.

M^* is obtained by adding the steady-state moments required to form elastic folds in all the tape springs. Half of the springs are bent in the opposite sense and the other half in the equal sense and hence the corresponding moments are respectively, as special cases of Eqs 35-36

$$(1+\nu)D\alpha \quad \text{and} \quad (1-\nu)D\alpha$$

where

$$D = \frac{Et^3}{12(1-\nu^2)}$$

Considering both tape springs

$$M^* = 2D\alpha \tag{39}$$

The maximum moment, M_{max}, that can be carried by a tape spring hinge without snapping could be estimated by considering the critical stress of the compression tape spring, assumed to be straight up to the point of buckling. However, this estimate is not very accurate. A better estimate is obtained from an analysis that includes the effect of the eccentricity of P with respect to the axis of the tape spring.

This analysis is based upon the following observations, made using the bending apparatus shown in Fig. 13.

1. When two end moments are applied to a tape spring hinge, the tension spring immediately forms edge buckles, which grow in amplitude with increasing moment. The edges of the tension spring remain essentially unstressed: all the tension load is carried by a central strip of width approximately equal to the width of the end fittings.
2. The tension spring is essentially straight. The compression spring bends very significantly; its shape is very smooth and the edges seem perfectly stable, suggesting that they are in tension, although the spring is overall in compression. Shortly before buckling a central bowing amplitude of 2-3 mm is measured in the compression tape spring.
3. The buckling of the compression spring is very sudden and is accompanied by a large reduction of the load carried by the spring.
4. An analysis of the buckling of the compression spring as an Euler strut is inaccurate for the range of spring lengths that are practical but becomes quite accurate over 200 mm. An alternative analysis based on a cylindrical panel that is simply supported on all edges vastly overestimates the buckling load. Also, it predicts a constant buckling load for tape springs of different lengths; this disagrees with both finite element and experimental analyses, which show a clear reduction in buckling load when the length increases. The buckling of an edge panel, free on one edge, significantly underestimates the buckling load. It is believed that this type of buckling would be stable, whereas unstable buckling is observed.

The pronounced curvature that is observed in the compression tape spring a long way before it buckles clearly indicates that this element is in bending as well as compression. This is due to the fact that the end blocks are offset from the centroid of the tape spring.

Hence, to obtain an accurate estimate of P_{cr}, the stress at points 1 and 2, defined in Fig. 23, is compared with the local shell-buckling values, which are determined as follows.

Local shell-buckling at point 1: Metallic tape springs are typically flat near the edges. Hence the expression for the critical stress in a rectangular

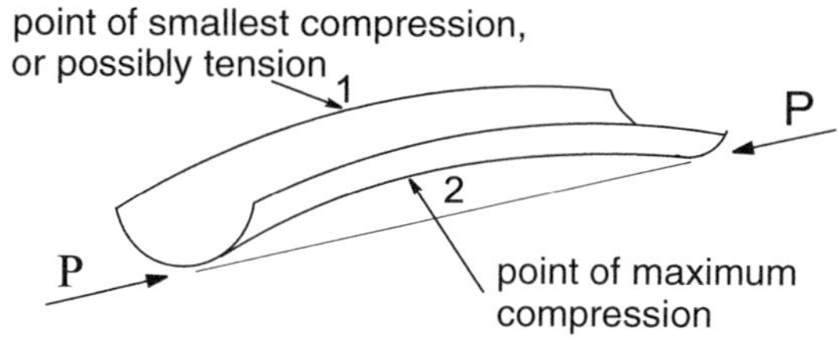

Figure 23: Buckling of compression tape spring.

plate is applicable (Allen and Bulson, 1980):

$$\sigma_{cr} = \frac{k\pi^2 E}{12(1-\nu^2)}\left(\frac{t}{b}\right)^2 \tag{40}$$

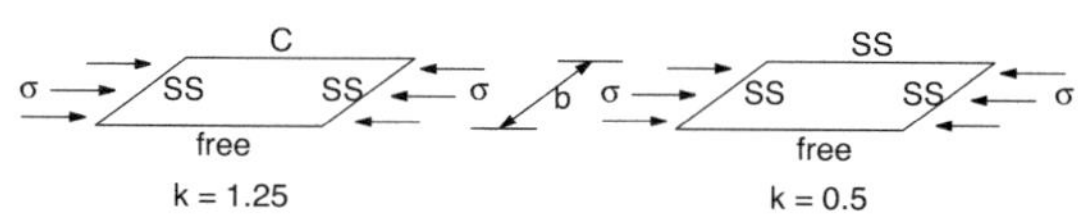

Figure 24: Values of k for different boundary conditions (from Allen and Bulson (1980)).

Two different values of k can be considered, see Fig. 24. The smallest value for b is the width of the flat edge of the cross-section and the maximum value is half the arc-length of the tape spring. The flat edge has a free side and therefore $k = 0.5$ whereas for half the arc length $k = 1.25$.

Local shell-buckling values at point 2: Timoshenko and Gere (1961) state that a simply supported cylindrical panel buckles under the compressive load:

$$\sigma_{cr} = 0.6\frac{Et}{r} \tag{41}$$

Stress at points 1 and 2: In general,

$$\sigma = P\left(\frac{1}{A} + \frac{wy}{I}\right) \tag{42}$$

Where y is the distance from the neutral axis and w is the eccentricity of P, defined in Fig. 25. w is the central deflection of an eccentrically loaded strut, see Fig. 25 (Bazant and Cedolin, 1991):

$$w = e\ \sec\left(\frac{\pi}{2}\sqrt{\frac{P}{P_E}}\right) = \frac{e}{\cos\sqrt{\frac{PL^2}{4EI}}} \tag{43}$$

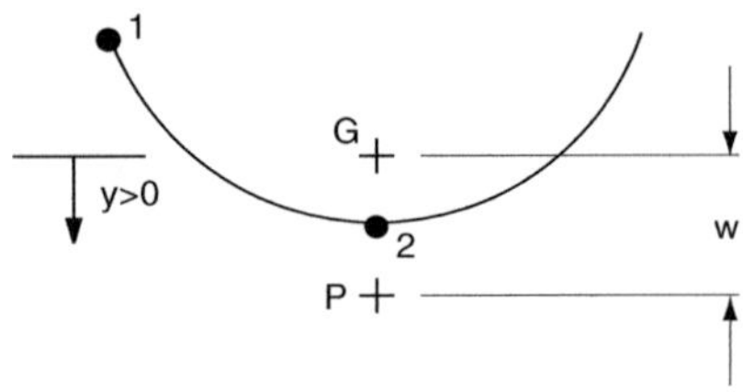

Figure 25: Definition of y.

where e is the eccentricity of P at the ends of the strut, see Fig. 26.

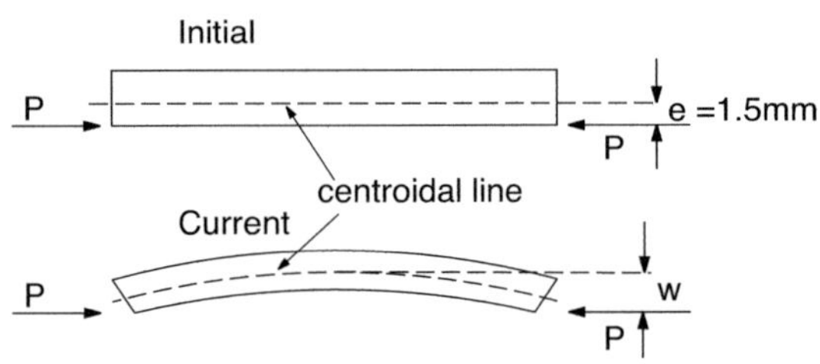

Figure 26: Eccentricity.

It turns out that for the range of hinge properties of practical interest σ_{cr} at point 1 is negative, i.e. tensile, and hence the critical buckling stress cannot be reached at this point.

The stress at point 2 is compressive and therefore, combining Eqs 42 and 43 and noting that $y = e$ for point 2, the buckling condition becomes:

$$\sigma_{cr2} = P_{cr} \left(\frac{1}{A} + \frac{e^2}{I \cos \sqrt{P_{cr} L^2/(4EI)}} \right) \tag{44}$$

Since σ_{cr2} can be found from Eq. 41, the only unknown in this equation is P_{cr}. For the particular tape-spring properties that have been considered above, Eq. 44 becomes

$$P_{cr} \left(\frac{1}{2.55} + \frac{1.5^2}{4.67 \cos \sqrt{P_{cr} L^2/(4 \times 210000 \times 4.67)}} \right) = 840 \tag{45}$$

This equation has been solved iteratively for different values of L, using Matlab, and the results have been multiplied by $h = 18$ mm and then plotted in Figure 27, assuming a knock-down factor on the critical buckling load of 0.8.

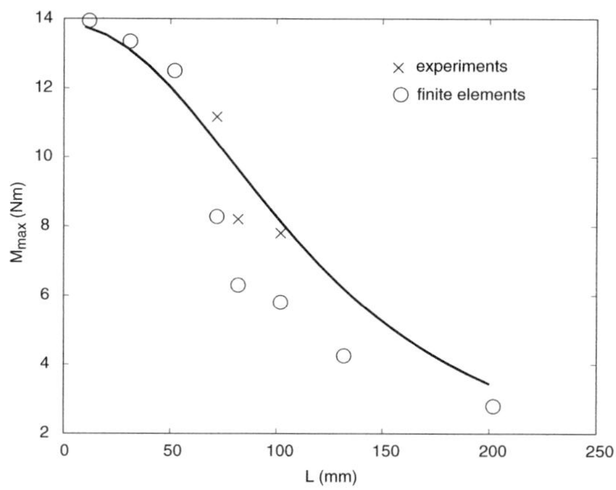

Figure 27: Comparison of Eqs 38 and 45 with experiments and finite-element simulation results.

It can be seen that: (i) increasing the length of the tape springs within the hinge quickly reduces the value of M_{max} and (ii) predicting M_{max} from Equations 38 and 45 is very accurate.

As a final test of the design equations that have been derived in this section, note that from Eq. 38 there is a linear relationship between M^{max} and the tape spring separation h. This relationship has been plotted in Fig. 28 along with a series of finite-element predictions and an experimental result.

5 Folding and Deployment Simulations

This section presents detailed numerical simulations techniques for thin-shell deployable structures, based on Mallikarachchi and Pellegrino (2011). The particular example that is studied is a thin-walled tube made of carbon fiber reinforced plastic with an elastically foldable central section made by cutting two longitudinal slots with round ends, to form a tape-spring hinge as shown in Figure 29. A variant of this hinge design, with three slots, was analyzed with the implicit finite element code ABAQUS/Standard by Yee and Pellegrino (2005) and Soykasap (2009). Also each folding section of the booms for the Mars Express spacecraft booms (Mobrem and Adams, 2009) is in fact a tape-spring hinge with two slots with enlarged round ends.

A particular hinge design has been studied by carrying out both quasi-static deployment tests using a bending rig as well as folding and deployment

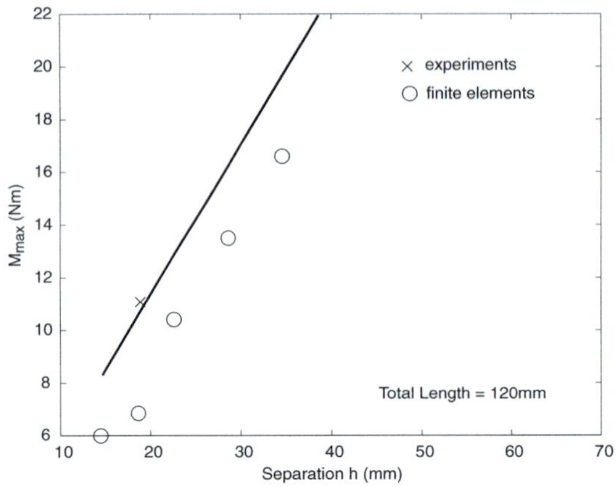

Figure 28: Further comparison of Eqs 38 and 45 with experiments and finite-element simulation results.

simulations with the commercial finite element package ABAQUS/Explicit (Dassault Systemes, 2013).

Figure 29: Photos of carbon-fiber reinforced plastic tape-spring hinge in three configurations: deployed, partially folded, fully folded.

5.1 Tape-Spring Hinge Finite Element Model

A finite element model of the tape-spring hinge was set up in ABAQUS/Explicit, see Figure 30. To simulate the deployment behavior one needs to compute first the folded configuration of the hinge and this requires that two cross-sections of the tape-spring hinge be rotated in opposite directions until the fully-folded configuration is reached. It is important to note that the boundary conditions imposed at these cross-sections may significantly affect the results of the simulations. For example, defining two rigid rings and rotating them would prevent the cross-sections from deforming when the tape-spring hinge is folded. To avoid this effect the actual stiffness of the holders used in the experiment was incorporated in the model and the rotations were imposed through these elastic holders.

The tape-spring hinge model consisted of 2546 nodes and 2412 shell elements (S4) with a minimum element length of around 3 mm. The model of each holder consisted of 182 nodes and 156 S4 shell elements with an approximate element length of 5 mm. The nodes of the holder models were rigidly attached to the nodes of the tape-spring hinge model using **Tie* constraints.

The elastic properties of the shell elements were defined in a cylindrical coordinate system (with directions x', y', z respectively longitudinal, circumferential and radial) by assigning the ABD matrix the laminate with the **Shell General Section* parameter.

In order to simulate the equal end moment conditions imposed during the deployment experiment, two reference nodes A and B were attached to a dummy node, C, using the ABAQUS command **Equation* and these reference nodes were rigidly attached to the nearest eight nodes of the holders.

The boundary conditions, see Figure 30, were defined as follows: all degrees of freedom at node A were restrained, except the rotation about the global X-axis. Node B was only allowed to translate along the global Z-axis and rotate about the X-axis. To keep the end moments equal, the rotational degrees of freedom of nodes A and B were restrained by prescribing the condition

$$\theta_X^A - \theta_X^B = \theta_X^C \tag{46}$$

where θ_X denotes a rotation about the global X-axis.

To simulate the folding process the folding angle θ_X^C was incremented from 0 deg to 170 deg (or in some cases only 160 deg) over a suitably chosen time interval; details are provided in Section 5.3.

To initiate the folding process the tape-spring hinge has to be pinched before one starts to rotate the ends. This process was simulated by defining two rigid plates connected by a single beam element which undergoes thermal contraction. The plates were restrained to the Y-Z plane, to maintain symmetry. The connections between the beam element and the rigid plates were modeled as fixed using CONN3D2 elements and *Weld* connector sections.

The definition of several contact surfaces is also required, as different parts of the tape-spring hinge come into contact with each other and also the rigid plates come into contact with the tape-spring hinge. The *General Contact* feature was assigned to the entire model by specifying *Contact Inclusions, All Exterior*. With this option ABAQUS/Explicit automatically defines potential contact surfaces around the whole hinge surface. However,

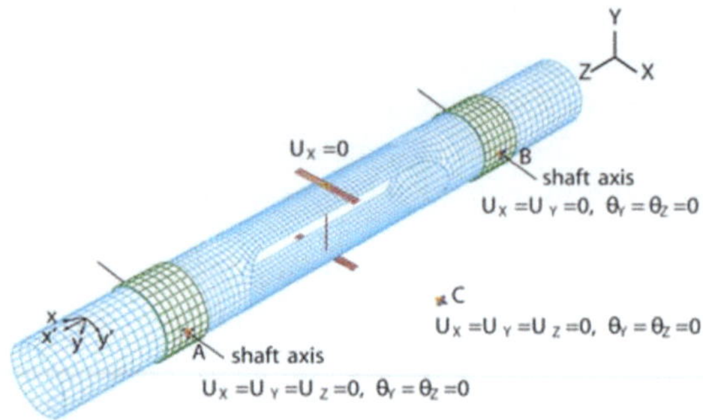

Figure 30: Finite element model of tape-spring hinge, showing global coordinate system X, Y, Z and intrinsic system x', y' in longitudinal and circumferential directions.

the beam element connecting the two plates used for pinching was allowed to go through the hinge by using the *Contact Exclusions* parameter.

5.2 ABAQUS/Explicit Simulation Techniques

Folding and deployment simulations of ultra-thin structures involve significant geometric changes that are associated with instabilities, dynamic snaps and extensive contact/sliding between different parts of the structure. The behavior that is of interest in the present study is quasi-static and hence inertia effects have to be minimized throughout the solution. In principle, this can be achieved by applying any loads sufficiently slowly and smoothly that the kinetic energy in the structure remains negligibly small but in reality there are many details that have to be fine-tuned to actually achieve this aim.

Three independent ways were used to control the analysis: integration time increment, loading rate and numerical damping. Next their effects and limitations when one attempts to optimize the related simulation parameters in folding/deployment studies are explained.

First, the integration time increment should be as large as possible, to reduce the number of increments to complete the simulation. However, explicit time integration is stable only if the Courant condition is satisfied (Belytschko et al., 2000; Geradin and Rixen, 1994). Essentially, the time increment cannot be larger than the time for a wave to travel between adjacent nodes in the finite element mesh. ABAQUS/Explicit includes damping effects and estimates the stable time increment limit at each time increment from the approximate relationship (Belytschko et al., 2000)

$$\Delta t = \alpha \left(\sqrt{1 + \xi^2} - \xi \right) \frac{l_{min}}{c_d} \tag{47}$$

where the wave speed is

$$c_d \approx \sqrt{\frac{E}{\rho}} \tag{48}$$

and ξ is the fraction of critical damping in the highest frequency mode of vibration.

Second, any loads should be applied as smoothly as possible and also the loading rate should be as high as possible, to minimize the number of integration increments required to complete the analysis, provided that the response of the structure remains quasi-static. The smoothness of the load application is achieved with the ABAQUS/Explicit command **Amplitude, Definition = Smooth Step*. This prescribes a fifth-order polynomial time variation with first and second time derivatives equal to zero at the beginning and end of the time interval. Regarding the overall loading rate, the key question is how short the simulation time can be made without exciting a significant dynamic response. The first mode that would get excited is the fundamental natural mode of the whole structure and, although its frequency and mode shape will change during folding/deployment, it is useful to obtain an approximate estimate at the beginning of the simulation. This can easily be done by an eigenvalue analysis of the structure in its initial configuration.

Third, numerical damping is introduced in the model to dissipate energy build-up at high frequencies, to avoid the sudden collapse of elements due to large out-balance forces that may develop at a few nodes, and to generally keep down the amount of kinetic energy in the structure. The amount of numerical damping should be as small as possible, to avoid affecting the results of the simulation. This is particularly critical for ultra-thin structures whose low bending stiffness could easily be swamped by damping effects, and also avoid decreasing the stable time increment (note that in Eq. 47 Δt decreases when ξ is increased).

ABAQUS/Explicit allows the inclusion of damping in several different ways, of which two —bulk viscosity and viscous pressure— are utilized; these two particular types of damping are explained next.

The first type of damping, bulk viscosity introduces an in-plane strain-rate dependent pressure

$$p_b = \xi \rho c_d l \dot{\epsilon}_v \tag{49}$$

and an in-plane curvature-rate dependent moment

$$m = \xi \frac{h^2}{12} \rho c_d l \dot{\kappa} \tag{50}$$

both distributed over all shell elements.

The second type of damping, viscous pressure, introduces a velocity-dependent normal pressure over all shell elements

$$p = -\ c_v \boldsymbol{v} \cdot \boldsymbol{n} \tag{51}$$

This normal pressure is very effective in quickly damping out dynamic effects and thus reach quasi-static equilibrium in a minimal number of increments. However, it is important to use an appropriate value of c_v; if it is too high the response of the structure will be overdamped and hence the simulation results may be incorrect. Note that, unlike bulk viscosity, introducing viscous pressure does not change the integration time increment.

There is an alternative technique, mass scaling, that is often used to speed up explicit analyses. This technique consists in artificially increasing the density of the material to $\beta\rho$, in order to increase the time increment from Eq. 47 to $\sqrt{\beta}\Delta t$. This technique would be useful in simulations involving rate-dependent materials, where the load-rate speed-up technique adopted in the present study could not be used, but offers no advantages in the present case.

The key test for the robustness of a particular analysis is to consider the history of the various energy terms and in particular the energy balance history. The energy balance, E_b, is defined as the difference between the energy stored in the structure and/or dissipated during the loading process, $E_i + E_v + E_k$, and the work of all external forces, E_w. In symbols:

$$E_b = E_i + E_v + E_k - E_w \tag{52}$$

where the internal energy E_i is equal to the sum of strain energy and artificial energy (due to hourglassing), E_v is the viscous dissipation, E_k is the kinetic energy. In the simulations presented here the artificial energy was kept negligibly small by using fully integrated elements and by avoiding any localized actions on the shells.

There are two main checks on the energies. First, at any particular time the kinetic energy has to be a small fraction ($< 1\%$) of the internal energy for the results at that time to be considered a valid quasi-static solution. In the case of multi-stable structures if the kinetic energy has reached high levels during an earlier part of the simulation, the possibility that the structure may have reached an alternative equilibrium configuration should be considered. Second, the energy balance should remain equal to the amount of external energy (e.g., thermal) introduced in the system.

In particular, Belytschko et al. (2000) discuss cases where an *arrested instability* in a structure, due to geometric and/or material softening, may result in the integrator losing stability, despite the use of Eq. 47 to compute the time increment. These authors have reported that all such instabilities can be detected by checking the energy balance, as discussed above.

5.3 Simulation of Tape-Spring Hinge

The fundamental natural period of vibration of the tape-spring hinge in the deployed configuration was computed by doing a frequency analysis in ABAQUS/Standard. The estimated period was 5 ms, hence the definition of the overall loading rate was initially set such that folding/deployment would occur over a time period of ten times the fundamental period, or 0.05 s. The entire simulation was run with the double precision solver and the energy components were recorded at every millisecond. The ratio between kinetic energy and internal energy was then monitored and the loading ratio was decreased until $E_k/E_i < 1\%$.

An example of a successful simulation is presented next, to provide a basis for discussion and as a starting point for various sensitivity studies.

The initial, unstressed configuration of the tape-spring hinge is shown in Fig. 50(a). At the beginning of the folding simulation the thermally controlled beam element connecting the two rigid plates was shortened by decreasing its temperature, which simulated the pinching of the tape-spring hinge and led to the configuration in Fig. 50(b). The next part of the folding simulation consisted in imposing a rotation $\theta_X^C = 170$ deg over 0.8 s (this simulation time was obtained by gradually increasing the initial value of 0.05 s estimated above). After 0.25 s, corresponding to $\theta_X^C \approx 16$ deg, see Fig. 50(c), contact between the rigid plates and the tape-spring hinge was disabled in order to avoid any spurious constraints on the folded shape. The rotation was then continued and the final outcome is shown in Fig. 50(d). Viscous pressure loading, discussed in Section 5.2, was used to maintain the energy ratio below 1%.

Figure 51 shows the variation of the energy terms during this simulation. Note that during the pinching phase there is an input of thermal energy that is not accounted for in Eq. 52 and this shows as an increase in energy balance. For the rest of the simulation the energy balance remained approximately constant. Throughout the simulation the kinetic energy remained much smaller than the internal energy, which indicates that the intended quasi-static behavior of the tape-spring hinge has been achieved.

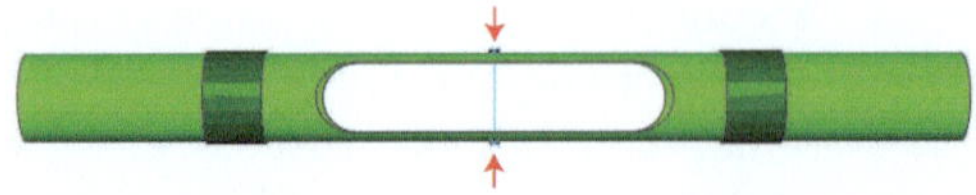

(a) Undeformed configuration

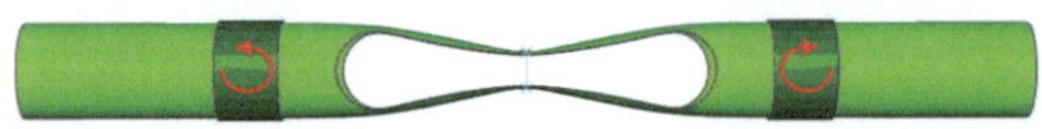

(b) Pinching

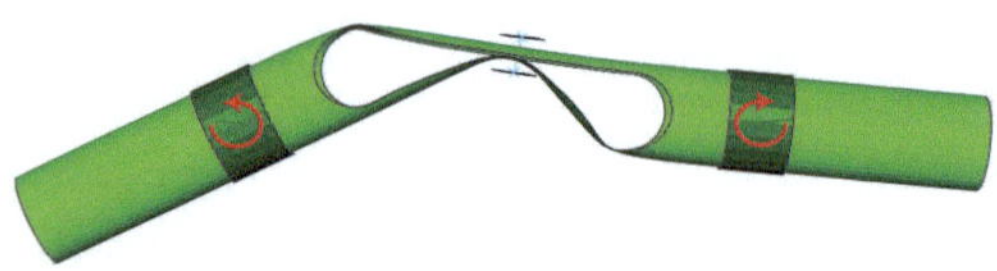

(c) Release of contact between hinge and rigid plates

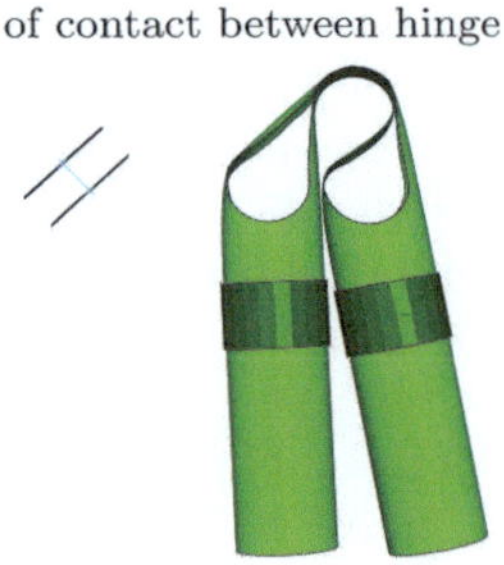

(d) Final folded configuration

Figure 31: Stages of folding simulation.

5.4 Selecting the Simulation Parameters

This section presents various sensitivity studies to determine a set of simulation parameters that provide accurate results and maximal speed up of the simulation. The studies were mainly focused on the deployment part of the simulation because this is the behavior that is of greatest interest for actual applications. However, it was important to first obtain an accurate representation of the folded configuration and hence two different folding simulations were carried out.

The first folding simulation used no bulk viscosity ($\xi = 0$) but only viscous pressure on both the outer and inner surfaces of the tape-spring

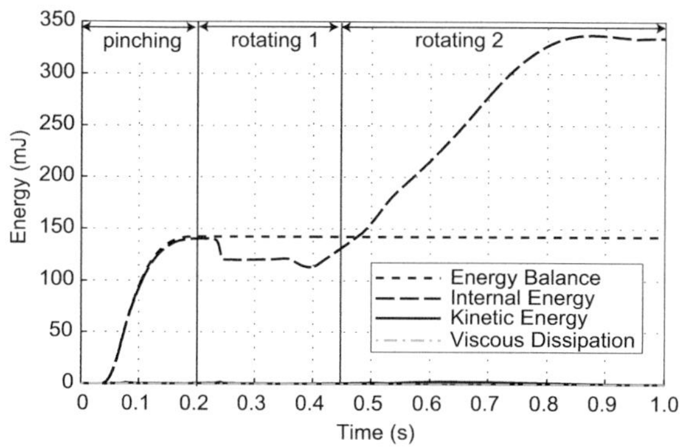

Figure 32: Energy histories for folding simulation with $\alpha = 0.8$, $\xi = 0$, $c_v = 2 \times 10^{-4} \rho c_d$; the maximum folding angle is $\theta_X^C = 170$ deg.

hinge during the rotational phase; the expression for the pressure is given by Eq. 51 with $c_v = 2 \times 10^{-4} \rho c_d$. The time increment factor in Eq. 47 was set to $\alpha = 0.8$. The results have been already presented in Section 5.3, in Fig. 51. Note that the viscous dissipation is zero because the energy absorbed by the viscous pressure is included in the external work. The energy balance remained constant during the rotation phase which indicates that the simulation was free of instabilities and hence the results are accurate. The kinetic energy was negligibly small at the end of the simulation, indicating that the folded configuration is in quasi-static equilibrium.

The second folding simulation used bulk viscosity ($\xi = 0.10$) to provide numerical damping. The energy histories are presented in Fig. 33; note the high level of viscous dissipation, representing 70% of the internal energy at the end of folding phase. Between 1.0 s and 1.1 s viscous pressure was applied (with $c_v = 2 \times 10^{-2} \rho c_d$) to decrease the kinetic energy. Reducing the time increment by setting α to less than 1.0 made little difference to these results.

The final configurations predicted by these two folding simulations are approximately equal, but the first simulation was chosen as a starting point for the deployment studies because it has lower kinetic energy and a smoother moment-rotation relationship.

A folded configuration with $\theta_X^C = 160$ deg (to avoid the effects of the contact between the ends of the tape-spring hinge) and $\alpha = 0.5$, $\xi = 0$, $c_v = 2 \times 10^{-4} \rho c_d$ was chosen as the starting point for several deployment

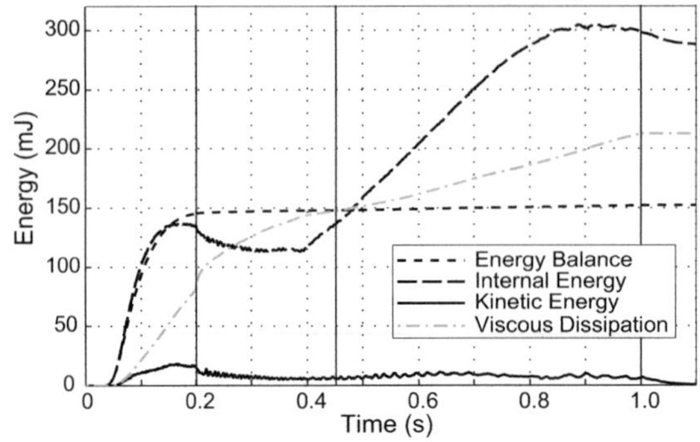

Figure 33: Energy histories for folding simulation with $\alpha = 0.8$, $\xi = 0.10$, $c_v = 0$ up to 1.0 s and then $c_v = \times 10^{-2} \rho c_d$.

simulations with different values of the three control parameters.

The sensitivity to α was examined by setting $\alpha = (0.25, 0.80)$ with $\xi = 0$ and $c_v = 0$. Figure 34 shows that in this case decreasing Δt decreases the change in energy balance.

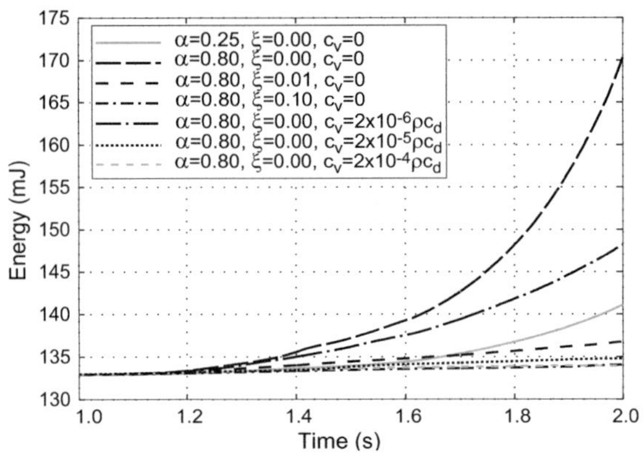

Figure 34: Energy balance histories for different values of α, ξ, c_v.

The sensitivity to ξ was examined by setting $\xi = (0, 0.01, 0.10)$. Figure 34 shows that increasing ξ leads to a decrease of the change in energy balance. Although increasing ξ leads to a small reduction in Δt, and hence to an increase in the number of simulation increments, it is much more effective in decreasing the energy balance than lowering α.

Despite this encouraging result, one should be cautious before relying on bulk viscosity to reduce the energy balance. The problem is that increasing ξ may prevent full deployment. Figure 35 shows the history of the viscous dissipation energy for the three cases that have been considered. It is interesting to note that the viscous dissipation during deployment is 13% and 17% of the internal energy in the fully folded configuration, respectively for ξ =0.01 and ξ=0.10.

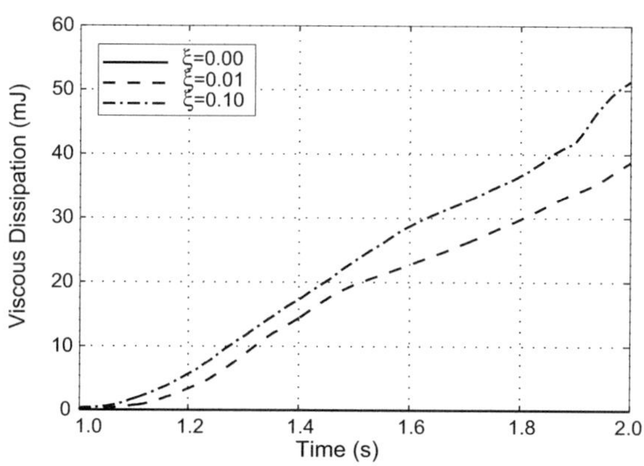

Figure 35: Variation of viscous dissipation for $\alpha = 0.80$, $c_v = 0$.

A feature common to the above simulations, in which bulk viscosity was introduced without viscous pressure, is that there was significant vibration, leading to a noisy moment-rotation deployment profile. This issue was addressed by examining the sensitivity to c_v. Four different values of c_v were considered, from 0 to $2 \times 10^{-4} \rho c_d$, while maintaining $\alpha = 0.8$ and $\xi = 0$ and the results are shown in Fig. 34 and Fig. 36. Figure 34 shows that the higher values of c_v are effective in decreasing the change in energy balance. Figure 36 shows that the particular value $c_v = 2 \times 10^{-5} \rho c_d$ suppresses most tape-spring hinge vibration from $\theta_X^C = 160$ deg to about 50 deg. However, increasing c_v by another order of magnitude radically changes the behavior of the structure.

In conclusion, the viscous pressure coefficient c_v is the most effective tool for keeping the change in energy balance small, but it is important to

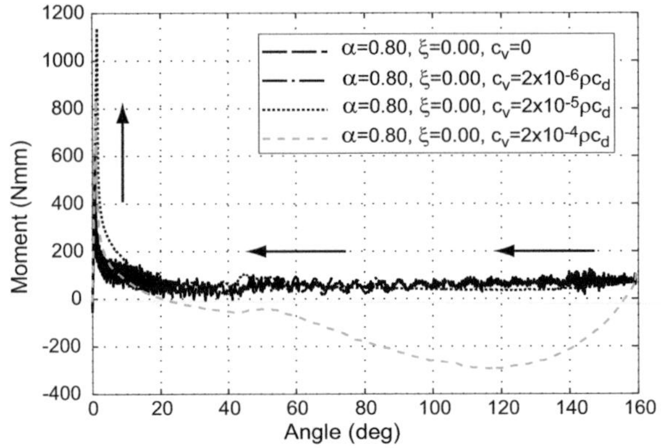

Figure 36: Sensitivity of moment-rotation response to c_v, for $\alpha = 0.8$ and $\xi = 0$. The line style matches Figure 34.

examine the effects of c_v on the structural response in order to avoid over-damping. Based on this study presented, it was decided that the parameters $\alpha = 0.8$, $\xi = 0$, $c_v = 2 \times 10^{-5} \rho c_d$ are best for deployment simulations of our particular hinge.

Once optimal deployment simulation parameters had been obtained, the overall simulation time was increased from 1 s to 3 s and Fig. 37 shows the energy history for this final simulation. Note that this particular simulation started from a folded configuration with $\theta_X^C = 170$ deg. These results are of very good quality, as the energy balance remains constant throughout the simulation, and show a sudden reduction in internal energy at 3.77 s, accompanied by a small spike in kinetic energy. These features correspond to the dynamic snap back of the hinge into its fully deployed configuration. This jump releases practically all the internal energy in the hinge.

5.5 Moment-Rotation Relationship

Next, a complete simulation of the process of folding and deploying the tape-spring hinge was carried out, but this time without initiating the folding process by pinching. The total simulation time was 6 s: 3 s to fold and 3 s to deploy. The simulation parameters were $\alpha = 0.8$, $\xi = 0$ and $c_v = 2 \times 10^{-5} \rho c_d$.

The energy histories have been plotted in Fig. 38. The energy balance remained approximately constant throughout the simulation and hence the

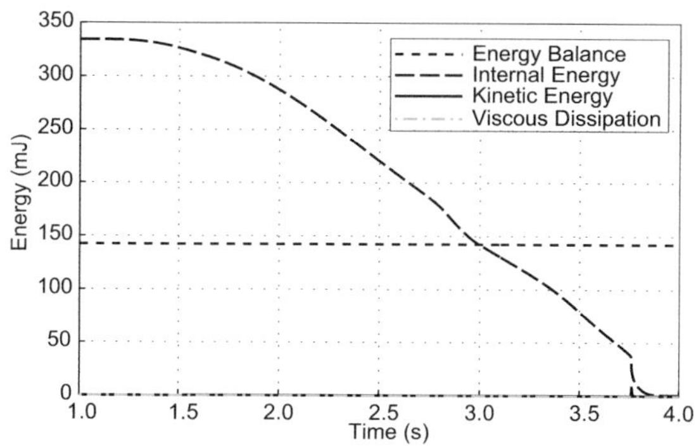

Figure 37: Energy histories for 3 s deployment with $\alpha = 0.8$, $\xi = 0$, $c_v = 2 \times 10^{-5} \rho c_d$.

results are accurate. Note that the internal energy profile is mostly symmetric but shows a spike during folding, at 0.32 s, that is not matched by a corresponding spike during deployment. In fact, during the initial stages of folding the strain energy increases at a considerably faster rate than it decreases during the corresponding stages of deployment. An important result is that the internal energy becomes approximately zero at the end of the deployment simulation, indicating that the unstressed, fully deployed configuration has been achieved. Note the kinetic energy spike at 0.32 s and the smaller spike at 5.76 s; they correspond to the hinge snapping during folding and snapping back, respectively. The viscous dissipation was, of course, zero throughout the entire simulation.

Next, turning to the moment-rotation profile, shown in Fig. 39, the most striking features are that the moment rises to a high peak of 4278 Nmm, snaps to about 67 Nmm and remains almost constant from an angle of approximately 20 deg all the way to the fully folded configuration. Note that there is a slight hump between 40 deg and 60 deg, associated with the snap back of the elastic fold at the center of the outer tape spring and an increase beyond 170 deg when a contact pressure develops between the two ends of the hinge. During deployment, the moment-rotation profile follows the folding profile down to 1.8 deg, then continues to rise gradually, with a final snap back to 2482 Nmm at 0.8 deg. It then joins the linear part of the folding response as the angle is further decreased.

This difference between the loading and unloading paths is characteristic of structures with an unstable post-buckling equilibrium path. Brush and

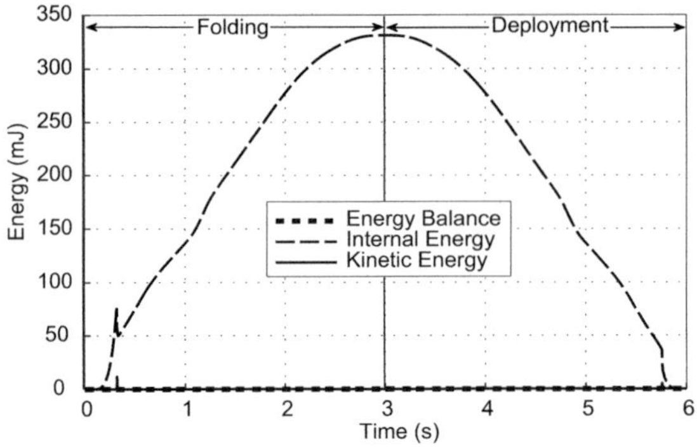

Figure 38: Energy histories for simulation of folding to $\theta_X^C = 170$ deg without pinching and deployment.

Almroth (1975) It was already know that a single tape spring behaves in this way (Seffen and Pellegrino, 1998; Warren et al., 2005), but having established that the post-buckling behavior of tape-spring hinges is also unstable is an interesting result with important implications.

Figure 39 includes snapshots of the deformed configuration just before and soon after snapping back. It is interesting to note that the mid-surface strains after the snap are almost zero, with a maximum strain of 0.01% near the transition between the straight and circular part of the slot.

5.6 Comparison of Experiment and Simulations

Figure 40 shows a comparison between the deformed configurations of the tape-spring hinge that were observed during deployment in the test rig, with snapshots from the finite element simulation. This qualitative comparison shows very good agreement in both the position of the regions of localized deformation and the overall geometric configurations of the tape-spring hinge.

Figure 41 shows a comparison between the deployment moment-rotation profile measured experimentally with the simulated response. Unlike the moment-rotation profile in Fig. 39, the particular simulation results shown here were obtained from a model that included two 5 mm long aluminum rods, to model the torsional stiffness of the holders. The torsional constant of the rod cross-section was 12 mm^4, to match the angle of snap back observed in the experiment, i.e. 2.3 deg.

Overall, the simulation has fully captured the experimentally observed

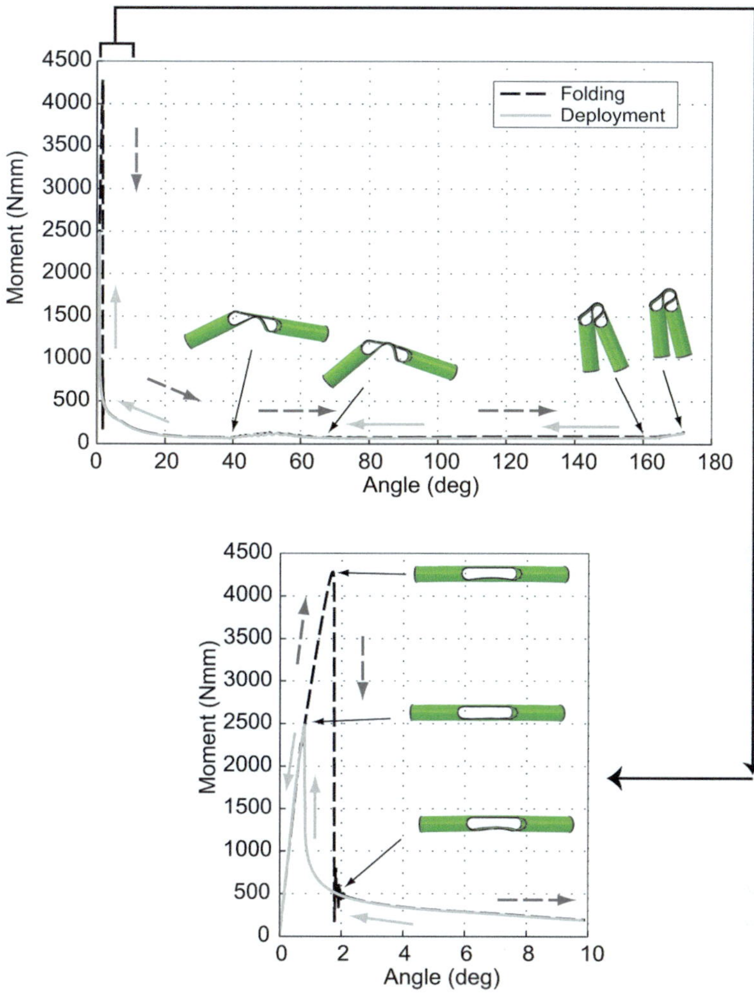

Figure 39: Moment-rotation profile for folding and deployment simulation (without pinching) up to $\theta_X^C = 170$ deg.

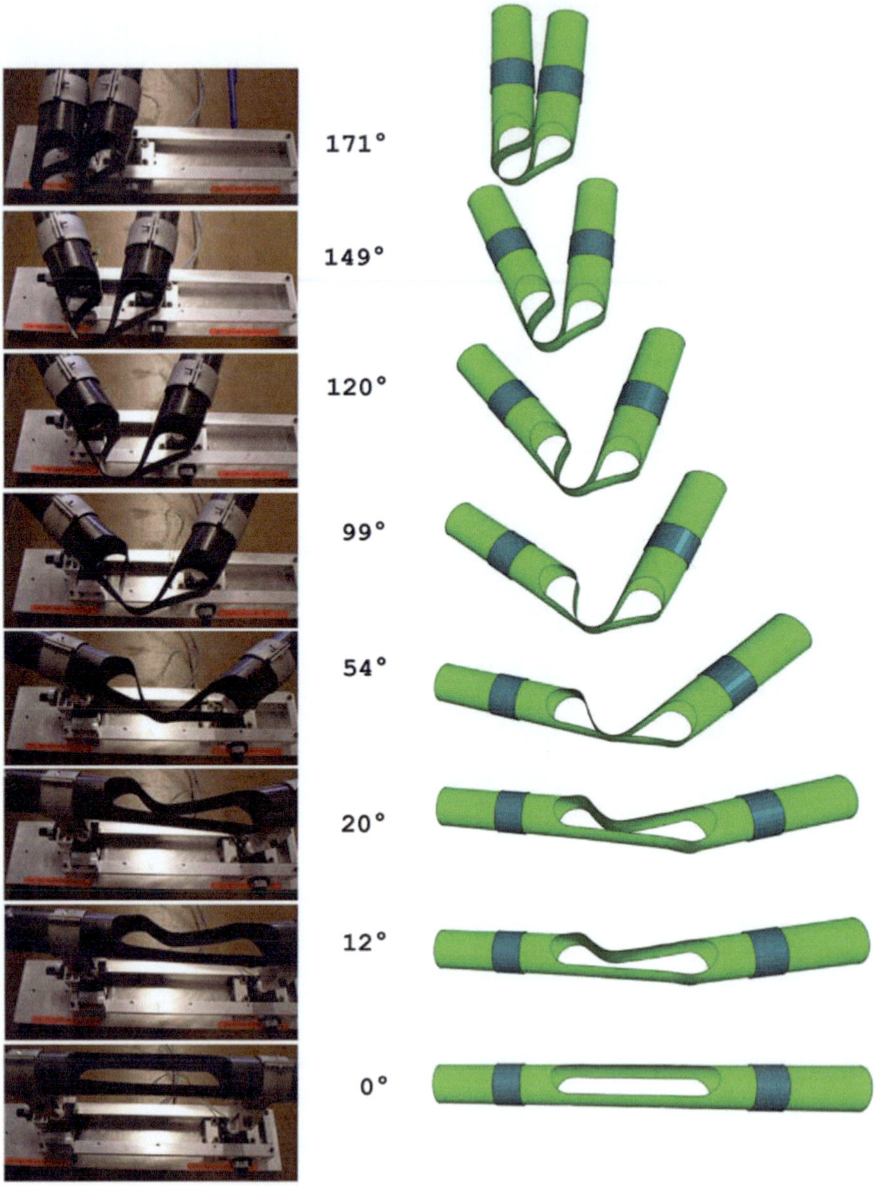

Figure 40: Comparison of hinge configurations during deployment.

behavior. However, the measured steady-state deployment moment had an average value of 80 Nmm whereas our simulation gave around 67 Nmm. Also the peak in the hump in the experiment was at ≈ 40 deg with a maximum moment of 104 Nmm, whereas in the simulation it was at ≈ 54 deg with a maximum of 117 Nmm. Also, the measured peak moment at snap back was 660 Nmm whereas it was significantly higher, 1078 Nmm, in the simulation.

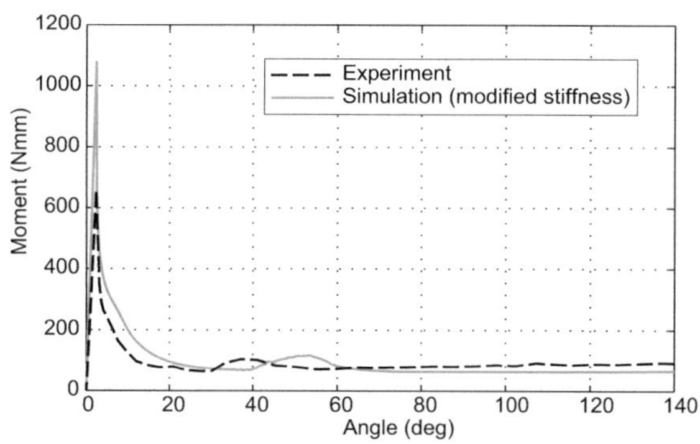

Figure 41: Comparison of moment-rotation profiles during deployment.

6 Deployment Dynamics

The folding simulation developed in the previous section, which mimicked the procedure to fold a tape-spring hinge, can be used to determine the initial conditions from which to initiate dynamic deployment.

6.1 Deployment Experiments

A deployable boom structure was constructed by integrating two parts, a composite tape-spring hinge and an aluminum tube, as shown in Fig. 42. The hinge section was made by cutting two slots into a 350 mm long, 0.2 mm thick thin-walled tube with internal diameter of 38.1 mm. It was made by laying two plies of plain-weave carbon fiber fabric (1k tows of T300 fibers) impregnated with HexPly 913 epoxy resin on a 38.1 mm outer diameter steel mandrel sprayed with Frekote 700-NC mold release agent. The weave directions were at 45 deg to the axis. The fabric was wrapped in PTFE release film and the assembly was inserted into a heat-shrinking sleeve, then the laminate was cured under vacuum for one hour at 125°C and 600 kPa.

After cooling, the mandrel was pulled out and two 140 mm long and 30 mm wide diametrically opposite parallel slots were cut with a Dremel 400 XPR high speed rotary tool. These cuts left two 25 mm wide and 110 mm long, longitudinally straight tape springs with transverse radius of curvature of 19.05 mm. Two nominally identical hinges were made with this process.

The tape-spring hinges were attached to a 0.9 mm thick aluminum tube with an outer diameter of 38.1 mm. The connection was made by inserting the aluminum tube into the tape-spring hinge to provide a 25 mm overlap which was covered with a heat shrinkable sleeve and then with a hose clamp tightened with a torque of 1.13 Nm. The complete structure was 1015 mm long including the 970 mm long boom and an additional length of 45 mm to provide a connection at the root of the boom, Fig. 42.

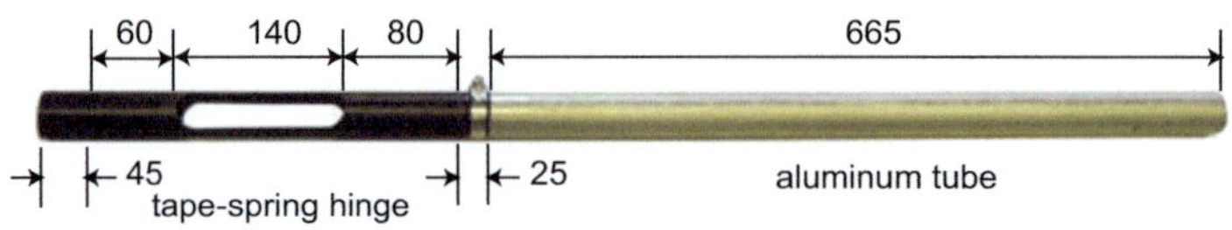

Figure 42: Geometry of test boom (all lengths are in mm).

Deployment tests were carried out on a rig that provided a single-point gravity offload through a string attached to the outer surface of the tube, at a point directly above the center of gravity of the aluminum-alloy tube. The other end of the string was run over a pulley located at a height of 4650 mm, directly above the center of the tape-spring hinge, and terminated at a fixed point; this constraint allowed the boom to move only horizontally. Figure 43 shows the experimental setup.

The root end of the boom was slid onto a 38 mm diameter 45 mm long solid aluminum cylindrical fitting, covered with a heat shrinkable sleeve, and tightened with two hose clamps, with a torque of 1.13 Nm. The fitting was attached to a 1 m by 1 m granite table for the deployment tests.

The deployment of the boom was recorded with a Phantom V12.1 high speed camera set at a rate of 400 frames per second. The camera was held directly above the folding part of the boom; its field of view included the tape-spring hinge, up to the hose clamp. A white sheet with black lines at 5 deg angles provided a horizontal background to measure the deployment angle.

The boom was folded 90 deg and then released three times, while recording the deployment with the high speed camera. After these tests the end fitting was turned over in order to repeat the test with the hinge folded in

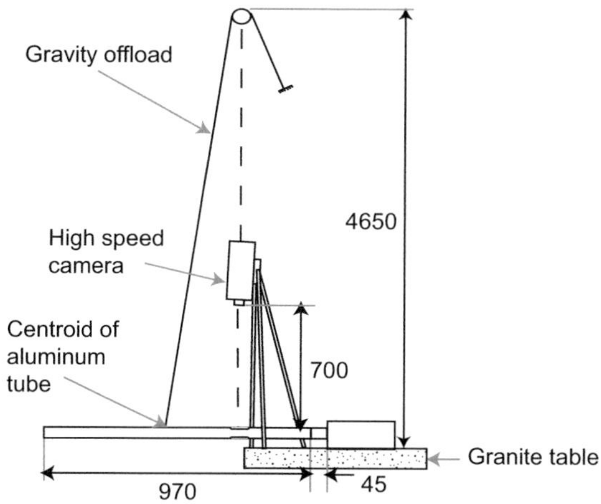

Figure 43: Experimental setup (lengths in mm).

the opposite sense. Three more tests were then carried out. Finally, the first tape-spring hinge was replaced with the second one and the whole set of tests was repeated. The 12 tests cases are defined in Table 3.

Test No.	Hinge	Fold direction	Trial No.
1	Hinge 1	clockwise	I
2			II
3			III
4		anti-clockwise	I
5			II
6			III
7	Hinge 2	clockwise	I
8			II
9			III
10		anti-clockwise	I
11			II
12			III

Table 3: Deployment experiments.

6.2 Experimental Results

A typical set of snapshots from a deployment test is presented in Fig. 44; note that the background has been removed for clarity. The time-series of the deployment angles for each test was derived by measuring the deployment angle from one frame out of every 25 frames (corresponding to a time step of 67.5 μs) using a Matlab program that computes the angle between two lines, drawn manually, that are parallel to the edges of the boom. The deployment angle vs. time relationship obtained in this way has been plotted in Fig. 45; note that the deployment angle is defined as 0 deg in the fully deployed configuration and its value is defined to be positive in the folded configuration.

In general, the deployment behavior of a boom can be divided into three phases: (i) main deployment; (ii) attempt to latch and further rotation of the boom beyond the fully deployed configuration (this may be repeated several times, see Fig. 45); and (iii) small amplitude vibration of the fully deployed boom. In the case shown in Fig. 45 the deployment phase lasted 1.49 s. The second phase lasted 2.48 s and involved two attempts to latch, reaching maximum overshoots of −36 deg and 9 deg. The vibration phase continued for about 20 s.

The behavior measured in the 12 tests is compared in Fig. 46. It can be seen that there is practically no difference in the main deployment phase, which takes an average of 1.49 s with a standard deviation of 0.02 s . In all cases there are two unsuccessful attempts to latch, followed by a third, successful attempt, but the duration of this second phase varies between 2.02 s and 2.55 s. This 25% variation is accompanied by a variation of the maximum overshoot angle between −28 deg and −40 deg.

Looking more closely at the deployment snapshots for the various tests, one finds that these differences are mainly associated with two effects. The first effect is the variability in the configuration of the hinge after the first latching attempt. In all tests apart from Test 11 the fold closest to the root of the boom is located on the outer tape spring, whereas in Test 11 this fold is located on the inner tape spring. These two configurations of the hinge, shown in Fig. 47, result in different effective lengths of the deploying boom. In the case of Test 11 the boom is effectively shorter and hence its moment of momentum after the latching attempt is lower and the moment of inertia about its center of rotation is also lower. Hence in this case the boom will rotate faster and may even achieve a smaller overshoot angle. The second effect is that the deployment angle does not completely identify the configuration of the boom but in some cases a significant degree of shearing takes place in the tape-spring hinge. Hence, although the deployment angle is the main parameter that describes the configuration of the boom, when

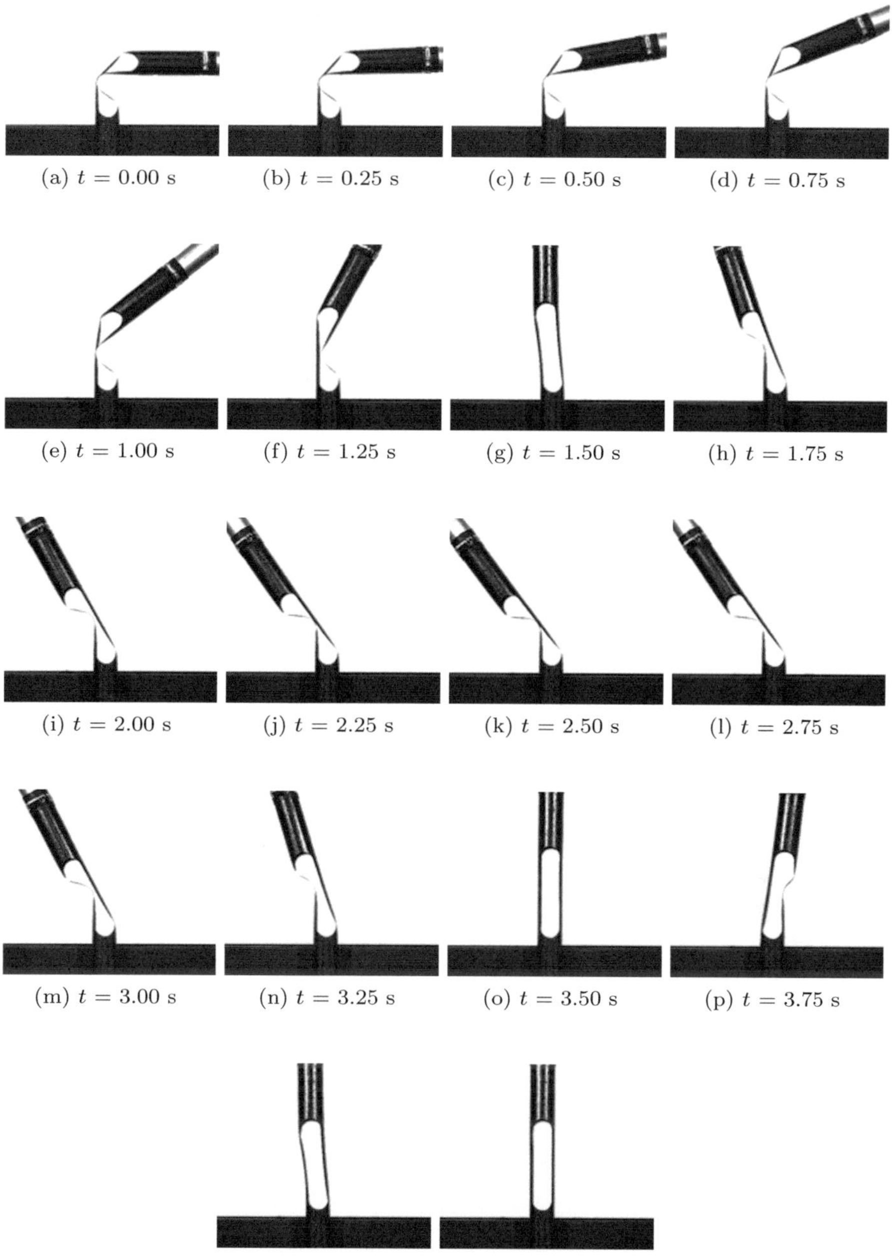

(a) $t = 0.00$ s (b) $t = 0.25$ s (c) $t = 0.50$ s (d) $t = 0.75$ s

(e) $t = 1.00$ s (f) $t = 1.25$ s (g) $t = 1.50$ s (h) $t = 1.75$ s

(i) $t = 2.00$ s (j) $t = 2.25$ s (k) $t = 2.50$ s (l) $t = 2.75$ s

(m) $t = 3.00$ s (n) $t = 3.25$ s (o) $t = 3.50$ s (p) $t = 3.75$ s

(q) $t = 4.00$ s (r) $t = 4.25$ s

Figure 44: Deployment snapshots at interval of 0.25 s.

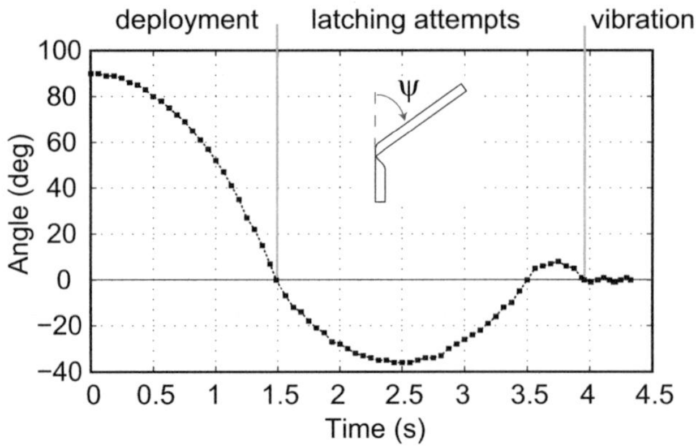

Figure 45: Angle-time relationship for test 1.

this angle is close to zero the boom may not be exactly in the fully deployed configuration, as shown in Fig. 48.

6.3 Numerical Simulations

The simulation techniques were based on Section 5. The geometry of the boom was defined in the deployed, i.e. unstressed, configuration. A complete finite-element model was defined, Fig. 49, consisting of 3919 nodes and 3787 four node fully integrated shell elements (S4). A fine mesh was defined for the tape-spring hinge region, in order to resolve the higher deformation gradients that occur in this region; the average element size was 2 mm and the smallest size 1.5 mm. A coarser mesh with average element size of 5 mm was defined for the rest of the boom.

The material properties of the tape-spring hinge were provided in the form of the ABD matrix using the command **Shell General Section*, which defines a general elastic shell with arbitrary cross-section and for which a user-defined stiffness matrix can be specified. The aluminum tube was defined as isotropic, with Young's modulus of 70 GPa and Poisson's ratio of 0.33. The mass of the connectors and the overlap between tube and hinge was applied at a set of nodes lying in the middle of the overlap region; these nodes are denoted as folding node-set in Fig. 49 as they will be used to drive the folding process. The breakdown of the mass distribution is presented in Table 4.

The gravity offload system was modeled by means of a single 4675 mm

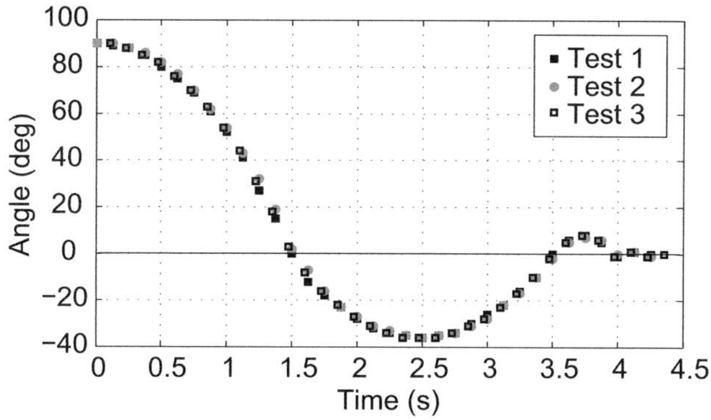

(a) Tests 1 to 3

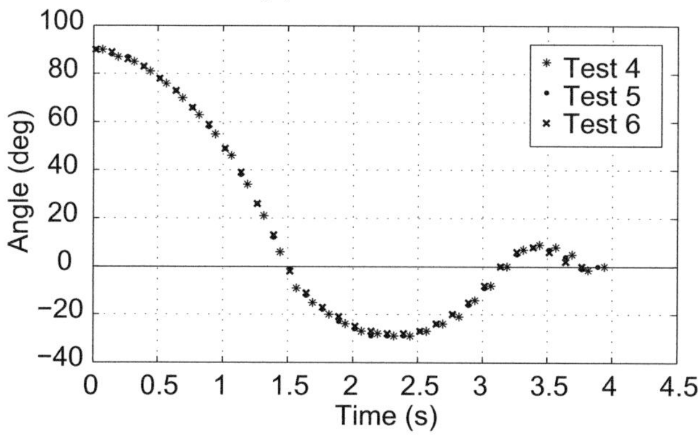

(b) Tests 4 to 6

Figure 46: Comparison of deployment angle vs. time responses for different tests.

long truss element (T3D2) with a Young's modulus of 100 GPa and a cross-sectional area of 0.031 mm^2. This truss element was connected at one end to the center of mass of the tube and at the other end to a fixed point with coordinates (4650 mm, 0, 130 mm). All translational degrees of freedom were connected between the end node of the truss element and the corresponding node of the tube, using CONN3D2 elements and *Weld* connectors (a weld connector ties all six degrees of freedom of the two specified nodes). A fixed-end condition was defined for the boom, by restraining all six degrees of freedoms of each node lying on the end section of the tape-spring region, Fig. 49. Gravity loading was defined for the whole boom, acting in the negative Y-direction.

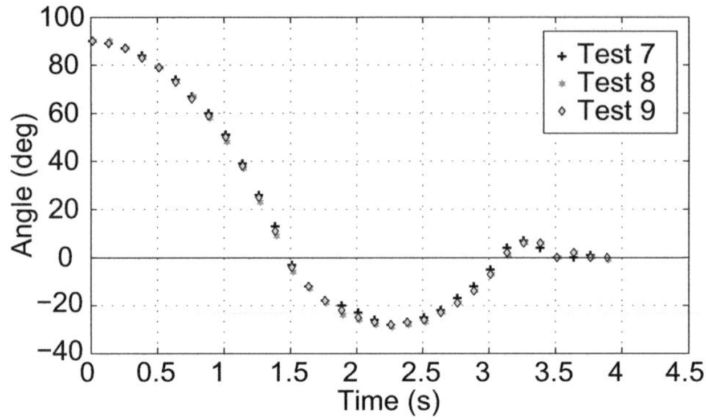

(c) Tests 7 to 9

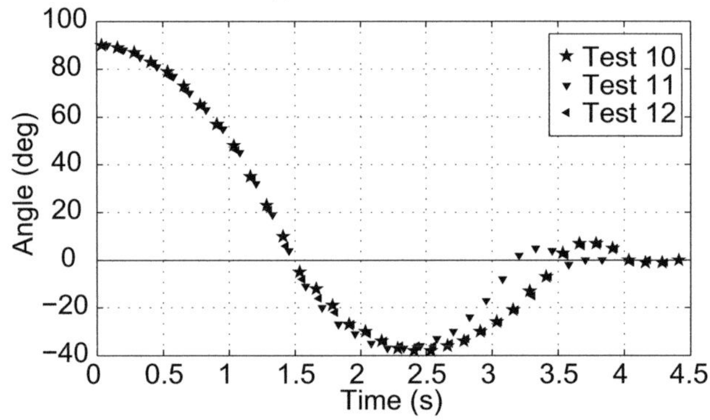

(d) Tests 10 to 12

Figure 46: Comparison of deployment angle vs. time responses for different tests (contd).

A simulation consists of two parts, the first part generates the folded configuration of the boom while the second part simulates its dynamic deployment, including all attempts by the tape-spring hinge to latch and the final post-latching vibration.

Regarding the first part, it should be noted that the detailed sequence of steps to obtain the folded configuration is less important than the final folded configuration itself, provided that the kinetic energy in the folded configuration is sufficiently small. Hence, starting from the fully deployed configuration the first step in the folding process is to pinch the hinge in the middle. This was done by defining two rigid plates, each attached to a reference node. These two reference nodes were restrained to the Y-Z plane and

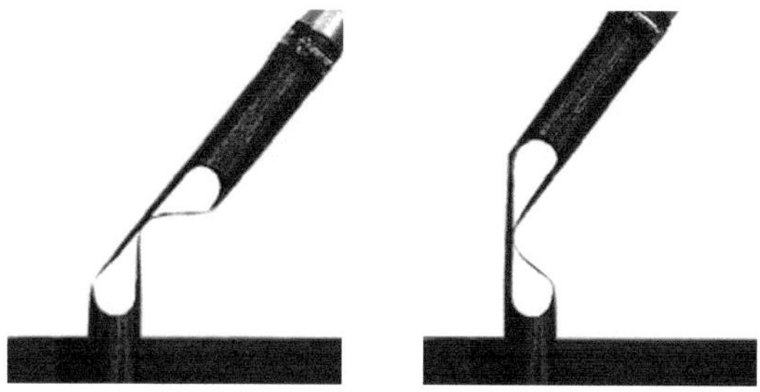

(a) Test 10 (b) Test 11

Figure 47: Two different configurations of tape-spring hinge at $t = 2.5$ s.

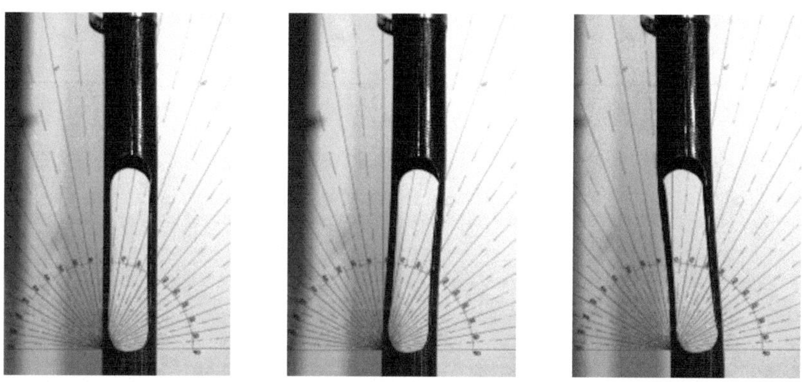

(a) Fully deployed configuration (b) Hinge sheared to the right (c) Hinge sheared to the left

Figure 48: Different configurations of tape-spring hinge when the deployment angle is zero.

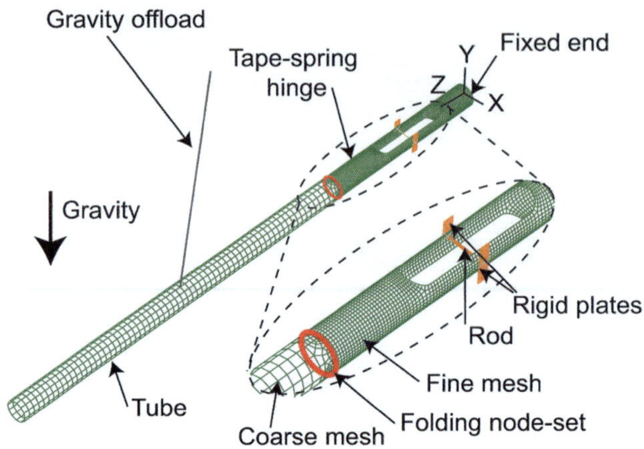

Figure 49: Finite element model of boom and definition of coordinate system.

Table 4: Mass distribution of finite element model in Fig. 49.

Region	Mass distribution
$0 < Z < 292.5$ mm	318 g/m^2
$Z = 292.5$ mm	25.79 g
$292.5 < Z < 970$ mm	2400 g/m^2

were connected by means of a vertical beam element. The beam element and the rigid plates were connected in all translational and rotational degrees of freedom using CONN3D2 elements and *Weld* connectors. Several frictionless contact surfaces were defined, to avoid that different parts of the boom or the rigid plates can penetrate into the boom. The **General Contact* feature, which defines all potential contact surfaces, was used to simulate this behavior. The setting *Contact Inclusions, All Exterior* was used to define all element-based surface facets as the set of potential self-contact surfaces. At the beginning of the folding simulation the beam element connecting the two rigid plates was shortened by decreasing its temperature and this caused the hinge to be pinched as shown in Figs 50(a) and 50(b). The next part of the folding simulation consisted in rotating the tape-spring hinge. The nodes in the folding node-set were rotated through 90 deg about the Y-axis over a time interval of 0.4 s; to minimize the dynamic effects associated with this process, the rotation was defined as a fifth-order polynomial function of time (*Smooth Step* in ABAQUS/Explicit).

In Section 5.4 it has been shown that a viscous pressure load can be used to further dampen this process and thus decrease the duration of the simulation. Hence, a viscous pressure load was applied over the entire surface of the boom throughout the folding simulation; the value of this pressure was defined by the pressure coefficient, c_v set equal to 2×10^{-5} times the density times the dilatational wave speed through the material, ρc_d. The dimensionless linear bulk viscosity factor, ξ, was set to zero.

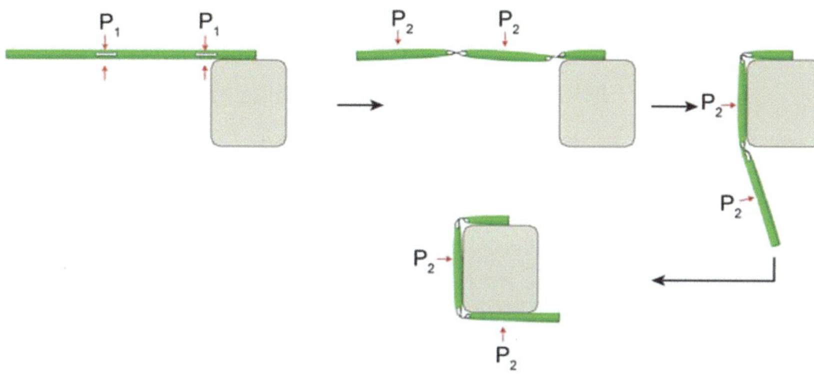

Figure 50: Stages of folding simulation.

At the end of the folding sequence, an additional damping step lasting 0.4 s was carried out with the viscous pressure load set 100 times higher, in order to quickly damp out any remaining kinetic energy while maintaining the nodes in the folding-node set fixed. This additional step had the effect of decreasing the kinetic energy of the boom to practically zero, see Fig. 51, and thus obtain the fully-folded static configuration of the boom. At the beginning of this step, contact between the rigid plates and the boom was disabled, in order to avoid spurious interference. Figure 50(c) shows the final folded configuration of the hinge region. Lastly, the viscous pressure loading was removed by decreasing c_v to zero in a simulation step lasting 0.10 s.

At this point the dynamic deployment simulation was started. Unlike the folding simulation, where the simulation time had no physical meaning but was simply chosen such that the kinetic energy would be negligible at the end, in the deployment analysis the simulation time is the actual time over

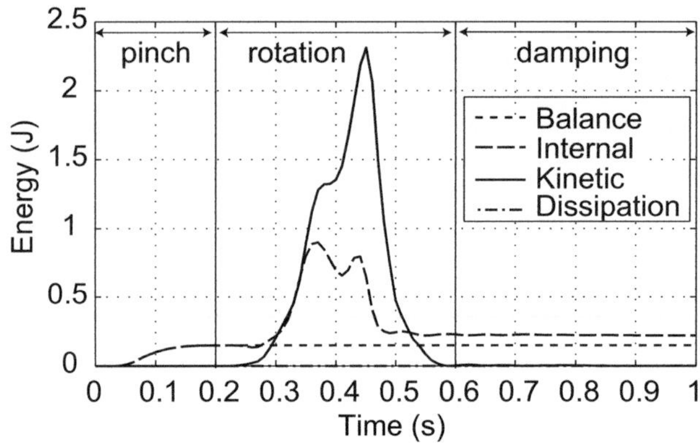

Figure 51: Energy variation during folding simulation.

which the motion occurs and so it is important to switch to a realistic model of the actual damping in the structure. Hence, the bulk viscosity ξ was set to 0.10 to damp out high frequency oscillations; this value was chosen on the basis of Section 5.4 and a further sensitivity study that considered the three values $\xi = [0, 0.01, 0.10]$.

Deployment was triggered by releasing all constraints on the folding-node set and the deployment simulation consisted in computing the response of the structure over a specified period of time, without any external actions being applied, apart from gravity loading.

Figures 52 shows the energy variation during the deployment simulation. Note that the internal and kinetic energies are respectively smallest and largest each time the boom attempts to latch. Also note that there is no energy dissipation during the deployment phase, but each latching attempt is accompanied by the dissipation of 10-20% of the total energy in the system. Finally, note that the energy dissipation remains larger after the boom has attempted to latch and continues its rigid-body rotation, as the kinetic energy associated with high-frequency vibration of the boom gets dissipated through viscous damping.

Looking again at Figs 51, 52, note that the energy balance term (defined as the sum of internal energy, kinetic energy and viscous dissipation, minus the external work done) remains fairly constant during both rotation/balance in Fig. 51 and deployment in Fig. 52, which indicates that the simulation is free of instabilities. The increase in the energy balance when the hinge is pinched, Fig. 51, is due to the thermal energy associated with

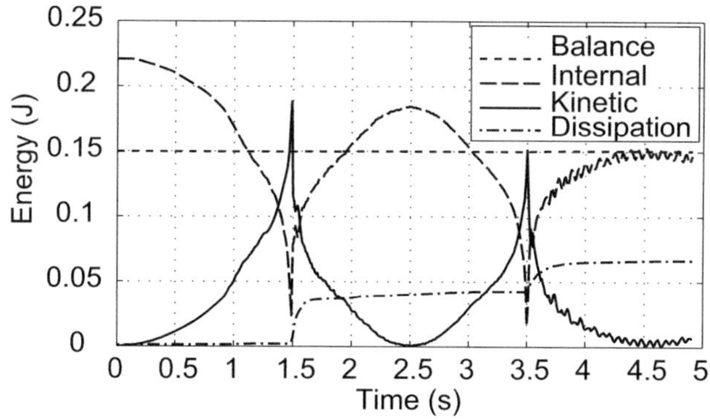

Figure 52: Energy variation during deployment simulation.

the decrease in the temperature of the beam element connecting the two rigid plates.

6.4 Sensitivity of Deployment Behavior

The sensitivity of key simulation results to different settings of the simulation parameters and also to changes in some modeling features is considered in the following sections.

Initial Conditions Figure 53(a) compares the angle-time relationships of two deployment simulations that were started from the same folded configuration. The final step in the folding simulation, consisting in the removal of the viscous pressure load, was extended from 0.10 s (case #1) to 0.15 s (case #2). This slight change results in a small increase of the kinetic energy at the beginning of deployment, from 0.01 mJ to 0.02 mJ, which in turn results in the tape spring hinge taking up two different configurations, Fig. 53(b), after the first latching attempt. In case #1 the fold in the *inner* tape spring is located closer to the root of the hinge, whereas in case #2 it is the fold in the *outer* tape spring that is closer to the root. This situation was observed in the experiments, Section 6.1, where the implications of this switch on the dynamics of the structure were also discussed.

Figure 53(c) compares the time-histories of the internal energy and the kinetic energy in the boom for the two cases. No difference in the energy responses can be seen before the first latching attempt, however afterwards there is a significant speeding up of the motion in the second case.

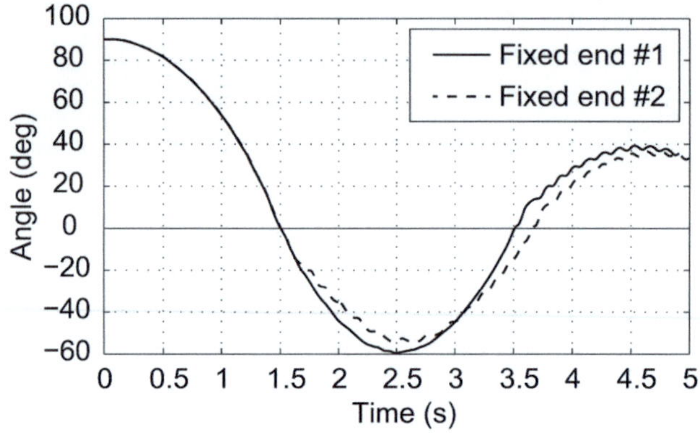

(a) Angle-time variation from simulations with fixed end and different initial conditions (case #1 and case #2).

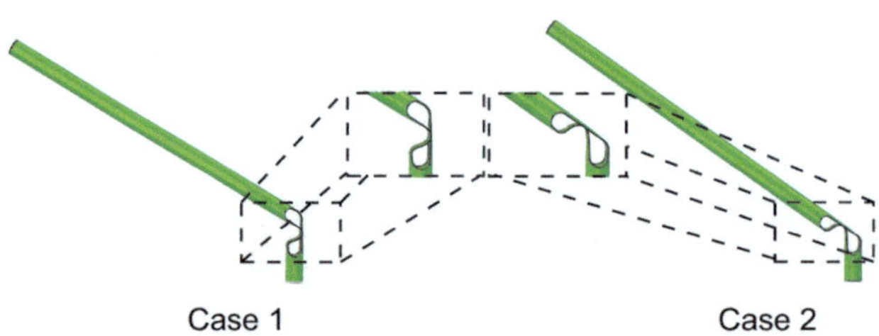

(b) Deformed configuration at point of maximum overshoot.

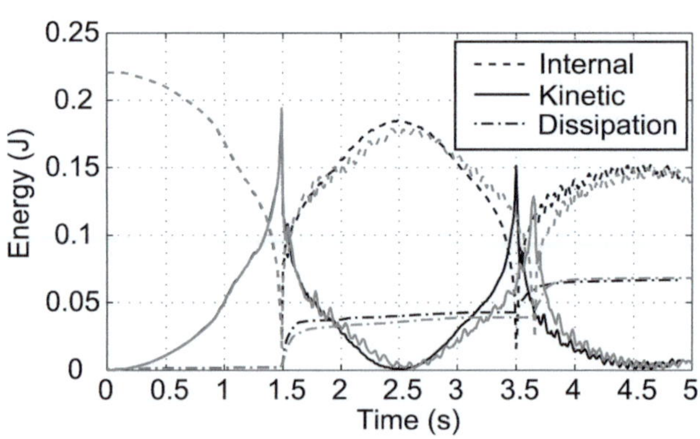

(c) Energy variation black lines; case #1, black lines; case #2 gray lines.

Figure 53: Deployment simulations with rigid end condition and two sets of slightly different initial conditions[58]

Boundary Conditions A comparison of the angle-time responses obtained from the simulations with a fixed end, in Fig. 53(a), show significantly larger overshoot angles than the experimental results, Fig. 46. Since this discrepancy relates to the amount of energy dissipated during the latching attempts, it seems likely that it may be addressed by refining the boundary conditions. Hence, to capture the compliance of the root, infinite elements were introduced in the model of the foundation.

Abaqus/Explicit (Dassault Systemes, 2013) provides first- and second-order infinite elements (Lysmer and Kuhlemyer, 1969), which act as remote dashpots that provide distributed damping along a boundary. These elements are used in conjunction with standard solid elements, which model the area around the region of interest, whereas the infinite elements model the far-field region.

A layer consisting of 100 eight node CIN3D8 infinite elements, with dimensions of $4 \times 4 \times 4$ mm^3 was defined. Because the infinite elements cannot be directly connected to the shell elements used to model the structure, the root of the boom was connected to three layers of elastic solid elements C3D8 through the "shell to solid" coupling feature, and these solid elements were attached to the infinite elements, Fig. 54(a). Both the solid elements and the infinite elements were defined as isotropic with Young's modulus of 70 GPa and Poisson's ratio of 0.33. Because the infinite elements allow the entire boom and its foundation pad to rotate as a rigid body, the (small) rigid-body rotation effects were subtracted from the boom response.

The angle-time response obtained from the simulation with infinite elements is shown in Fig. 54(b). Until the first latching attempt occurs, it is practically identical to the response obtained for a fixed end, but afterwards there are significant differences. The overshoot angles are much smaller (-21.5 deg instead of -54.6 deg after the first latching attempt), indicating that a much larger energy dissipation has occurred during the latching attempt. Thus, according to this new simulation the boom becomes latched at the second attempt.

It is interesting to examine the energy variation, plotted in Fig. 54(c), for the two latching attempts at $t \approx 1.5$ s and $t \approx 2.8$ s. The internal (strain) energy decreases as the boom approaches the fully deployed configuration and drops to practically zero as the boom attempts to latch, while the kinetic energy increases correspondingly. Large amounts of energy are dissipated during the two latching attempts, ≈ 0.08 J and ≈ 0.12 J respectively. The main difference is that during the second (successful) latching attempt practically all internal and kinetic energies are dissipated, whereas only 50% of the total energy is dissipated during the first (unsuccessful) attempt. A comparison of this plot with the energy variation plots for the

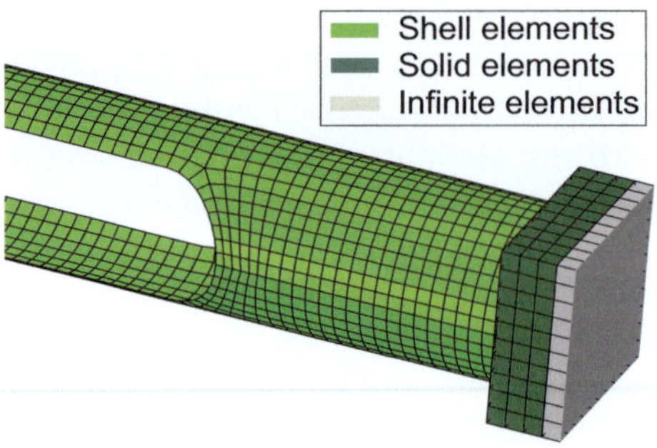

(a) Finite element model.

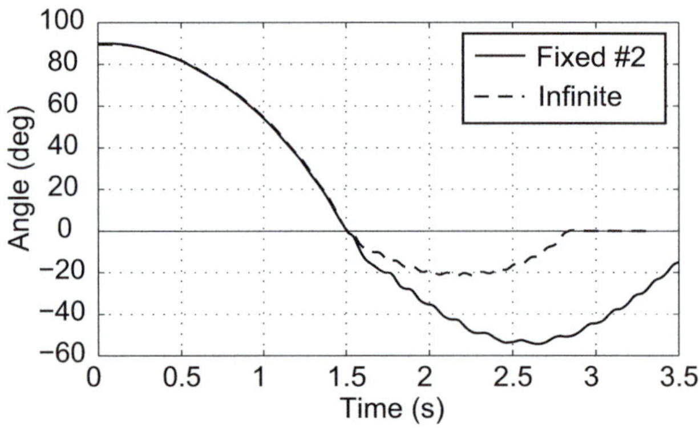

(b) Comparison of angle-time responses.

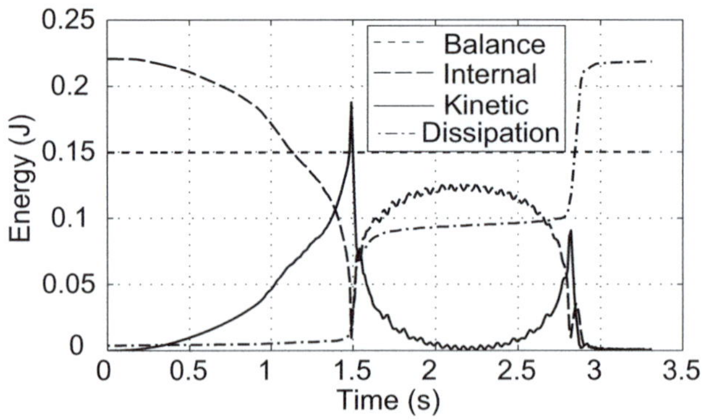

(c) Energy variation.

Figure 54: Deployment simulations with infinite elements.

rigid foundation models, in Figs 52 and 53, shows that the predicted energy dissipation during each latching attempt is much larger for the infinite foundation model, which explains why this simulation leads to complete latching after a smaller number of attempts.

Simulation Parameters In the previous two sections it has been shown that two extreme sets of angle-time responses for the boom can be obtained by prescribing either a fixed-end boundary condition with very little numerical damping or by extending the boundary to infinity. This section considers specific values of the simulation parameters, in order to capture more closely the angle-time response obtained in the experiments. The key issue is how to increase the damping in the simulations with fixed-end boundary condition.

The simulation with fixed-end boundary condition and $\xi = 0.10$ was selected as a reference and the sensitivity of the predicted behavior to changes in ξ and c_v was investigated. First, the sensitivity to ξ was considered. Three values of ξ, 0.10, 0.30 and 0.50, for deployment starting from the same folded configuration were considered and the resulting change in the predicted angle-time response was practically negligible.

The alternative to varying ξ is to vary c_v, but this time the effects were too severe because increasing the viscous pressure loading over the entire boom quickly overdamped the whole deployment. Specifically, the initial deployment behavior (which had been predicted correctly using a rather small value of c_v) became too slow. It was found that more realistic results are obtained by applying viscous pressure loading only to the tape-spring hinge region and this was done in two different ways. Figure 55 compares the angle-time responses obtained by varying c_v from 0 to $2 \times 10^{-5} \rho c_d$ during deployment; in this case the viscous pressure had been applied through the whole deployment. Note that the larger values of c_v have the effect of decreasing the number of latching attempts, which leads to better agreement with the experimental behavior in Fig. 46. However, increasing c_v also had the unwanted effect of delaying the initial deployment phase.

Comparison of these latest results with the previous simulations, in Figs 52 and 54(c), suggested that the energy dissipation mechanism should be introduced only for the duration of a latching attempt. Hence the viscous pressure loading was applied to the hinge region just before the snap-back of the tape-springs and it was then removed 0.2 s after snap back. It was found that setting $c_v = 2 \times 10^{-5} \rho c_d$ leads to a more realistic variation of the energy dissipation. Figure 56(a) compares the angle-time response obtained in this way with the results of Test 1 and Fig. 56(b) shows the corresponding energy variation during deployment.

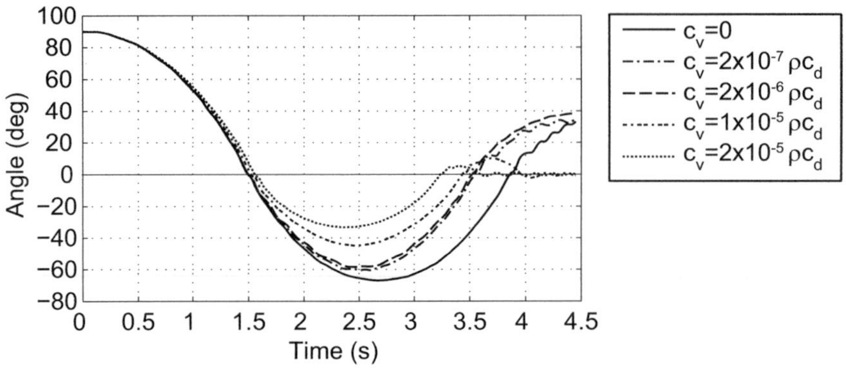

Figure 55: Comparison of angle-time responses obtained by varying c_v.

6.5 Comparison of Simulations and Experiments

Figure 57 presents a comparison between a set of four representative deployment experiments, selected from the 12 presented in Section 6.1, with the two "extreme" simulations and the "tuned" simulation developed in Section 6.4.

Focusing first on the deployment phase, it can be seen that the measured angle-time response is consistent among the four experiments, that all three simulations give practically the same results, and also that there is good agreement between experiments and simulations. It can thus be concluded that the initial deployment phase is repeatable and in excellent agreement with the simulations.

Next, consider the behavior of the boom as it attempts to latch. The first thing to note is that after the first latching attempt the experimental results show considerable scatter, with a wide range of overshoot angles (−28 deg to −38 deg) and time delay between the first and second latching attempt (1.6 to 2.0 s). The spread becomes smaller between the second and third latching attempts (5 deg to 9 deg and 0.4 s to 0.6 s) and in all experiments the boom successfully latched at the third attempt. Turning to the simulations, note that the two "extreme" simulations obtained by assuming either infinite or rigid foundation models provide lower and upper bounds, respectively, for the full range of measured experimental behavior. The bounds on the overshoot angles are of greatest interest to estimate the potential of interference with neighboring structures; they are 21 deg, or 25% lower than the smallest measured value and 51 deg, or 40% higher

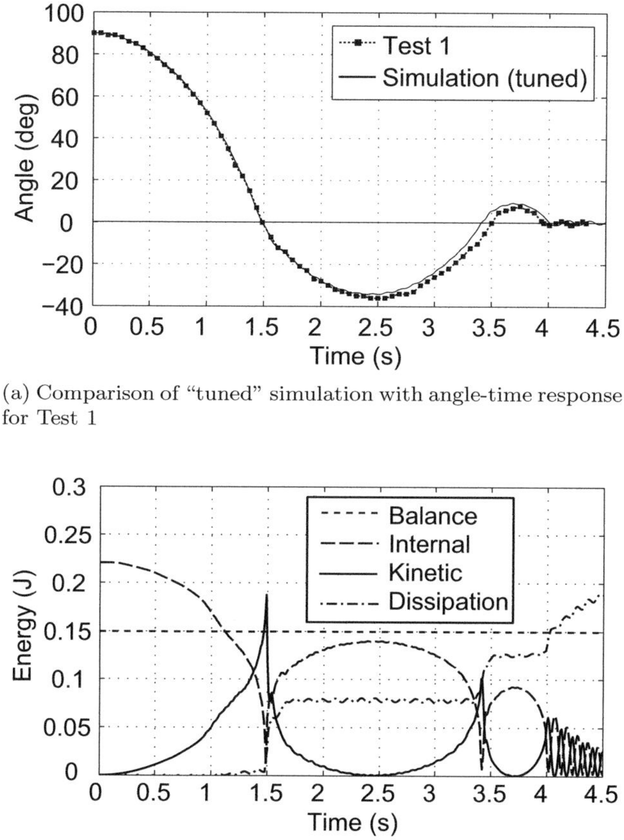

(a) Comparison of "tuned" simulation with angle-time response for Test 1

(b) Energy variation

Figure 56: "Tuned" deployment simulation.

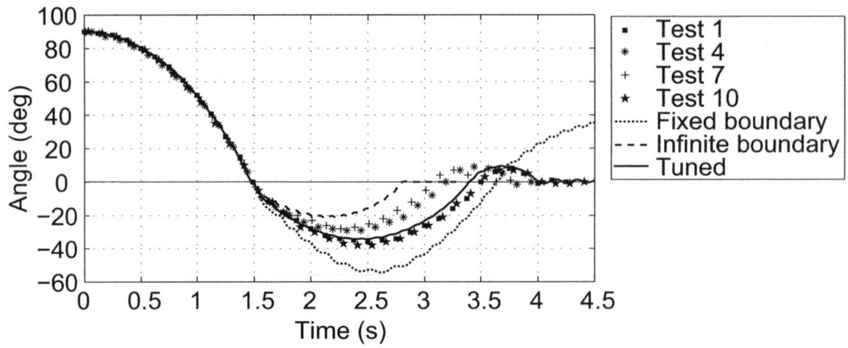

Figure 57: Comparison of four measured angle-time responses with three different simulations.

than the largest measured value.

The third simulation presented in Fig. 57 (labelled "tuned") was obtained with a rigid foundation model, but applying a viscous pressure loading with $c_v = 2 \times 10^{-5} \rho c_d$ to the tape spring region of the boom during each latching attempt. This simulation predicts the correct number of latching attempts, with first and second overshoot angles of 34.2 deg and 9.5 deg respectively. It produces results that fit among the measured response curves, although somewhat closer to the larger amplitude behavior.

The final phase of the deployment behavior consists in a small-angle oscillation of the latched boom. This phase is both repeatable and predictable, with a 1 deg amplitude and 0.2 s natural period of vibration.

It was explained in Section 6.2 that the rotation angle of the boom does not fully describe a partially deployed configuration of the boom, as the same rotation angle may correspond to boom configurations that are sheared by different amounts. The easiest way of confirming the accuracy of the results obtained from the third simulation is to compare snapshots of the tape-spring region of the boom with images obtained from the simulation, see Fig. 58.

7 Optimized Designs of Tape-Spring Hinges

The main problem, depicted in Fig. 59, is the design of a tubular boom rigidly connected to a spacecraft bus, that can be wrapped around the spacecraft. Upon release, the boom self-deploys and then becomes fully

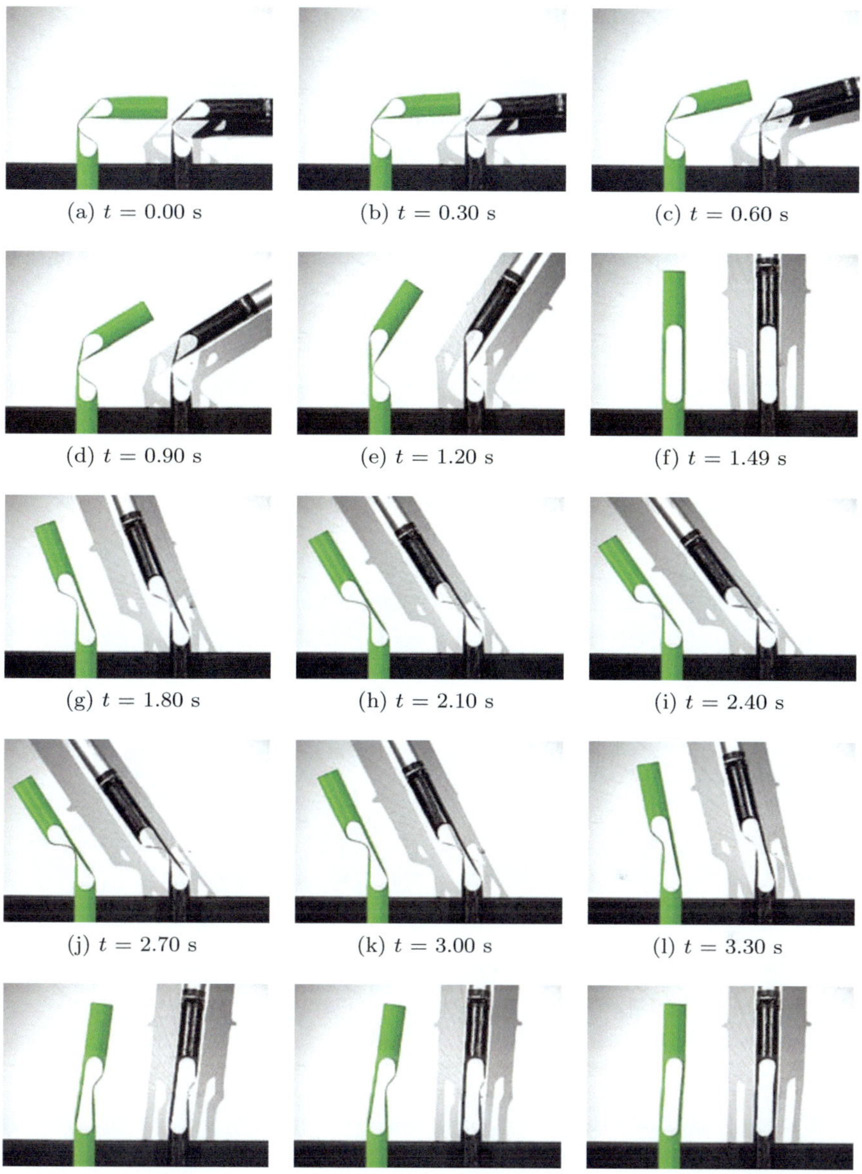

(a) $t = 0.00$ s (b) $t = 0.30$ s (c) $t = 0.60$ s

(d) $t = 0.90$ s (e) $t = 1.20$ s (f) $t = 1.49$ s

(g) $t = 1.80$ s (h) $t = 2.10$ s (i) $t = 2.40$ s

(j) $t = 2.70$ s (k) $t = 3.00$ s (l) $t = 3.30$ s

(m) $t = 3.60$ s (n) $t = 3.90$ s (o) $t = 4.26$ s

Figure 58: Comparison of snapshots from Test 1 and simulation with rigid foundation and $c_v = 2 \times 10^{-5} \rho c_d$.

latched; overshooting will occur for certain boom designs and sets of initial conditions. Overshooting is caused by buckling of the tape springs after a latching attempt, and it is characterized by the rotational velocity maintaining the same sign during the latching attempt. This behavior minimizes the potential for damage of the spacecraft or other payloads. The boom is 1 m long, has diameter of 38 mm and is made of a 0.22 mm thick two-ply plain weave $[\pm 45]_s$ laminate, as described in Section 2.1; the spacecraft has dimensions of 360 mm by 300 mm by 310 mm; the boom is rigidly attached to the spacecraft side as shown in Fig. 59. Two tape-spring hinges are formed in the boom, at distances of 200 mm and 600 mm from the root, allowing it to be wrapped around the spacecraft, as shown in the figure. The chosen laminate has a number of advantages, e.g. small thickness and a fiber orientation that avoids sharp folds at the end end of the hinge slots, but the lack of longitudinal fibers will lead to boom designs with somewhat reduced axial and bending stiffness. The chosen laminate may also be potentially sensitive to long-term stowage effects (Kwok and Pellegrino, 2012, 2013).

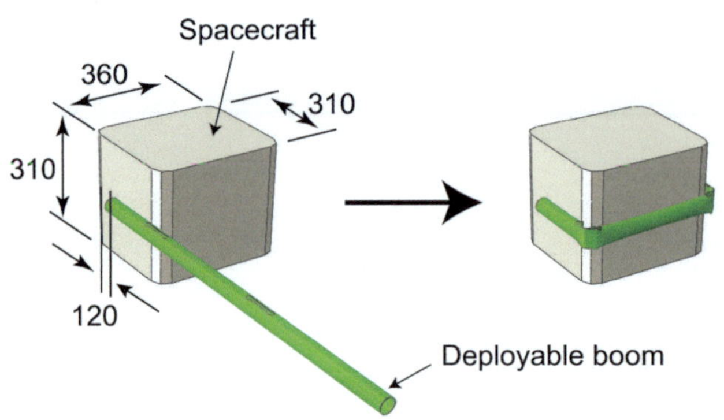

Figure 59: Deployable boom mounted on small spacecraft (dimensions in mm).

Because this main design problem is computationally quite challenging, as a full folding and deployment analysis is required for each trial design of the tape spring hinges in the boom, a simpler, preliminary design problem is considered first. This preliminary design problem consists in finding a good design for a single tape-spring hinge that can be folded 180 deg, without considering the deployment behavior.

To carry out such a design study for tape spring hinges, the basic configuration shown in Fig. 60 was chosen and the effects of varying three parameters, the slot length, L, the slot width, SW, and the diameter of the end circles, D, were explored. These three parameters fully define the dimensions of a solid that extrudes the slot from a cylindrical tube.

Based on the observed behavior of the tape-spring hinge in Section 6.4 the three parameters were varied in the ranges $L \in (45, 100)$ mm, $SW \in (8, 12)$ mm and $D \in (10, 18)$ mm.

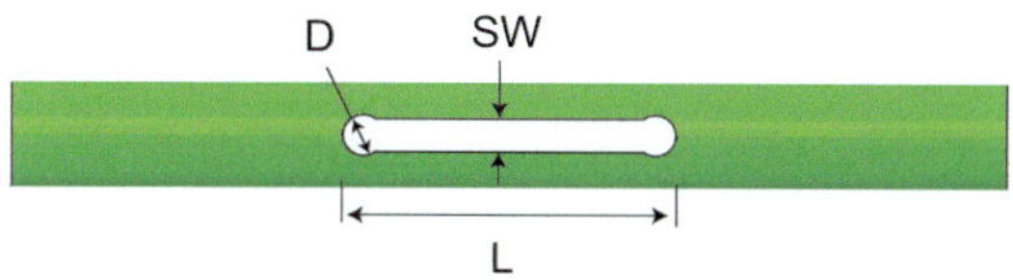

Figure 60: General geometry chosen for tape-spring hinge optimization study.

7.1 Tape-Spring Hinge Design

This part of the study considered 180 mm long tape-spring hinges. The hinges were not connected to any other structures. The hinges were modeled as discussed in Section 6.3; the finite element model is shown in Fig. 61.

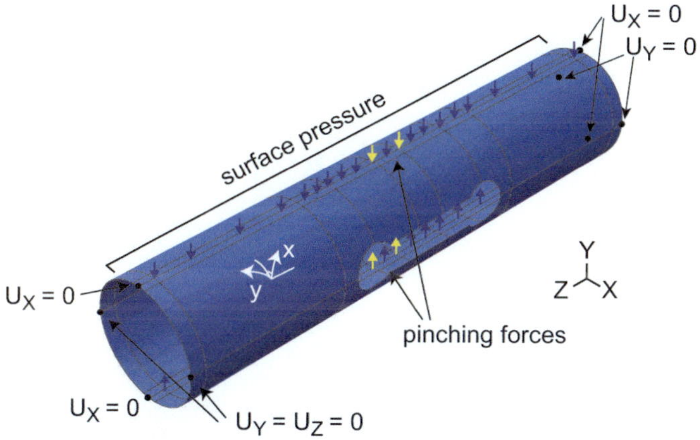

Figure 61: Finite element model of tape-spring hinge.

The single-hinge design problem consists in finding good hinge designs that can be folded 180 deg and this requires that a complete folding simulation is carried out for each trial design. These simulations were carried out in two stages. First, the hinge was pinched and flattened in the middle, Figs 62(a) and 62(b) and then it was folded by applying equal and opposite rotations on the sets of end nodes. The acceleration at any nodes in the finite-element model were minimized by defining a smooth fifth-order polynomial function of time for the applied rotations. Once the tape-spring hinge had been folded completely, the pinching and flattening forces were slowly removed (again, using fifth-order polynomial functions of time) to arrive at the final, fully folded configuration shown in Fig. 62(c).

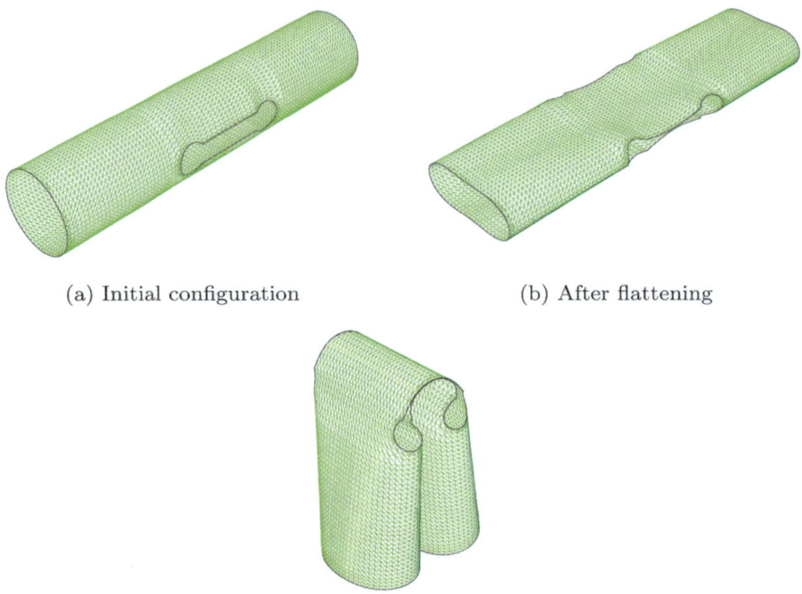

(a) Initial configuration (b) After flattening

(c) Folded configuration

Figure 62: Folding sequence of tape-spring hinge.

A sample plot of the energy variation corresponding to a folding simulation of a tape-spring hinge with design parameters SW=10 mm, D=15 mm and L=60 mm is shown in Fig. 63. Note that the kinetic energy is negligibly small compared to the internal energy, which confirms that the response is quasi-static throughout the simulation. Also the energy balance term remains approximately zero throughout the simulation, which confirms that the simulation is free of numerical instabilities, see Section 5.2.

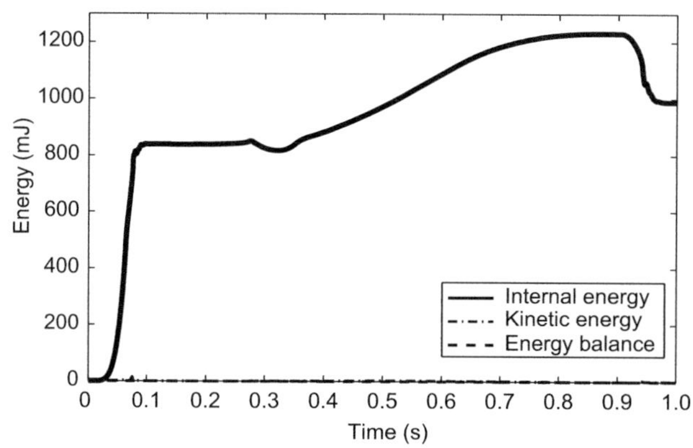

Figure 63: Energy variation during folding simulation of hinge with SW=10 mm, D=15 mm and L=60 mm.

The failure criterion of Section 2.3 was applied to the stress-resultants in the fully-folded configuration, by computing the values of FI-1, FI-2 and FI-3 at every integration point. The average over each element was then computed. These three indices should be less than 1 for a design to be safe, although in practice a value well below 1 should be used to provide an adequate margin. The relative magnitudes of the three indices are useful to determine the dominant loading condition at each point of the hinge. A sample set of results is shown in Fig. 64, where it should be noted that the most critical region of the hinge is in the end circle, near the inner edge of the slot.

Sensitivity to Slot Length The first design parameter to be considered was the slot length. The sensitivity of the failure indices were calculated for changes in the value of L in the range 45 mm to 90 mm, at 5 mm intervals, while maintaining SW = 10 mm and D = 15 mm. Table 5 presents the maximum values of all three failure indices obtained for each case. Note that in most cases FI-1 remains below 0.50, whereas FI-2 and FI-3 are generally much larger and in several cases even larger than 1.0. Hence, the dominant failure modes are bending and in-plane loading coupled with bending.

Based on these results, the two cases L = 60 mm and L = 90 mm were

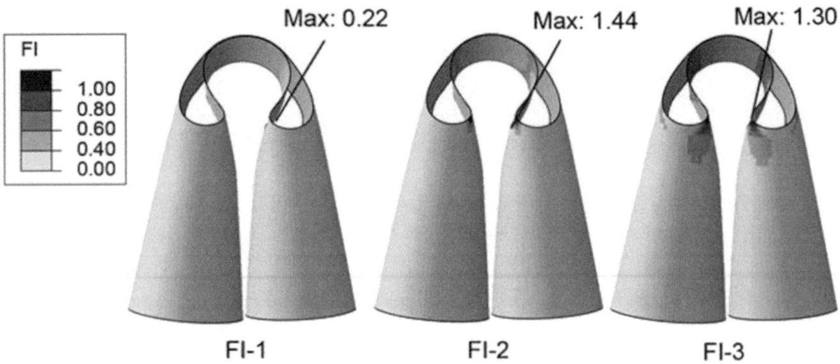

Figure 64: Distribution of three failure indices in fully folded configuration for hinge with $L = 60$ mm, $SW = 10$ mm and $D = 15$ mm.

selected for further investigation. The case $L = 90$ mm is clearly a safe choice; the case $L = 60$ mm was also considered to explore the potential of decreasing the failure indices by modifying SW and D.

Hinge Design Parameters			FI-1	FI-2	FI-3
L (mm)	SW (mm)	D (mm)			
45	10	15	0.54	1.87	1.29
50	10	15	0.22	1.80	1.25
55	10	15	0.19	1.43	1.13
60	10	15	0.22	1.44	1.30
65	10	15	0.23	1.01	1.06
75	10	15	0.32	0.79	0.91
80	10	15	0.36	0.67	0.87
85	10	15	0.37	0.54	0.78
90	10	15	0.26	0.51	0.85
100	10	15	0.48	0.72	0.96

Table 5: Failure indices for hinge designs with different values of L.

Sensitivity to Slot Width and End Circle Diameter Next, the sensitivity of the hinge design to SW and D was considered. D was initially set equal to 15 mm and SW was varied from 8 mm to 12 mm, in steps of

2 mm. The sensitivity to D was then studied by varying D from 13 mm to 17 mm with SW fixed at 10 mm. The maximum failure indices for all of these cases are presented in Table 6, showing that in all cases FI-2 and FI-3 exceed the value 1.0. From these results it was concluded that the design with $L = 60$ mm cannot be improved significantly by varying SW or D and so the study focused on the alternative value $L = 90$ mm.

Hinge Design Parameters				FI-1	FI-2	FI-3
L (mm)	SW (mm)	D (mm)	$\approx SW/D$			
60	8	15	1/2	0.17	1.15	1.07
60	10	15	2/3	0.22	1.44	1.30
60	12	15	4/5	0.20	1.10	1.06
60	10	13	3/4	0.25	1.18	1.12
60	10	17	3/5	0.20	1.10	1.03

Table 6: Failure indices for hinge designs with different values of SW and D, for $L = 60$ mm.

It has been seen in Table 5 that the design with $L = 90$ mm, $SW = 10$ mm and $D = 15$ mm can be folded safely. This design was taken as a starting point to search for designs with greater safety margins; in this case the sensitivity to SW and D was investigated by varying SW from 8 mm to 12 mm while maintaining the ratio SW/D equal to 1/2, 2/3 or 1. The results of this study are presented in Table 7. The lowest value of the maximum failure index, FI-3=0.69, corresponds to the design with $SW = D = 12$ mm which, however, may be problematic to build as it is difficult to manufacture a perfectly smooth transition between a straight edged slot and the end circle.

Designs with $D \leq SW$ lead to sharp transitions in the shape of the slot ends and can result in unexpected stress concentrations. To avoid potential problems such designs were avoided.

Comparison with Intuitive Hinge Design This part of the study considered longer (220 mm long) tape-spring hinges.

It is interesting to compare several of the hinge designs considered above with the intuitive hinge design considered in Sections 4 and 6. The intuitive design had two tape-springs separated by slots with $SW = D = 30$ mm and $L = 140$ mm. The main concerns with this design were its low torsional stiffness and the fact that the the folds could move along the tape springs, resulting in loss of symmetry and uncertainty in the configuration of the folded hinge.

Table 7: Failure indices for hinge designs with different values of SW and D, for $L = 90$ mm.

Hinge Design Parameters				FI-1	FI-2	FI-3
L (mm)	SW (mm)	D (mm)	$\approx SW/D$			
90	8	15	1/2	0.26	0.51	0.78
90	10	10	1	0.53	0.46	0.82
90	10	15	2/3	0.33	0.60	0.85
90	12	12	1	0.26	0.34	0.69
90	12	18	2/3	0.26	0.53	0.80
90	12	24	1/2	0.13	0.56	0.75

Figure 65 compares the moment-rotation plot for the top hinge design presented in Table 7, for which all three failure indices are < 1, with the intuitive hinge design. Both plots were obtained from quasi-static simulations in which the nodes on each end section of the hinge were constrained to remain in a plane, and the rotation of the cross-section was monotonically increased. The values of the three failure indices were calculated at each step of the simulation and it was confirmed that all three indices are < 1 throughout the simulation. Hence, the linear-elastic material model assumed in the simulation is valid.

The new hinge design has much wider and shorter tape-springs, and hence can sustain much larger moments before snapping. Note that the elastic snap-through limit is much higher for the new hinge design.

The stiffness properties of the new hinge design are also superior to the old, intuitive design. Table 8 shows that the torsional stiffness is increased by a factor of approximately six, the peak moment by over four, the axial stiffness by a factor of approximately two and the bending stiffness by approximately one quarter.

7.2 Boom Design

This section turns to the main design problem, which requires simulations of folding and deployment of the complete boom, folded according to the scheme in Fig. 59.

The tape-spring hinge design study in Section 7.1 considered hinges whose end sections were unconstrained at either end and so were allowed to freely ovalize. Also, those hinges were required to fold 180 deg. The present situation is different in both respects, because the hinges are now part of a boom that prevents any significant deformation of the tubular end sections of the hinge regions. In fact, for the root hinge no deformation is allowed on

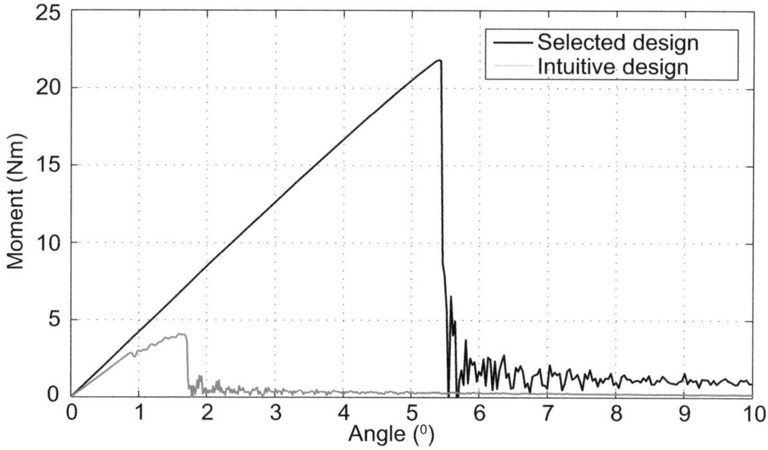

Figure 65: Comparison of moment-rotation plot for new hinge design with previous (intuitive) design.

Design	Tape-spring width (mm)	Axial stiffness (N/mm)	Bending stiffness (Nm/deg)	Torsional stiffness (Nmm/deg)	Peak moment (Nm)
Intuitive	25	597	3.30	97	4.97
Selected	56	1151	4.17	576	21.83

Table 8: Comparison of previous and optimized designs.

the side connected to the spacecraft. A further difference is that the hinges of the boom are required to fold only 90 deg.

To restrict the range of options, three potential hinge designs were selected among those previously considered. Their design parameters are presented in Table 9. For each hinge design, the last column in this table presents the maximum failure index that had been obtained in Section 7.1.

Finite Element Model and Simulation Techniques The finite-element model of the complete boom consisted of approximately 18000 nodes and 17500 four node fully integrated shell elements, S4; the number of elements varies between different hinge designs, due to changes in the slot width. The

Design #	Slot parameters			Max FI from Section 7.1
	L (mm)	SW (mm)	D (mm)	
1	90	10	15	0.85
2	85	10	15	0.78
3	90	8	15	0.78

Table 9: Three sets of hinge design parameters considered for boom designs.

finite element mesh was finer over the hinge regions, to capture the details of the complex deformation that occurs in these regions. The spacecraft was modeled with 112 reduced integration eight node brick elements, C3D8R, which were defined to form a single rigid body connected to a reference node. Fixed boundary conditions were specified at the root of the boom and also at the reference node of the rigid body representing the spacecraft. Contact between the boom and the spacecraft and self contact between different parts of the boom was defined as frictionless (a parametric study of friction-related effects showed them to be small) by means of the **General Contact* definition. Figure 66 shows the complete model.

The folded configuration of the boom, see Fig. 67, was generated during the first part of the simulation. Folding of both hinges was started simultaneously, by pinching each hinge in the middle by applying sets of equal and opposite forces P_1, at the center of each hinge and also on either side at a distance of 5 mm. The maximum value of P_1 was 30 N. Once P_1 had reached its maximum value, two additional forces, P_2, were applied normal to the boom in the direction of folding (i.e., the negative Y-direction in the initial configuration of the boom) and at distances of 400 mm and 800 mm from the root of the boom. All of these forces were defined to remain normal to the boom throughout the folding simulation (follower forces). P_2 was slowly increased up to a maximum value of 10 N, at which point the boom had become wrapped around the spacecraft. A viscous pressure with the magnitude provided in Table 10 was applied over the surface of the boom throughout the entire folding simulation.

Deployment of the boom was initiated by the sudden removal of the forces P_1 and P_2. For this part of the simulation the bulk viscosity factor was set to 0.1 and a viscous pressure loading with the value provided in Table 10 was applied only to the hinge regions and only just before each hinge had become fully latched, to suppress artificial high frequency oscillations. The use of viscous pressure during the deployment simulations was minimized, to avoid overdamping the structural response.

A sample energy variation plot for a complete simulation is provided

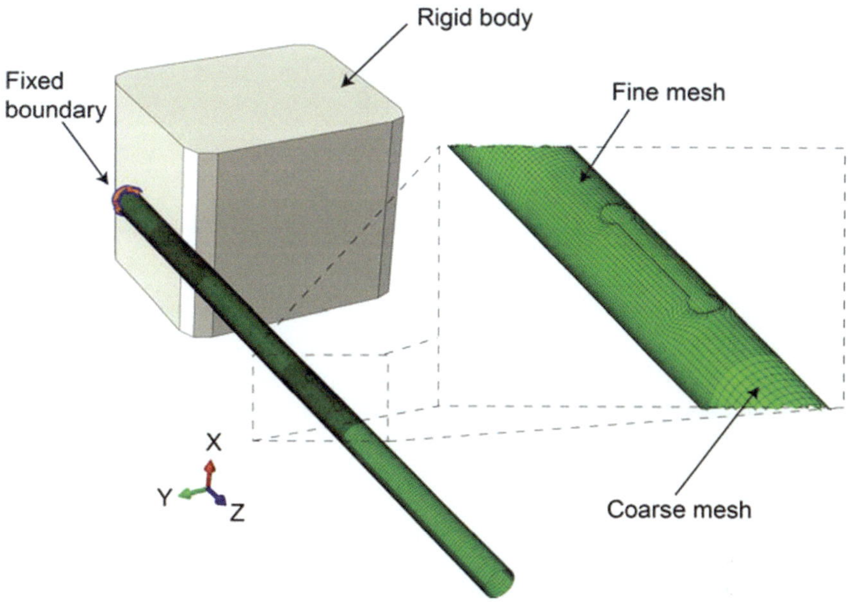

Figure 66: Finite element model of spacecraft and boom.

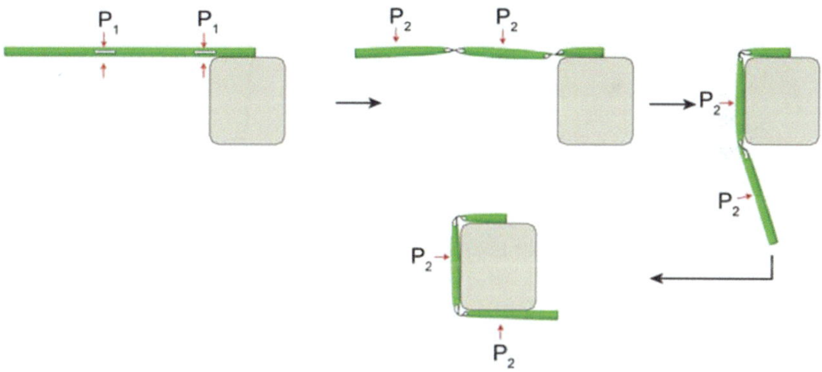

Figure 67: Boom folding sequence.

Table 10: Simulation parameters

Property	*Value*
time scaling factor	0.98
mass scaling factor	1.00
linear bulk viscosity factor	0.10
viscous pressure coefficient for:	
• quasi-static simulation	$2{\times}10^{-4}\rho c_d$
• dynamic simulation [a]	$2{\times}10^{-5}\rho c_d$

[a] Pressure applied only over tape springs, during snapping.

in Fig. 68. Note that the kinetic energy is practically zero at the beginning of the deployment simulation, indicating that the initial conditions for boom deployment were static and the energy balance term remained almost constant throughout the entire simulation, indicating that no numerical instabilities had been encountered. Also note that a complete simulation of folding and dynamic deployment typically lasted 23 h on a workstation with two 6-core 3.3 GHz Intel Xeon processors.

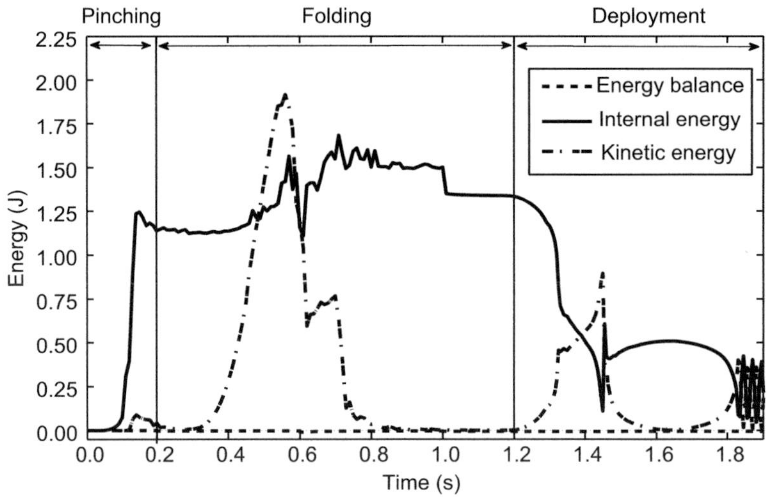

Figure 68: Energy variation for folding and deployment simulation of boom with hinges with $L = 90$ mm, $SW = 10$ mm and $D = 15$ mm.

Analysis of Boom Design #1 A deployment simulation of this boom design gave the results shown in Fig. 69, where θ_1, θ_2 are the total rotations of the first and second segments of the boom as defined in the figure. The simulation shows that during the initial part of the deployment, between $t = 0$ s and $t = 0.12$ s, θ_1 remains unchanged while θ_2 decreases from 180 deg to 90 deg. This corresponds to the tip segment of the boom rotating and the root segment not moving. At $t \approx 0.12$ s the second hinge latches and the root segment of the boom begins to move as a whole. At $t \approx 0.25$ s the first latching attempt of the root hinge occurs, but it is unsuccessful and hence the boom continues to rotate in the same sense although its angular velocity is reduced. The boom continues to rotate until it reaches at $t \approx 0.45$ s a maximum rotation $\theta_1 = \theta_2 \approx 35$ deg, then the boom reverses its sense of rotation and makes a second, successful latching attempt at $t \approx 0.63$ s. Afterwards the deployed boom oscillates about the fully deployed configuration.

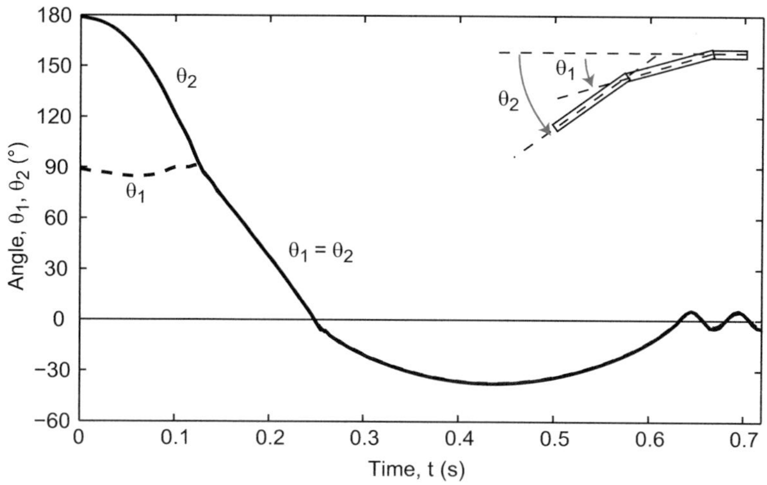

Figure 69: Deployment of boom with hinges based on design #1.

The procedure to estimate if damage is likely to occur during folding and deployment of the boom is analogous to Section 7.1 but this time the values of the three failure indices are considered both in the fully-folded (static) configuration and in the full set of (dynamic) configurations that occur during deployment.

The results of the failure check for the fully folded configuration of the present boom design are shown in terms of contours of the three failure indices, in Fig. 70. The top set of pictures represents the root hinge and the bottom set represents the tip hinge. The plots show that the root hinge is subjected to higher stress resultants than the tip hinge; this happens because the fixed connection at the root of the boom restricts deformation of the root cross-section whereas the second hinge is less constrained. Note that the largest failure index is FI-3, which indicates that this boom design is governed by the interaction between axial and bending stress resultants.

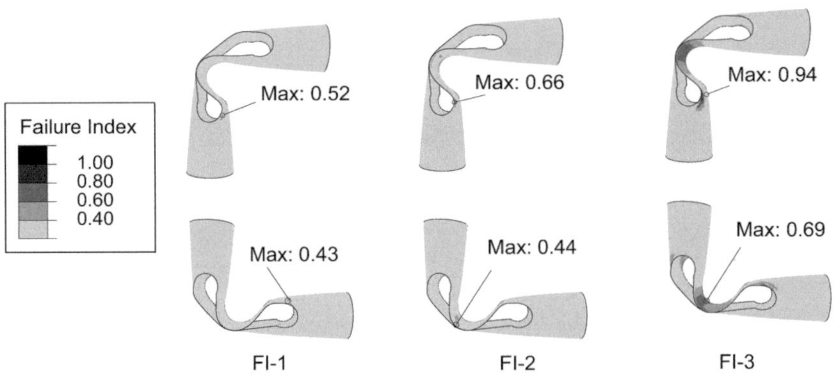

Figure 70: Maximum values of three failure indices for root hinge (top) and tip hinge (bottom) in fully folded configuration of boom design #1.

This analysis has been repeated for the configurations obtained from the deployment simulation. The stress-resultants every 0.005 s were obtained from ABAQUS/Explicit and the three failure indices were calculated. It was found that the most critical conditions are reached at the root hinge during the latching of the tip hinge; FI-3 goes from 1.13, just before the tip hinge becomes latched, to 1.15, just after it has latched. The corresponding configurations of the boom are shown in Fig. 71. The value of FI-3 is generally smaller than 1.0 at other stages of the deployment.

Selection of Boom Design The unsuccessful latching attempt that was predicted earlier for the boom design #1 is undesirable for several reasons. First, the boom could become damaged and second it was noted in Mallikarachchi and Pellegrino (2011) that the deployment dynamics become less predictable after the first latching attempt and hence the motion

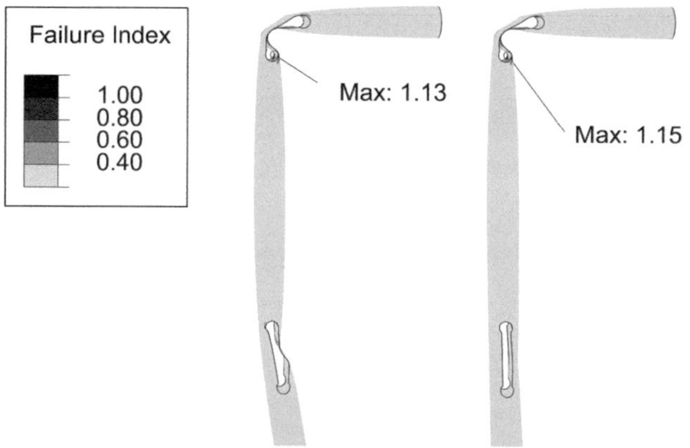

Figure 71: Maximum failure index at three stages of deployment, for boom design #1.

envelope for negative values of θ_i cannot be estimated reliably.

The analysis presented above has been repeated for the boom designs #2 and #3, and the maximum values of the failure indices have been presented in Table 11, both for the fully folded configuration and during deployment. The critical failure index is always FI-3, hence in all cases the dominant loading condition is the interaction between axial and bending stress resultants. Note that for all three designs the largest failure index during deployment is FI-3 and the maximum values are always reached at the root hinge. The specific location where each maximum value occurs has been shown in Fig. 72.

Design #	Fully folded	During deployment
1	0.94	1.15
2	0.81	1.05
3	0.88	0.95

Table 11: Values of maximum failure index (FI-3) for three boom designs.

Figure 73 presents a comparison of the variation of the tip hinge angle vs. time for the three boom designs. The rotation of the root hinge has not been plotted because it has been seen in Fig. 69 that it remains initially unchanged and then follows the tip hinge. Note that design #1 is the only one for which the first latching attempt is unsuccessful, whereas both

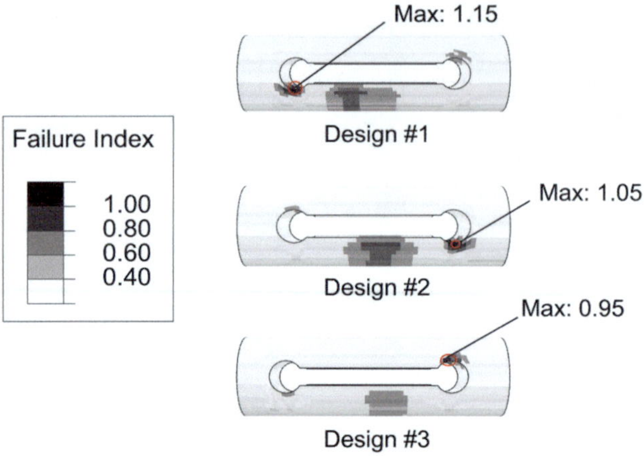

Figure 72: Contour plots (on undeformed hinge configuration) of maximum values reached by failure index FI-3 during deployment, for each boom design.

design #2 and design #3 latch straight away. Considering the results of the failure analysis in Table 11, together with these latest results, design #3 was selected as it is the only one for which the failure indices are always < 1.0 and also complete latching occurs at the first attempt.

7.3 Experimental Verification Studies

Verification of Hinge Design The failure criterion of Section 2.3 was tested by building and testing two of the tape-spring hinges that had been analyzed in Section 7.1.

The design parameters for the first hinge were $L = 60$ mm, $SW = 10$ mm and $D = 15$ mm. According to Table 7 this hinge reaches a maximum failure index FI-2=1.44 when it is folded 180 deg. A contour plot of the maximum failure index in the fully folded configuration has been plotted in Fig. 74(a) on the undeformed configuration of this hinge; note that the maximum value exceeds 1.0 in the regions highlighted in black. The same regions were seen to have become damaged when an actual hinge based on this design was folded. The enlarged detail in Fig. 74(b) shows one of the regions that had cracked.

The design parameters for second hinge were $L = 90$ mm, $SW = 8$ mm and $D = 15$ mm. According to Table 7 the maximum failure index reaches

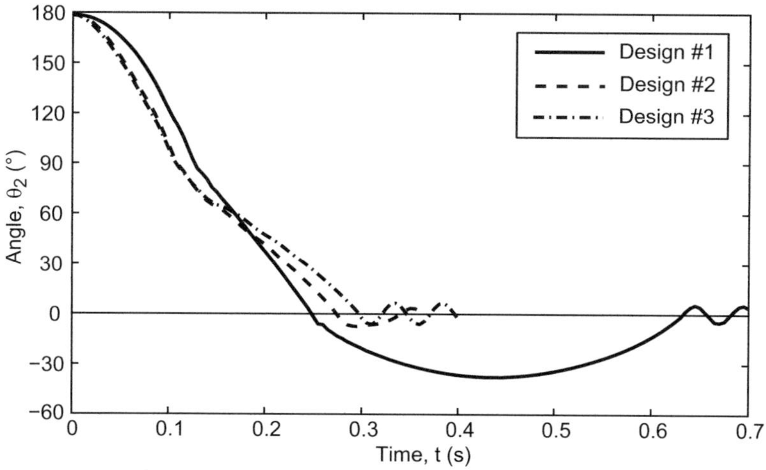

Figure 73: Deployment comparison for three boom designs.

the value FI-3=0.78 when this hinge is folded 180 deg, and hence this hinge is expected to remain undamaged when it is folded. This was confirmed by building the hinge and folding it several times; afterwards the hinge was inspected and no damage was observed.

Deployment of Boom Design #3 A boom based on design #3 was built and tested to verify the accuracy of the simulation results and failure analysis on which the selection of this design had been based.

The manufacturing procedure for the boom was similar to that of the tape-spring hinges described in Section 6.1. The complete boom is shown in Fig. 75. The root end of the boom was slid onto a 50 mm long cylindrical mandrel with 37.8 mm diameter, forming part of a solid aluminum-alloy fixture. The overlap region between the boom and the fixture was wrapped with plastic tape and then clamped with a hose clamp tightened with a torque of 1.13 Nm. The fixture was bolted to a granite table.

The boom is able to self-deploy while carrying its own self weight, however, to minimize gravitational effects, a single-point off-load was provided through a cord attached to the outer surface of the boom at a distance of 400 mm from the root. The other end of the cord was run over a pulley placed directly above the root hinge, at a height of 4.65 m.

A 400 mm long, 360 mm wide, and 310 mm tall wood box was placed

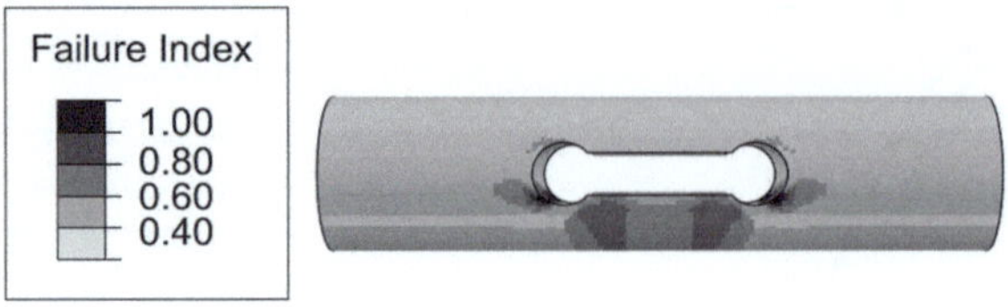

(a) Maximum failure index plotted on undeformed configuration

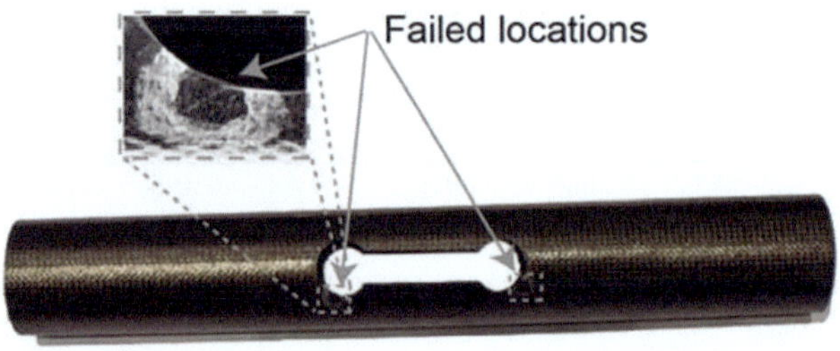

(b) Experiment

Figure 74: Experimental evidence of failure critical hinge design (L = 60 mm, SW = 10 mm and D = 15 mm).

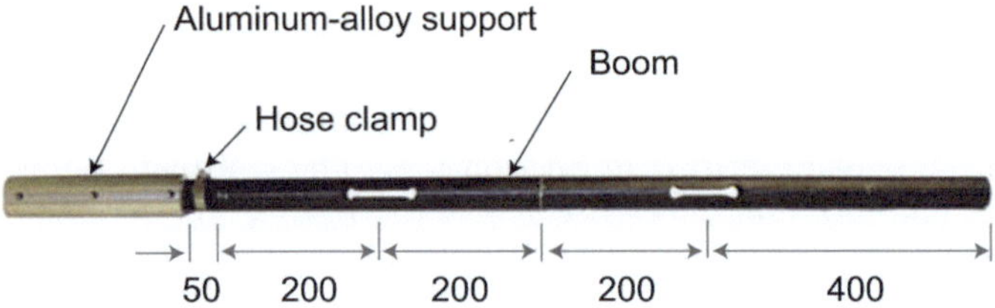

Figure 75: Complete boom (units: mm).

next to the boom to simulate the spacecraft. Non-contact measurement techniques are preferred for ultrathin structures, hence the motion of the boom was recorded with two high speed cameras. A Phantom V12.1 high speed camera with a wide angle lens was held directly above the root hinge to capture the detailed behavior. A Casio EX-FH20 high speed camera was placed 4.8 m above the boom to capture the deployment sequence. Figure 76 shows a schematic of the experimental setup.

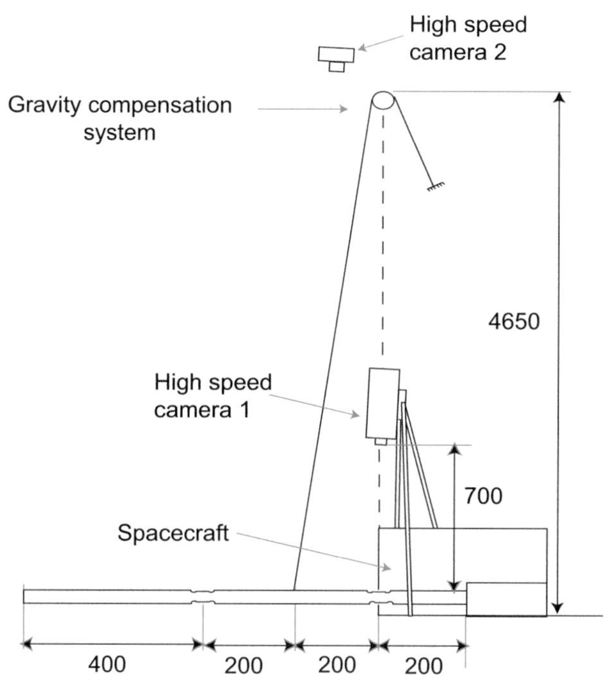

Figure 76: Deployment experiment setup (units: mm)

Great care was needed to fold the boom, specially the root hinge, because the boom design was known to have only a small margin against failure in the folded configuration. First the root hinge was pinched and folded 90 deg and then the same procedure was followed to fold the tip hinge, Fig. 77. The fully folded boom was held folded by hand, near the tip, and was released while recording the deployment with the two cameras, at rates of 1000 and 420 frames per second, respectively.

Snapshots of the deployment were extracted from the videos recorded

(a) Pinching root hinge (b) Folding

(c) Fully folded configuration

Figure 77: Setting up deployment experiment.

by the two cameras and the hinge angles were measured using a specially written Matlab program that computes the angle between two lines manually drawn by an operator. This was done by loading selected pictures, manually defining the center lines of different parts of the boom, and then computing the angle between these lines.

Both simulation and experiment showed the boom reach its fully deployed configuration in about 0.3 s, become fully latched, and then oscillate around the deployed configuration. The experiment also confirmed that the root hinge remains folded until the tip hinge has deployed completely. Figure 78 shows snapshots of the deploying boom, taken by the second high speed camera.

Figure 79 compares the measured angle vs. time response of the root hinge with the predictions obtained from the Abqus/Explicit simulations.

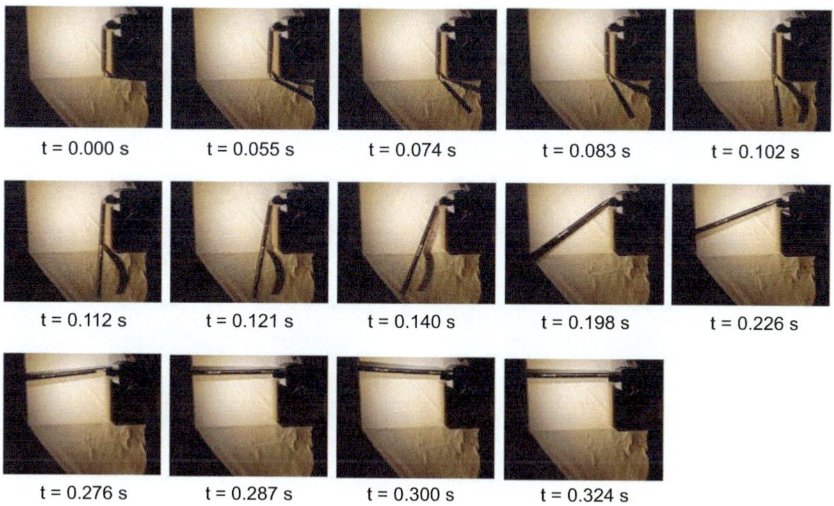

Figure 78: Snapshots of boom deployment experiment.

Note that the root hinge remains fully folded for the initial 0.1 s of the deployment. During this period the tip hinge deploys and latches; it then remains latched throughout the rest of the deployment. The experimental and simulated responses show excellent agreement. In the experiment the boom reached its fully deployed state at 0.296 s whereas the simulation predicts latching after 0.297 s. The final vibration portion of the response is also well predicted, with a maximum predicted amplitude of 4.8 deg vs. the measured value of 3.5 deg the predicted period of vibration is practically identical to the measured value.

Figure 80 compares the localized deformation of the root hinge. Note that the simulation results (on the left of each image) show that between $t = 0.105$ s and $t = 0.130$ s the fold locations in the inner tape-spring should be swapped, but the experimental results (on the right of each image) show that this change did not happen. It was seen in Section 6.2 that folded tape-spring hinges can flip between two different asymmetric configurations. When the tip hinge latches a dynamic wave travels through the boom and in the simulation this wave flips the root hinge configuration. This did not happen in the experiment, but the overall behavior of the boom was not affected significantly.

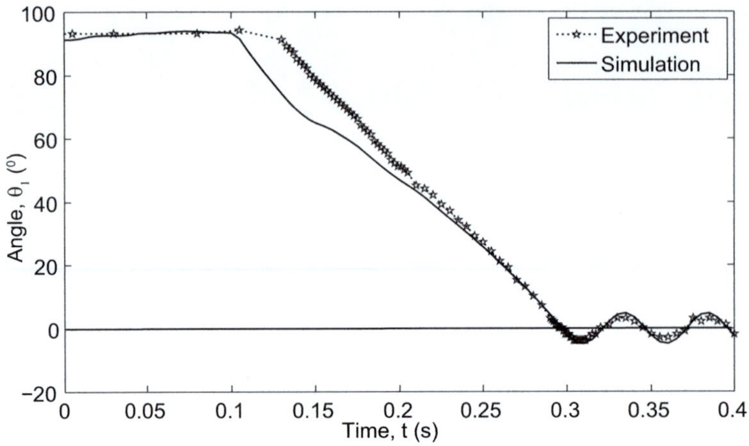

Figure 79: Comparison of root hinge angle time variation.

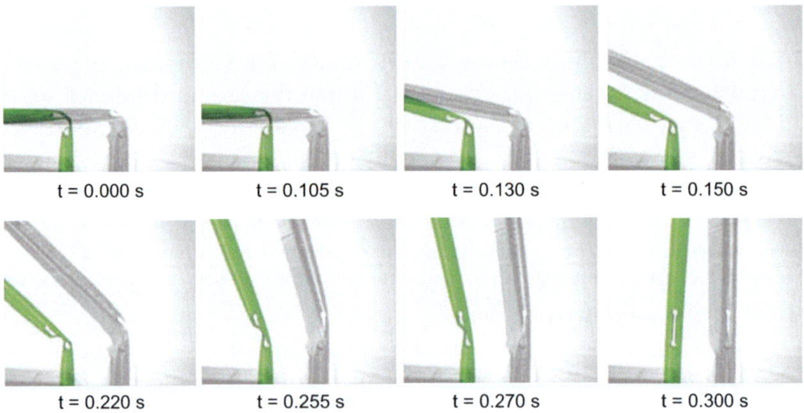

Figure 80: Comparison of local deformation of the root hinge; the simulations are shown on the left and the experimental images on the right.

Acknowledgement Financial support from AFOSR award no. FA9550-13-1-0061 (program manager Dr David Stargel) is acknowledged.

Bibliography

D.S. Adams and M. Mobrem. Lenticular jointed antenna deployment anomaly and resolution onboard the Mars Express spacecraft, *Journal of Spacecraft and Rockets*. Vol. 46, 403-410, 2009.

H.G. Allen and P.S. Bulson. *Background to Buckling*. McGraw-Hill, London, 1980.

Z.P. Bazant and L. Cedolin. *Stability of Structures*. Oxford University Press, Oxford, 1991.

T. Belytschko, W.K. Liu, and B. Moran. *Nonlinear Finite Elements for Continua and Structures*, J. Wiley & Sons., Chichester, 2000.

D.O. Brush and B.O. Almroth. *Buckling of Bars, Plates and Shells*, McGraw-Hill, 1975.

C.R. Calladine. *Theory of Shell Structures*. Cambridge University Press, Cambridge, 1983.

C.R. Calladine, C. R. The theory of thin shell structures 1888-1988. Proceedings of the Institution of Mechanical Engineers, Vol. 202(42), 1-9, 1988.

I.M. Daniel and O. Ishai. *Engineering Mechanics of Composite Materials, Second Edition*. Oxford University Press, New York, 2006.

Dassault Systemes Simulia Corp., ABAQUS/Explicit Version 6.11, Providence, RI, 2013.

M. Geradin and D. Rixen. *Mechanical Vibrations: Theory and Application to Structural Dynamics*, Wiley, Chichester, 1994.

Hexcel Corporation, *Technical Data Sheet HexPly913*, (http://www.hexcel.com/NR/rdonlyres/D2E16C53-1991-4DD5-8EA2-277ED3323147/0/HexPly_913_eu.pdf), 2007.

R.L. Karkkainen and B.V. Sankar. A direct micromechanics method for analysis of failure initiation of plain weave textile composites, *Composites Science and Technology*, Vol. 66, 137–150, 2006.

A.B.H. Kueh and S. Pellegrino. ABD matrix of single-ply triaxial weave fabric composites, *48th AIAA/ASME/ASCE/AHS/ASC Structures, Structural Dynamics, and Materials Conference*, Honolulu, Hawaii, AIAA-2007-2161, 2007.

K. Kwok and S. Pellegrino. Micromechanical modeling of deployment and shape recovery of thin-walled viscoelastic composite space structures, *53rd AIAA/ASME/ASCE/AHS/ASC Structures, Structural Dynamics and Materials Conference*, Honolulu, Hawaii, AIAA-2012-1910, 2012.

K. Kwok and S. Pellegrino. Folding, stowage and deployment of viscoelastic tape springs, *AIAA Journal*, Vol. 51(8), 1908-1918, 2013.

J. Lysmer and R.L. Kuhlemeyer. Finite dynamic model for infinite media, *Journal of the Engineering Mechanics Division, Proceedings of the American Society of Civil Engineers*, Vol. 95, No. 4, 859–878, 1969.

H.J. Luhmann, C.C. Etzeler, and R. Wagner. Design and verification of mechanisms for a large foldable antenna. *Proceedings 23rd Aerospace Mechanisms Symposium*, Marshall Space Flight Center, Huntsville, Alabama pp 113-126. NASA-CP-3032, 1989.

H.M.Y.C. Mallikarachchi and S. Pellegrino. Quasi-static folding and deployment of ultra-thin composite tape-spring hinges, *Journal of Spacecraft and Rockets*, Vol. 48, 187-198, 2011.

H.M.Y.C. Mallikarachchi and S. Pellegrino. Failure criterion for two-ply plain weave CFRP laminates, *Journal of Composite Materials*, Vol. 47(11), 1357-1375, 2013

H.M.Y.C. Mallikarachchi, and S. Pellegrino. Deployment dynamics of ultrathin composite booms with tape-spring hinges. *Journal of Spacecraft and Rockets*, 51(2): 604-613, 2014.

H.M.Y.C. Mallikarachchi and S. Pellegrino. Design of ultrathin composite self-deployable booms. *Journal of Spacecraft and Rockets*, Vol. 51(6), 1811-1821, 2014.

M. Mobrem and D.S. Adams. Deployment analysis of lenticular jointed antennas onboard the Mars Express spacecraft, *Journal of Spacecraft and Rockets*, Vol. 46, 394-402, 2009.

Y.G. Panovko and I. I. Gubanova. *Stabilities and oscillations of elastic systems: paradoxes, fallacies and new concepts.* Consultants Bureau, New York, 1965.

S. Pellegrino. Large retractable appendages in spacecraft. *Journal of Spacecraft and Rockets*, Vol. 32, 1006-1014, 1995.

S. Pellegrino, C. Green, S.D. Guest, and A.M. Watt. *SAR Advanced Deployable Structure.* Department of Engineering, University of Cambridge, CUED/D-STRUCT/TR191, 2000.

F.P.J. Rimrott. Storable tubular extendible members. *Engineering Digest*, September 1966.

F.P.J. Rimrott. Self-extension of an open section tube. In W. T. Koiter & G. K. Mikhailov, editors, *Theory of Shells*, 495–508. North-Holland Publishing Co. 1980.

K. Schulgasser. Configuration of a bent tape of curved cross-section. *Transactions of the ASME, Journal of Applied Mechanics*, 59, 692-693, 1992.

K.A. Seffen and S. Pellegrino. Deployment dynamics of tape springs, *Proceedings of the Royal Society of London Series A*, Vol. 455, 1003–1048, 1999.

C. Sickinger, L. Herbeck, T. Strohelin, and J. Torrez-Torres. Lightweight Deployable Booms: Design, Manufacture, Verification, and Smart Materials Application, *55th International Astronautical Congress, IAF/IAA/IISL*, Vancouver, Canada, 2004.

O. Soykasap, A. Watt and S. Pellegrino. New deployable reflector concept, *45th AIAA/ASME/ASCE/AHS/ASC Structures, Structural Dynamics, and Materials Conference*, Palm Springs, California, AIAA 2004-1574, 2004.

O. Soykasap. Micromechanical models for bending behaviour of woven composites, *Journal of Spacecraft and Rockets*, Vol. 43, 1093-1100, 2006.

O. Soykasap, S. Pellegrino, P. Howard, and M. Notter. Folding large antenna tape spring, *Journal of Spacecraft and Rockets*, Vol. 45(3), 560-67, 2008.

O. Soykasap. Deployment analysis of a self-deployable composite boom, *Composite Structures*, Vol. 89, 374-381, 2009.

W. Szyszkowski, K. Fielden, and D.W. Johnson. Self-locking satellite boom with flexure-mode joints. *Applied Mechanics Reviews*, Vol. 50, S225-S231, 1997.

S.P. Timoshenko and J.H. Gere. *Theory of Elastic Stability.* McGraw-Hill, New York, 1961.

The MathWorks Inc., Matlab Version R2010a, Natick, MA.

Torayca, *Technical Data Sheet No. CFA-001, T300 Data Sheet* (http://www.toraycfa.com/pdfs/T300DataSheet.pdf), 2015.

P.A. Warren. *Foldable member*, US Patent 6,374,565 B1, 2002.

P.A. Warren, B.J. Dobson, J.D. Hinkle, and M. Silver. Experimental characterization of lightweight strain energy deployment hinges. *46th AIAA/ASME/ASCE/AHS/ASC Structures, Structural Dynamics & Materials Conference*, Austin, Texas, AIAA 2005-1809, 2005.

W. Wuest. Einige Anvendungen der Theorie der Zylinderschale. *Z. angew. Math. Mech.* Vol. 34, 444-454, 1954.

J.C.H. Yee, O. Soykasap, and S. Pellegrino. Carbon fibre reinforced plastic tape springs. *45th AIAA/ASME/ASCE/AHS/ASC Structures, Structural Dynamics and Materials Conference*, Palm Springs, CA, AIAA 2004-1819, 2004.

J.C.H. Yee and S. Pellegrino. Composite tube hinges. *Journal of Aerospace Engineering*, Vol.18(October), 224-231, 2005.

A short introduction to morphoelasticity: the mechanics of growing elastic tissues

A. Erlich, Th. Lessinnes, D. E. Moulton, and A. Goriely
Mathematical Institute, University of Oxford, UK

Abstract Growth is a key process in the life and development of all biological organisms and depends on a number of genetic, biochemical, environmental, and mechanical factors. In particular, growth can be affected by mechanical stresses and, in turn, generate new stresses to modify shape, create patterns, and tune the overall response of the tissue. From a mathematical perspective, the modelling of growth processes and in particular its interplay with mechanics is particularly challenging since, unlike traditional mechanical systems, the reference state where key physical quantities need to be evaluated evolve with time. An extra difficulty comes from the fact that the geometry of the object also evolves in time due to the addition of mass. From a modelling perspective, it is particularly important to isolate these effects as they are generated by different processes. In this short introduction, we first give a general overview of the problem of biological growth. Second, the mathematical problem of growth modelling for biological system is considered and illustrated on a number of examples starting with simple one-dimensional systems. Third, we present the general framework of nonlinear morphoelasticity to describe the mechanical response of growing elastic tissues and the remodelling of material properties. The first few introductory pages of these lecture notes are reproduced from (Goriely and Moulton, 2010)

1 Introduction

Growth is the process by which a body increases in size through the addition of mass. In biological systems, growth can serve a number of different purposes, and occurs in many different forms. Growth may be restricted to particular locations on the body. In particular, tip growth is often found in microscopic filamentary systems (Gooday and Trinci, 1980; Howard and Valent, 1996; Goriely and Tabor, 2003). Surface growth and accretion refers to the deposition of material on the surface of a body - this type of growth

D. Bigoni (Ed.), *Extremely Deformable Structures*,
CISM International Centre for Mechanical Sciences
DOI 10.1007/ 978-3-7091-1877-1_6 © CISM Udine 2015

is found in the formation of horns, teeth, and seashells (Skalak and Hoger, 1997; Thompson, 1992). In volumetric growth, on the other hand, growth occurs throughout the bulk of the body. This is common in the growth of hearts, tumors, arteries, the brain and airways (Taber and Humphrey, 2001; Cowin, 2004; Budday et al., 2014; Moulton and Goriely, 2011).

Continuum mechanics and nonlinear elasticity provide a natural framework to study growth. Of foremost importance is capturing the correct relationship between growth and elasticity. The basic idea is that in a growth process, the deformation of a body can be due to both a change of mass, and an elastic response. This concept, first put on mathematical terms by Rodriguez et al. Rodriguez et al. (1994), states that the deformation tensor can be decomposed into a growth tensor instructing different parts of the body how to add mass locally and an elastic tensor which captures the elastic reorganization necessary to ensure integrity of the body.

A key benefit to studying growth through continuum mechanics is that it provides direct access to stresses. The relationship between stress and growth is a key concept in the theory of elastic growth. Stress may induce growth, for instance a body may grow until a homeostatic "target" stress is reached. On the other hand, growth can induce residual stress, in particular when the body is inhomogeneous or growing differentially. Residual stresses are stresses that persist in the body even after all body forces have been removed. They are known to play an important role in the functioning of a body. For instance, residual stress is key in the regulation of the transmural stress in arteries (Humphrey, 2003), and is also known to occur in tree and plant tissues (Vandiver and Goriely, 2008), leading to tension or compression woods. A particularly spectacular example of these internal stresses is found in the "Wapas" trees in French Guiana. Due to residual stress, these trees explode when cut, giving them the nickname "killing trees" (Détienne and Thiel, 1988). More generally, the generation of residual stress is connected to the question of how growth can alter the properties of a material (Goriely and Ben Amar, 2005; Ben Amar and Goriely, 2005).

These lecture notes are structured as follows. We begin with a 1D theory of growth, providing several examples to illustrate the basic concepts and some of the questions that arise in morphoelasticity (a complete treatment of one-dimensional growth is presented in (Moulton et al., 2012; Lessines et al., 2014)). Then, we develop the full 3D theory from first principles. In particular we discuss in details issues related with the thermodynamics of growing materials.

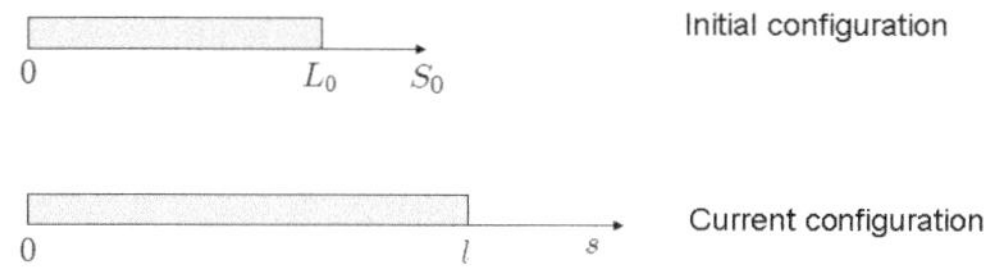

Figure 1. One-dimensional deformation of a rod.

2 1D growth

In this section we develop the theory of elastic growth in one dimension. Consider a rod of length L_0 in its initial reference configuration which undergoes a deformation so that it is of length l in the current configuration. Let S_0 describe position along the rod in the initial configuration and s position in the current configuration - see Figure 1. We define the geometric stretch

$$\lambda = \frac{\partial s}{\partial S_0}. \tag{1}$$

In general $\lambda = \lambda(S_0)$, and λ is constant for a uniform stretch. Also, $\lambda > 1$ corresponds to elongation and $\lambda < 1$ to a reduction. We next consider several classes of deformation.

2.1 Pure elastic deformation

In a pure elastic deformation, $\lambda = \alpha$, where $\alpha = \frac{\partial s}{\partial S}$ is the *elastic stretch*. Letting σ denote the stress, the characteristics of the deformation are determined by a constitutive relationship between σ and α. For example, "Hooke's Law" for a linear material gives $\sigma = E(\alpha - 1)$, where E is the Young's modulus. Solving for α and inserting in (1), the deformation is described by

$$s = \left(\frac{\sigma}{E} + 1\right) S_0, \tag{2}$$

from which we can write $l = \left(\frac{\sigma}{E} + 1\right) L_0$. In this case, the length increases linearly with the stress. A more accurate description is given by the nonlinear neo-Hookean relationship $\sigma = \mu(\alpha^2 - \alpha^{-1})$, where $\mu > 0$ is the elastic modulus. This relationship is derived in the contribution by Ogden and Saccomandi in this volume.

2.2 Pure growth

A pure growth is characterized as having no elastic response. Here, $\lambda = \gamma$, where $\gamma = \frac{\partial S}{\partial S_0}$ describes the growth. Whereas in the previous section α describes a pure elastic *stretch*, γ describes the *addition* or *removal* of material. If $\gamma > 1$ there is growth while $\gamma < 1$ indicates shrinking or resorption of material. In general, the material may undergo growth or resorption at different times and at different rates, and a constitutive relation is needed. For pure growth, the constitutive law is of the form

$$\frac{\partial \gamma}{\partial t} = G(\gamma, s, S_0). \tag{3}$$

In the simple case of uniform growth, $G = 1$, which gives $\gamma = t$ and $s = S_0 t$. Exponential growth is captured by $G = k\gamma$. Then $s = S_0 e^{kt}$ and the length of the rod grows as $l(t) = L_0 e^{kt}$. Alternatively, suppose that growth only occurs towards the tip. For an exponentially growing tip, we take

$$\frac{\partial \gamma}{\partial t} = G(\gamma, s) = \begin{cases} k\gamma & 0 < s < a \\ 0 & a < s < l \end{cases} \tag{4}$$

Equation (4) is coupled with $\frac{\partial s}{\partial S_0} = \gamma$ and initial conditions $s(0) = S_0$ and $\gamma(0) = 1$. The difficulty computationally is that $s = s(S_0, t)$ is a moving interface. To understand the behavior of the growing tip, consider a discrete numeric process. At time t_i, the rod will grow according to $\gamma_i = e^{kt_i}$. How much does the rod grow in one time step, from t_i to t_{i+1}? Letting s_n be the current configuration at the nth step, we have

$$e^{kt_i} = \gamma_i = \frac{\partial s_{i+1}}{\partial S_0} = \frac{\partial s_{i+1}}{\partial s_i}\frac{\partial s_i}{\partial S_0} = \frac{\partial s_{i+1}}{\partial s_i} e^{kt_{i-1}}. \tag{5}$$

This implies that $\frac{\partial s_{i+1}}{\partial s_i} = e^{k(t_i - t_{i-1})} = e^{k\Delta t}$. Thus the region $0 < s_i < a$ grows by a factor $e^{k\Delta t}$, meaning that the rod extends by an amount $\Delta l = a(e^{k\Delta t} - 1)$. The important thing to note is that this is independent of the actual time, *i.e.* the rod extends by the same amount at every time step. Even though the tip is growing exponentially, the length of the rod is only increasing *linearly.*

2.3 Example: the kinematics of tumour growth

As a second example, we present a toy model for tumor growth. Our aim is to write a model of tumour growth for a three dimensional tumour considered as a growing spheroid. Experimentally, cancer cells can be cultured

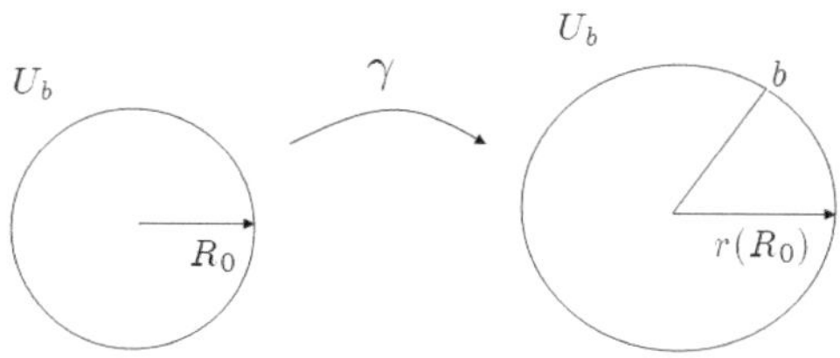

Figure 2. Setup for the tumor growth toy model.

in vitro to grow on a gel or in a pool of nutrients and form a sphere-like shape, the spheroid Sutherland (1988); Freyer et al. (1990). These spheroids are of great interest as they can be used in a variety of settings to understand the response of cancer cells under different physical, chemical, and genetic treatments Santini and Rainaldi (2000); Helmlinger et al. (1997). Mathematically, the spherical symmetry of these aggregates allows for a reduction of the problem difficulty by only considering the evolution of the radius (Franko and Freedman, 1984; Adam and Maggelakis, 1990; Byrne and Chaplain, 1995; Friedman, 2006; Roose et al., 2007).

The setup is depicted in Figure 2. The tumor is surrounded by a bath of nutrient, which is transmitted to the tumor through diffusion with uptake. We first write a one dimensional model. And consider the problem of the growth of a line due to nutrient intake. The centre of tumour is represented by the side of the line at $S_0 = 0$. Nutrient diffuses in from the tip at $S_0 = L_0$. We first assume that the tumour is allowed to growth without being compressed by surrounding healthy tissues. The line concentration of nutrient is labelled $u(s,t)$. The concentration of nutrient in healthy tissues is labelled u_0. Its initial reference length is parameterised by S_0. Any region of the tumour (and in particular the one which is between 0 and S) grows proportionally to its nutrient intake:

$$\partial_t S(S_0, t) = \kappa \int_0^{s(S_0,t)} u(x,t)\mathrm{d}x, \tag{6}$$

where S parameterised the reference arc-length of the tumour and s its current arc length. Taking one partial derivative of (6) by S_0 leads to the local law:

$$\partial_t \gamma = \kappa\, \alpha\, \gamma\, u. \tag{7}$$

The nutrient dynamics is assumed to follow Fick's law of diffusion with a sink term Q accounting for the nutrient consumed by the tumour:

$$u_t = Du_{ss} - Q. \tag{8}$$

We distinguish Phase 1 during which the nutrient fully permeates the tumour from Phase 2 in which the tumour has grown so large that the nutrient no longer reaches its core. As a result, a necrotic core forms and only the periphery of the tumour receives nutrient. We do not model a later phase where vascularisation occurs.

In Phase 1, the boundary conditions are $u(l) = u_0$, and $u_x(0) = 0$ accounting the no-flux condition through the core of the tumour. Assuming that the characteristic time scale for diffusion is much shorter than that of growth, we solve (8) independently of (7) and find

$$u(s,t) = \begin{cases} u_0 & \text{if } s > l, \\ \frac{u_l}{a^2}\left(s^2 + a^2 - l^2\right) & \text{if } s \leq l, \end{cases} \tag{9}$$

where l is the position of the front of the tumour and $a = \sqrt{\frac{2Du_0}{Q}}$ is the permeation length of the nutrient.

Substituting (9) in (6) leads to

$$\begin{aligned} \partial_t S &= \frac{\kappa\, u_0}{a^2} \int_0^S \left(x^2 + a^2 - l^2\right) \mathrm{d}x \\ &= \frac{\kappa\, u_0}{a^2} \left(\frac{S^3}{3} + (a^2 - l^2)S\right). \end{aligned} \tag{10}$$

Specialising (10) to the tumour front ($S = l$) gives

$$\partial_t l = \kappa u_0 l \left(1 - \frac{2}{3}\frac{l^2}{a^2}\right). \tag{11}$$

Eq. (11) indicates that growth is exponential in Phase 1.

When $l > a$, the nutrient does not diffuse to the inner most part of the tumour and a necrotic core forms. The point $l = a$ marks the onset of Phase 2. At this point, a new concentration profile occurs defined by the boundary conditions $u(l) = u_0$ and $u(l - a) = 0$. The concentration profile becomes

$$u = \begin{cases} u_0 & \text{if } x > l, \\ \frac{u_0}{a^2}\left(x - (l - a)\right)^2 & \text{if } l - a \leq x < l, \\ 0 & \text{if } x < l - a, \end{cases} \tag{12}$$

where l and a are defined as before and are once again $u(x)$ is C^1 by construction. Also note that the profiles (9) and (12) are identical when $l = a$.

Substituting (12) in (6) and specialising to the tumour edge ($S = l$) leads to

$$\begin{aligned} \partial_t l &= \frac{\kappa\, u_0}{a^2} \int_{l-a}^{l} \Big(x - (l-a) \Big)^2 \,\mathrm{d}x \\ &= \frac{\kappa u_0 a}{3}, \end{aligned} \tag{13}$$

which is explicitly solved by

$$l(t) = a + \frac{\kappa\, u_0 a}{3}(t - t_{\mathrm{cr}}), \tag{14}$$

where t_{cr} is the time at which the exponential growth satisfying (11) reaches $l = a$ whence subsequent growth obeys the linear law (14).

Although the simple model (7,8) correctly predicts the apparition of a necrotic core it can easily be improved. We now show how to modify (7,8) to take into account the three-dimensional shape of the tumour. Referring back to the sketch 2, we define $r(R_0, t)$ as the radial position for a solid sphere in the current configuration, where R_0 is the radial position in the initial configuration, and $b(t)$ as the radius in the current configuration. We assume that growth is exponential but proportional to the concentration of the nutrient. Hence, growth is modelled by

$$\begin{aligned} \frac{\partial \gamma}{\partial t} &= k\gamma u(r) \\ \frac{\partial r}{\partial R_0} &= \gamma. \end{aligned} \tag{15}$$

The three dimensional Fick's law reads in spherical coordinates

$$\frac{\partial u}{\partial t} = \frac{D}{r^2}\frac{\partial}{\partial r}\left(r^2 \frac{\partial u}{\partial r} \right) - Q, \tag{16}$$

where D is the diffusion coefficient and Q is the nutrient uptake (assumed to be constant). The outer boundary condition for $u(r,t)$ is $u(b,t) = U_b$, that is the bath provides a constant nutrient supply to the outer surface. The other boundary condition, and the growth behavior, depends on the size of the tumor. If b is small enough so that $u(r)$ can diffuse all the way to the center of the sphere, the boundary condition is $0 \le u(0) < \infty$. For some b_{cr}, the diffusion satisfies $u(0) = 0$. For $b > b_{cr}$, the boundary condition is $u(b - a) = 0$, where $a > 0$ is the penetration length of the nutrient. This is shown schematically in Figure 3.

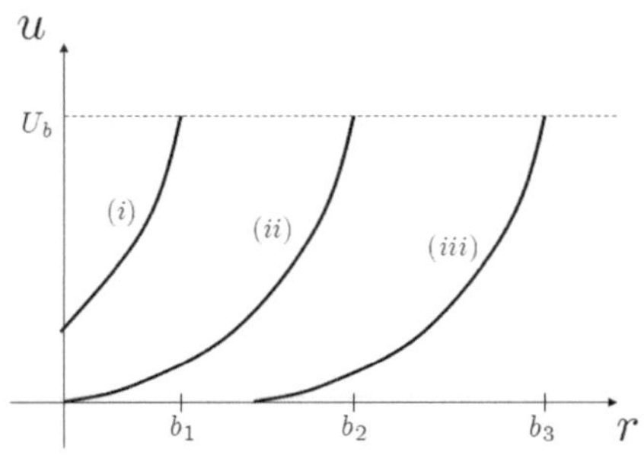

Figure 3. The concentration of the nutrient $u(r)$ as a function of r for different sized tumors. In (i), the tumor radius b_1 is small enough for the nutrient to diffuse all the way to the center, and growth is exponential. In (ii), the radius is a critical value b_{cr} such that $u(0) = 0$. In (iii), $b > b_{cr}$ – here growth only occurs in the region where $u(r) > 0$ and so the growth is linear.

For $b < b_c$, *i.e.* when the tumor is "small", the entire sphere grows and growth is therefore exponential. After this initial phase, growth is restricted to a spherical shell and so is equivalent to the tip growth problem discussed above. Thus, as a function of time, the size of the tumor $b(t)$ increases exponentially at first and then transitions to a linear rate.

Apart from the effect of nutrient, it is believed that mechanical stress has also an effect on growth as high stress may limit the ability of cells to divide through contact inhibition Helmlinger et al. (1997); Hamilton (1998). We will consider this effect in the next Section.

2.4 Growth with elastic response

In an elastic body, growth is subject to an elastic response. The fundamental assumption of morphoelasticity, to be justified later, is that the geometric stretch $\lambda = \frac{\partial s}{\partial S_0}$ is the product of an elastic term α and a growth term γ. That is,

$$\lambda = \alpha\gamma, \tag{17}$$

where α satisfies some constitutive relationship with the stress, $H(\alpha, \sigma) = 0$, and γ satisfies a growth equation $\partial_t \gamma = G(\alpha, \gamma, \sigma)$.

To give an example, consider a rod growing between two walls. We take the growth to be linear and uniform, so that $\gamma = 1 + t$, and use the linear

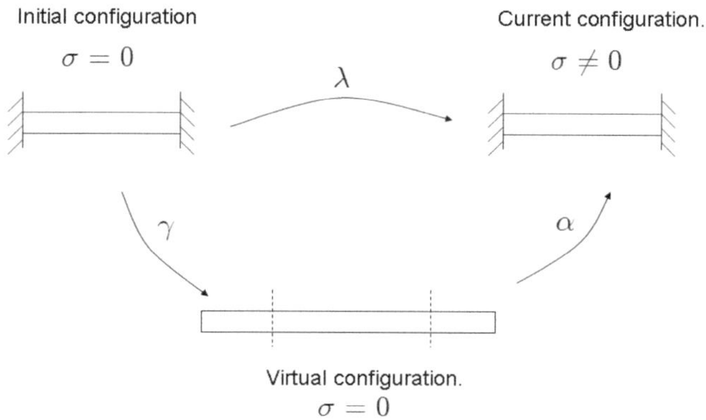

Figure 4. Decomposition of the deformation into a growth step which leads to an incompatible, virtual configuration, and an elastic response step, which fits the rod back into a physically compatible state and introduces stress.

Hookean model for the elastic relationship: $\sigma = E(\alpha - 1)$. Due to the wall constraint, we must have $\lambda = 1$. This implies $\alpha = 1/\gamma$, from which we obtain

$$\sigma = -E\left(\frac{t}{1+t}\right). \tag{18}$$

Essentially, as the rod grows, a compressive stress builds up to contain the rod in the fixed space. Observe from (18) that as $t \to \infty$, $\sigma \to -E$. This is problematic in that it states that only a finite stress is developed in compressing an infinite rod. The problem is due to the linearity assumption. If instead the non-linear neo-Hookean relationship $\sigma = \mu\left(\alpha^2 + \alpha^{-1}\right)$ is used, we obtain

$$\sigma = \mu\left(\frac{1}{1+t^2} - t + 1\right). \tag{19}$$

For the non-linear model, there is nothing that slows down the growth and $\sigma \to -\infty$ as $t \to \infty$.

To better understand the relationship between growth and elasticity, it is useful to consider a two-step process, pictured in Figure 4. Starting from the initial, stress-free configuration, in Step 1 the rod *grows* to a configuration which is stress free but not compatible because it does not fit within the walls. Hence this configuration is called a virtual configuration. Step 2 is the elastic response, which may be thought of as the problem of fitting the rod back in between the walls. This step introduces stress, and maps the rod from the virtual configuration to the current configuration.

In many situations, the growth of a body depends on the stress level within the body. For instance, the body may function best at a particular stress, and grow until this target stress is reached. This situation can be modelled by

$$\frac{\partial \gamma}{\partial t} = k\gamma(\sigma - \sigma^*) \tag{20}$$

where σ^* is the target stress or *homeostatic stress.* Consider the rod between two walls geometry and a neo-Hookean material with $\sigma = \mu\left(\alpha^2 - \alpha^{-1}\right)$. As before, $\gamma = 1/\alpha$, and (20) may be recast in terms of α as

$$\frac{\partial \alpha}{\partial t} = \alpha\sigma^* - \mu(\alpha^3 - 1). \tag{21}$$

Setting the right hand side to zero, define α^* as the solution to $\alpha\sigma^* = \mu(\alpha^3 - 1)$, the equilibrium elastic stretch.

Assuming that the initial configuration is stress free, $\alpha(0) = 1$. If $\sigma^* > 0$, $\alpha^* > 1$ and so α will increase up to α^* as $t \to \infty$. This corresponds to γ decreasing from 1 to $\gamma^* = 1/\alpha^*$. In this case, the target stress is tensile, and the rod contracts until the target stress is reached. On the other hand, if $\sigma^* < 0$, the target stress is compressive. Here, $\alpha^* < 1$, and so α decreases to α^*; that is, the rod grows until the proper compression is reached.

2.5 Example: the mechanics of tumour growth

We now revisit the one-dimensional model of tumour growth proposed in Sec. 2.3. In a one dimensional model, we assume that the centre of the tumour is at $s = 0$. In Phase 2 a necrotic core extends until $s = l - a$ followed by growing tissues covering the interval $s \in [l - a, l]$. These two layers compress the healthy tissue trapped between the tumour and the external medium modelled as a fixed wall at distance $d > l$ from the centre of the tumour. The setup is similar to Phase 1 but there are now only two layers since the necrotic core has not yet formed. We allow for necrotic, growing, and healthy tissues to have different mechanical properties. In 1D and in the absence of distributed loads, the stress σ is constant along the length of the structure. We modify the growth law (7) of the active layer to account for the effect of stress on growth:

$$\partial_t \gamma = \kappa\, \alpha\, \gamma\, u \left(1 - \frac{\sigma}{\sigma^\star}\right) H\left(1 - \frac{\sigma}{\sigma^\star}\right), \tag{22}$$

where H is the Heaviside function and $\sigma^\star$ is a critical homeostatic level of stress beyond which growth stops.

The only constitutive assumption needed is that σ is a monotonic function of α. Since necrotic, growing, and healthy tissues may have different constitutive laws, we consider three functions: $\alpha_n(\sigma)$ (in the necrotic tissue), $\alpha_t(\sigma)$ (in the tumour) and $\alpha_h(\sigma)$ (in the healthy tissue).

We restate the problem in dimensionless variables according to

$$\begin{aligned} t &= \frac{t^\star}{\kappa u_0}; \qquad \ell = a\ell^\star; \qquad L_0 = L_0^\star a; \qquad L = aL^\star; \\ l &= al^\star; \quad \sigma = n\,|\sigma^\star|; \quad u = u_0\, u^\star \end{aligned} \tag{23}$$

and drop the *s after rescaling. For instance (22) simplifies to

$$\partial_t \gamma = \alpha\, \gamma\, u\,(1+n)\, H\,(1+n)\,. \tag{24}$$

In Phase 1, the non-dimensionnalised version of (9) is

$$u = \begin{cases} 1 & \text{if } x > l, \\ x^2 + 1 - l^2 & \text{if } x \le l, \end{cases} \tag{25}$$

Note that since the stretches α_t in the tumour and α_h in the healthy tissue are themselves independent of S, we have

$$d = \int_0^L \alpha_t \mathrm{d}S + \int_0^{d-L_0} \alpha_h \mathrm{d}S = \int_0^L \alpha_t \mathrm{d}S + \int_0^{L_h} \alpha_h \mathrm{d}S = L\,\alpha_t(n) + L_h\,\alpha_h(n), \tag{26}$$

which gives an implicit equation for n in function of L. We also have

$$\begin{aligned} \partial_t L &= \partial_t \int_0^{L_0} \gamma \mathrm{d}S_0, \\ &= H(1+n)\,(1+n) \int_0^{L_0} \alpha\gamma u \mathrm{d}S_0, \\ &= H(1+n)\,(1+n) \int_0^{l} u \mathrm{d}s, \\ &= H(1+n)\,(1+n) \int_0^{\alpha_t L} s^2 + 1 - (\alpha_t L)^2 \mathrm{d}s, \\ &= H(1+n)\,(1+n)\,\alpha_t L \left(1 - \frac{2}{3}\alpha_t^2\, L^2\right). \end{aligned} \tag{27}$$

Eqs. (26, 27) together with the constitutive laws for both healthy and tumour tissues form a closed system.

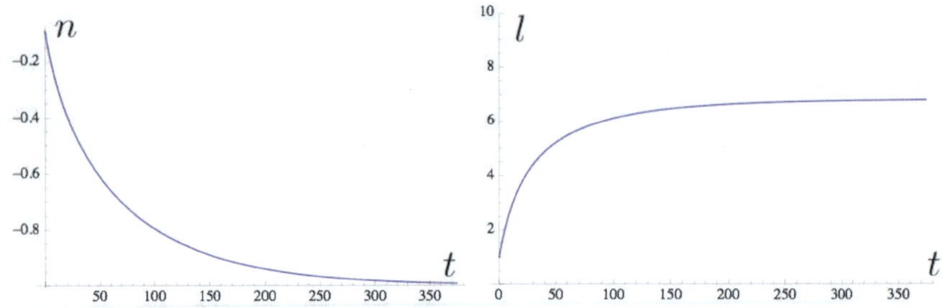

Figure 5. Stress $n[|\sigma^\star|]$ (left) and position of the front $l[a]$ (right) in function of time. For a tumour growing towards an homeostatic stress $\sigma^\star < 0$ when all three tissues are assumed to be Neo-Hookean: $\sigma = \mu_i/3(\alpha_i^2 - 1/\alpha_i)$ where i takes value in $\{n, t, h\}$ and we assumed as an example that $\mu_n = \mu_h = \mu_t = -\sigma^\star$.

In Phase 2 the non-dimensional nutrient profile is

$$u = \begin{cases} 1 & \text{if } x > l, \\ \left(x + 1 - l\right)^2 & \text{if } l - 1 \leq x < l, \\ 0 & \text{if } x < l - 1, \end{cases} \tag{28}$$

so that

$$\begin{aligned} \partial_t(L_n + L) &= H(1+n)\,(1+n)\int_0^l u\mathrm{d}s, \\ &= H(1+n)\,(1+n)\int_{\alpha_n L_n}^{\alpha_n L_n + 1} (s - \alpha_n L_n)^2 \mathrm{d}s, \\ &= H(1+n)\,\frac{1+n}{3}. \end{aligned} \tag{29}$$

Eq. (26) becomes

$$d = L_n \alpha_n + 1 + \alpha_h L_h, \tag{30}$$

since the growing tissue extends over a current length of

$$\alpha_t L_t = 1. \tag{31}$$

Eqs. (29-31) together with the constitutive laws for each subregions constitute a closed system of equations. The initial value for the differential equation (29) is $L_n(0) = 0$. A numerical solution of this system is given in Fig. 5.

For large times, the stress asymptotically reaches the (dimensional) value $\sigma^\star$. Accordingly, the healthy tissue is asymptotically stretched to $\alpha_h(\sigma^\star)$ and the front of the tumour reaches

$$l \overset{t\to\infty}{\to} d - \alpha_h(\sigma^\star)L_h. \tag{32}$$

We conclude that before vascularisation, the final size of the tumour is dictated by the available space d, the constitutive law of *the healthy tissue*, and the homeostatic stress.

3 3D growth

To appropriately model the 3D growth of soft biological tissues aspects of growth and elasticity should be incorporated into a theoretical framework. In their seminal paper, Rodriguez *et al.* (Rodriguez et al., 1994) proposed a multiplicative decomposition of the deformation gradient into growth and elastic parts.

Their work was based on similar ideas from plasticity: The Kröner-Lee decomposition (Kröner, 1960; Lee, 1969) which splits the deformation gradient into an elastic and a plastic part. In this Section, we formally introduce the theory of morphoelasticity, which is an formalisation of Rodriguez *et al.*'s idea. This introduction, much like a traditional derivation of basic equations of nonlinear solid mechanics, will include a part on kinematics (geometric deformation, *i.e.* the study of strains, see Section 3.1), balance laws (balance of forces and other physical quantities, see Section 3.2) and constitutive relationships (which relate strains to stresses, see Section 3.3).

A challenging aspect of morphoelasticity is that (as opposed to traditional nonlinear solid mechanics) morphoelastic systems can be open to fluxes of mass, energy, momentum, and entropy. These quantities must be accounted for carefully. This will be the subject of Section 3.2. It will result in a continuum formulation of the second law of thermodynamics known as the Clausius-Duhem inequality. This formalism will be used Section 3.5 to derive general statements about growth laws.

3.1 Kinematics

In traditional continuum mechanics it is common to consider the body in two different configurations, the *initial reference configuration* and the *current configuration.* The initial reference configuration describes the body in its initial state at $t = 0$ (*i.e.* before deformation) and is required to be

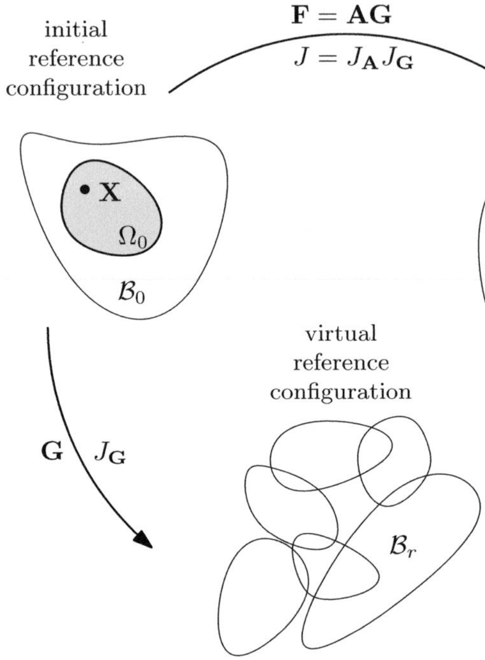

Figure 6. Deformation gradient decomposition. The fact that the virtual reference configuration $\mathcal{B}_r$ does not fit into Euclidean space is highlighted.

stress-free. The current configuration describes the deformed body at time t. Mathematically, they are subsets of the Euclidean space, denoted $\mathcal{B}_0 \in \mathbb{E}^3$ and $\mathcal{B}_t \in \mathbb{E}^3$, respectively. Material points in $\mathcal{B}_0$ are described by the the vector $\mathbf{X}$ which are mapped to $\mathbf{x}$ in $\mathcal{B}_t$. This deformation map is

$$\varphi : \ \mathcal{B}_0 \to \mathcal{B}_t \qquad \mathbf{X} \mapsto \mathbf{x} = \varphi(\mathbf{X}, t), \tag{33}$$

for which we assume the inverse φ^{-1} to exist. Strain, the geometric deformation of a body $\mathcal{B}_t$ with respect to its reference state in $\mathcal{B}_0$, is quantified through the deformation gradient $\mathbf{F}$. The latter is defined as the gradient of the deformation map with respect to initial reference coordinates

$$\mathbf{F}(\mathbf{X}, t) = \nabla_{\mathbf{X}} \varphi(\mathbf{X}, t) = \frac{\partial \mathbf{x}}{\partial \mathbf{X}}(\mathbf{X}, t). \tag{34}$$

As mentioned in the introduction, in morphoelasticity we decompose the deformation gradient into a growth deformation gradient $\mathbf{G}$ and an elastic deformation gradient $\mathbf{A}$:

$$\mathbf{F}(\mathbf{X},t) = \mathbf{AG}. \tag{35}$$

This decomposition requires the introduction of the virtual reference configuration $\mathcal{B}_r \in \mathbb{E}^3$. We define the determinants of the deformation gradients as $J := \det \mathbf{F}$, $J_A := \det \mathbf{A}$ and $J_G := \det \mathbf{G}$ and we have $J = J_A J_G$. If the deformation is isochoric (*i.e.* locally volume preserving), $J_A = 1$.

Formally[1], $\mathbf{G}$ maps vectors from $\mathcal{B}_0$ to $\mathcal{B}_r$ and $\mathbf{A}$ maps from $\mathcal{B}_r$ to $\mathcal{B}_t$ and their composition $\mathbf{F} = \mathbf{AG}$ maps vectors from $\mathcal{B}_0$ to $\mathcal{B}_t$. On a physical level, the decomposition can be interpreted as follows. Growth (geometrically speaking) is a map from a stress-free configuration $\mathcal{B}_0$ to another stress-free configuration $\mathcal{B}_r$, but the latter will in general not fit into Euclidean space as the grown material will overlap or (if locally $\det \mathbf{G} < 1$, *i.e.* resorption) have holes. That is, $\mathbf{G}$ can break the continuity of the tissue. The purpose of the elastic deformation is to restore continuity of the tissue, filling the holes at the expense of introducing tensile stress, and moving apart the overlapping regions at the expense of introducing compressive stress. For this reason, the current configuration $\mathcal{B}_t$ is no longer stress free, even when it is unloaded. Stress that remains even in the absence of loads is called *residual stress.*

The configuration $\mathcal{B}_r$ is not physically realised in a growing biological tissue, and for this reason we refer to it as the *virtual reference configuration.* Because of possible overlaps of tissue, a bijective deformation map of the kind (33) cannot be defined between $\mathcal{B}_0$ and $\mathcal{B}_r$ (*i.e.* it is impossible to keep track of material points if $\mathcal{B}_r$ is not Euclidean) and $\mathbf{G}$ cannot be defined as a gradient of a deformation map in the fashion of (34).

3.2 Balance laws

In this section we state the balance laws for mass, linear momentum, angular momentum, energy and the entropy imbalance. The energy balance and the entropy imbalance are continuum formulations of the first and second law of thermodynamics, which leads to the Clausius-Duhem formulation of the second law.

---

[1] In the main text, we follow a standard introduction of kinematics in morphoelasticity in which $\mathcal{B}_0$, $\mathcal{B}_r$ and $\mathcal{B}_t$ are Euclidean, which leads to restrictions like the lack of a bijective map between $\mathcal{B}_0$ and $\mathcal{B}_r$.

From a differential geometric perspective, $\mathcal{B}_0$, $\mathcal{B}_r$ and $\mathcal{B}_t$ are differentiable manifolds and $\mathbf{F}$, $\mathbf{A}$ and $\mathbf{G}$ are maps from tangent bundles to tangent bundles of the respective manifolds. From this perspective, $\mathcal{B}_r$ is perfectly compatible at the expense of being non-Euclidean. We will not go further into differential geometric aspects here. Detailed treatments can be found in (Marsden and Hughes, 1994) and (Yavari, 2010).

Generalising the conservation equations of traditional continuum mechanics to morphoelasticity and open systems requires careful balance and the correct interpretation of source terms. These equations have been studied by several authors and we will keep our notation relatively close to Epstein (Epstein, 2012).

Note that the overdot denotes derivatives with respect to the time variable. For some quantity a, we denote the time derivative $\dot{a} = \mathrm{d}a/\mathrm{d}t$. Also note that spatial differentiation (divergence, gradient) is always implied with respect to the current variable $\mathbf{x} \in \mathcal{B}_t$ (unless stated otherwise, like in 34).

Slow growth assumptions of morphoelasticity Having stated the kinematic assumptions of morphoelasticity in the previous section, we now introduce some mechanical assumptions.

1. Separation of timescales: The timescale of growth is much slower than that of elasticity.
2. Mass is added from the initial reference configuration $\mathcal{B}_0$ to the virtual reference configuration $\mathcal{B}_r$ and is constant throughout the elastic deformation (which maps from $\mathcal{B}_r$ to $\mathcal{B}_t$).
3. Mass is added through a volumetric source (*i.e.* no mass boundary flux).

The first assumption is typically valid for biological growth with time scale of the order of hours, days, or years, whereas the timescale of elasticity relates to wave propagation in elastic materials which is no larger than seconds.

We will now briefly explore a consequence of the second assumption and explore the third assumption when stating the mass balance.

We assume that mass is added in the growth process from $\mathcal{B}_0$ to $\mathcal{B}_r$, but no mass is added during the elastic deformation from $\mathcal{B}_r$ to $\mathcal{B}_t$. We define $\mathrm{d}M_r = \rho_r \mathrm{d}V_r$ as the mass element in the virtual reference configuration $\mathcal{B}_r$, where ρ_r is the virtual reference mass density and $\mathrm{d}V_r$ is a virtual reference volume element. Similarly, let $\mathrm{d}M_t = \rho \mathrm{d}V_t$ where ρ is the current density. Since no mass is added from $\mathcal{B}_r$ to $\mathcal{B}_t$, $\mathrm{d}M_t = \mathrm{d}M_r$, and with appropriate volume element transformation $\mathrm{d}V_t = J_A \mathrm{d}V_r$ we obtain

$$J_A \, \rho = \rho_r. \tag{36}$$

We now make the additional assumption that ρ_r is constant in time. This assumption implies that the grown mass has the same density (locally) as the surrounding pregrown tissue. This assumption allows for a simple coupling between the mechanics and geometry of growth.

Mass balance We think of a growing tissue as a collection of cells which take nutrients from the extracellular fluid. In our continuum modeling approach, the tissue has a density (mass per volume) $\rho(\mathbf{x}, t)$ in $\mathcal{B}_t$. Nutrient uptake contributes a volumetric growth rate function $\rho(\mathbf{x}, t)\,\gamma(\mathbf{x}, t)$ in $\mathcal{B}_t$. Physical laws are most naturally stated in the current configuration. We will present the mass balance for a volume $\Omega_t \subset \mathcal{B}_t$ on which the mass balance takes the form

$$\frac{\mathrm{d}}{\mathrm{d}t}\int_{\Omega_t} \rho \mathrm{d}V_t = \int_{\Omega_t} \rho\gamma \mathrm{d}V_t. \tag{37}$$

We now transform this balance equation which is integrated over Ω_t into a local, pointwise statement at $\mathbf{x}$, t. Since Ω_t is a region which evolves with time, it will prove useful to transform to the corresponding region $\Omega_0 \subset \mathcal{B}_0$ and evaluate time derivatives in this static region. Volume elements and area elements transform as

$$\int_{\Omega_t} \mathrm{d}V_t = \int_{\Omega_0} J \mathrm{d}V_0. \tag{38}$$

Evaluating time derivatives in $\mathcal{B}_0$ and taking into account that $J_A\,\rho$ is constant in time leads us to the local, pointwise statement

$$\gamma = \mathrm{tr}\left(\dot{\mathbf{G}}\mathbf{G}^{-1}\right). \tag{39}$$

Tht is, for constant mass density, the change in mass of a body to properties of the geometric deformation gradient.

Compatible and incompatible sources As we have discussed regarding the slow growth assumption, new mass (with the same density as the surrounding material) enters the tissue with the local growth rate $\rho\gamma$. This material will come with its own momentum, energy and entropy. One possibility is that the new material has the same properties as (locally) the surrounding tissue. We call these type of sources *compliant*, using the terminology of (Epstein, 2012), and they balance out and do not appear in the local statements of the laws. Alternatively, the new material may have sources that are different from the pregrown material, having for instance locally a higher energy density than the surrounding tissue, or due to microscopic organisation possibly a lower entropy. We call these sources *non-compliant* (denoted with an overbar), and they do appear in the local statements.

Linear momentum balance The linear momentum balance is the balance between rate of change of linear momentum and the sum of the forces. We introduce the volumetric body force density (force per unit current volume) $\mathbf{b}$, the surface traction (force per area) $\mathbf{t}$ which according to Cauchy's postulate can be decomposed into the Cauchy stress $\mathbf{T}$ times the surface normal $\mathbf{n}$, *i.e.* $\mathbf{t} = \mathbf{Tn}$. The non-compliant momentum source (force per volume) is $\overline{\mathbf{p}}$:

$$\frac{\mathrm{d}}{\mathrm{d}t}\int_{\Omega_t} \rho\mathbf{v}dV_t = \underbrace{\int_{\Omega_t} \mathbf{b}\mathrm{d}V_t + \int_{\partial\Omega_t} \mathbf{Tn}\mathrm{d}A_t}_{\text{body and traction forces}} + \underbrace{\int_{\Omega_t} \gamma\rho\mathbf{v}\mathrm{d}V_t}_{\text{compliant}} + \underbrace{\int_{\Omega_t} \bar{\mathbf{p}}\mathrm{d}V_t}_{\text{non-compliant}}\,. \tag{40}$$

After a standard transport and localisation procedure, the local (pointwise) equivalent statement is

$$\rho\dot{\mathbf{v}} = \mathbf{b} + \nabla\cdot\mathbf{T} + \bar{\mathbf{p}}. \tag{41}$$

Angular momentum balance The angular momentum balance is the balance between change in angular momentum and the moments (sometimes also called torques) that balance it. Without loss of generality, we state moments with respect to the coordinate system origin. Since the non-compliant linear momentum source $\overline{\mathbf{p}}$ enters the angular momentum balance, we require no additional non-compliant angular momentum source (it could be include into $\overline{\mathbf{p}}$) and we assume that there is no source of body couple (that is the system is non-polar). We have

$$\begin{aligned}\frac{\mathrm{d}}{\mathrm{d}t}\int_{\Omega_t} \rho\mathbf{x}\times\mathbf{v}\mathrm{d}V_t &= \underbrace{\int_{\Omega_t} \mathbf{x}\times\mathbf{b}\mathrm{d}V_t + \int_{\partial\Omega_t} \mathbf{x}\times\mathbf{Tn}\mathrm{d}A_t}_{\text{moments through body and traction forces}} \\ &+ \underbrace{\int_{\Omega_t} \gamma\rho\mathbf{x}\times\mathbf{v}\mathrm{d}V_t}_{\text{compliant}} + \underbrace{\int_{\Omega_t} \mathbf{x}\times\overline{\mathbf{p}}\mathrm{d}V_t}_{\text{non-compliant}}\,. \end{aligned}\tag{42}$$

Localising the angular momentum balance, leads to the symmetry of the Cauchy stress tensor $\mathbf{T} = \mathbf{T}^T$.

Energy balance (first law of thermodynamics) The total energy is made up of internal energy $\mathcal{E}$ (*i.e.* elastic and chemical energy) and kinetic energy $\mathcal{K}$. The mechanical power $\mathcal{P}$ is the work per time which the body is exposed to through body force forces $\mathbf{b}$ and surface traction $\mathbf{t}$, introduced in the context of linear momentum balance. The heating $\mathcal{Q}$ is composed of a volumetric heat source r (for radiation) and a surface flux $\mathbf{q}$ called the Fourier heat flux. Apart from $\overline{\mathbf{p}}$, the non-compliant energy source $\overline{\varepsilon}$ enters the balance equation:

$$\begin{aligned}\underbrace{\frac{\mathrm{d}}{\mathrm{d}t}\int_{\Omega_t}\rho\left(\varepsilon+\frac{1}{2}|\mathbf{v}|^2\right)\mathrm{d}V_t}_{\dot{\overline{\mathcal{E}(\Omega_t)+\mathcal{K}(\Omega_t)}}} = & \underbrace{\int_{\Omega_t}\mathbf{b}\cdot\mathbf{v}\mathrm{d}V_t+\int_{\partial\Omega_t}\mathbf{n}\cdot\mathbf{T}\mathbf{v}\mathrm{d}A_t}_{\mathcal{P}(\Omega_t)} \\ & +\underbrace{\int_{\Omega_t}\rho r\,\mathrm{d}V_t+\int_{\partial\Omega_t}\mathbf{q}\cdot\mathbf{n}\mathrm{d}A_t}_{\mathcal{Q}(\Omega_t)} \\ & +\underbrace{\int_{\Omega_t}\gamma\rho\left(\varepsilon+\frac{1}{2}|\mathbf{v}|^2\right)\mathrm{d}V_t}_{\text{compliant}} \\ & +\underbrace{\int_{\Omega_t}\overline{\varepsilon}+\overline{\mathbf{p}}\cdot\mathbf{v}\mathrm{d}V_t}_{\text{non-compliant}}\,.\end{aligned} \tag{43}$$

The local (pointwise) statement is

$$\rho\dot{\varepsilon}=\mathbf{T}:\dot{\mathbf{F}}\mathbf{F}^{-1}+\rho r-\nabla\cdot\mathbf{q}+\overline{\varepsilon}. \tag{44}$$

Note that in this derivation, the term $\mathbf{b}\cdot\mathbf{v}$ is be replaced through the linear momentum balance according to (41), which explains why $\overline{\mathbf{p}}$ does not appear in the local statement.

Entropy / free energy imbalances (second law of thermodynamics) According to the entropy imbalance, the change in entropy density η is greater than the heat supply (as defined in the energy balance) divided by the temperature density θ. The non-compliant energy source $\overline{\varepsilon}$ and the entropy source $\overline{\eta}$ also enter the imbalance.

$$\frac{\mathrm{d}}{\mathrm{d}t}\int_{\Omega_t}\rho\eta\mathrm{d}V_t\geq\underbrace{\int_{\Omega_t}\frac{\rho r}{\theta}\mathrm{d}V_t-\int_{\partial\Omega_t}\frac{\mathbf{q}\cdot\mathbf{n}}{\theta}\mathrm{d}A_t}_{\mathcal{Q}(\Omega_t)/\Theta(\Omega_t)}+\underbrace{\int_{\Omega_t}\gamma\rho\eta\mathrm{d}V_t}_{\text{compliant}}+\underbrace{\int_{\Omega_t}\frac{\overline{\varepsilon}-\overline{\eta}}{\theta}\mathrm{d}V_t}_{\text{non-compliant}}, \tag{45}$$

which locally takes the form

$$\rho\dot{\eta}\geq\frac{\rho r}{\theta}-\nabla\cdot\left(\frac{\mathbf{q}}{\theta}\right)+\frac{\overline{\varepsilon}-\overline{\eta}}{\theta}. \tag{46}$$

For our analysis of constitutive relationships, it is useful to apply a Legendre transform to the free energy density $\psi = \varepsilon - \theta\eta$. After applying the time derivative and substituting (44) for the internal energy, we obtain the free energy imbalance

$$\rho\dot{\psi} \leq \mathbf{T} : \dot{\mathbf{F}}\mathbf{F}^{-1} - \rho\dot{\theta}\eta - \frac{\mathbf{q} \cdot \nabla_{\mathbf{x}}\theta}{\theta} + \overline{\eta}. \tag{47}$$

This formulation of the second law of thermodynamics in terms of the free energy density is known as the Clausius-Duhem inequality.

3.3 Constitutive relationships

In the context of nonlinear elasticity, a constitutive law or constitutive relationship relates strains to stresses.[2]. For a hyperelastic material, stresses can be derived from a scalar field called the strain-energy function W. Its independent variables depend on the modeling assumptions at hand. Typical assumptions in the context of modelling biological tissues are:

1. Hyperelastic and isothermal materials, *i.e.* strain energy density depends only on the elastic deformation gradient, $W = W(\mathbf{A})$.
2. Incompressible materials, *i.e.* $\det(\mathbf{A}) = 1$.

For incompressible, isothermal, hyperelastic materials the Cauchy stress $\mathbf{T}$ can be derived from the strain energy density as

$$\mathbf{T}(\mathbf{A}) = \frac{\partial W}{\partial \mathbf{A}}\mathbf{A}^T - p\mathbf{1}, \tag{48}$$

where p is a Lagrange multiplier which enforces incompressibility. Incompressibility is a typical assumption for soft biological tissues.

There is generally very little information on constitutive relationships in soft biological tissues (Taber, 1995), and neo-Hookean materials have proven very useful as modeling starting points. The strain-energy density function of a neo-Hookean solid in three dimensions is $W(\mathbf{A}) = \mu(I_1 - 3)$ where μ is a material parameter and I_1 is the first principal invariant of the right Cauchy-Green strain tensor $\mathbf{C} = \mathbf{A}^T\mathbf{A}$ for which $I_1 = \operatorname{tr}\mathbf{C} = \alpha_1^2 + \alpha_2^2 + \alpha_3^2$ where α_i, $i = 1, 2, 3$ are the three principal elastic stretches. We obtain

$$W(\alpha_1, \alpha_2, \alpha_3) = \mu\left(\alpha_1^2 + \alpha_2^2 + \alpha_3^2 - 3\right). \tag{49}$$

3.4 Growth laws and literature review

In morphoelasticity, the evolution of growth is described as a differential equation in time

[2] Note that in the context of rational thermodynamics, the notion of constitutive relationships carries a wider meaning. This will be discussed in Section 3.5

$$\dot{\mathbf{G}} = \mathcal{G}\left(\mathbf{T}, \mathbf{F}, \mathbf{G}, \boldsymbol{\mu}; t, \mathbf{x}, \mathbf{X}\right),$$

where the dot denotes time differentiation. Here, $\mathbf{F}$ and $\mathbf{G}$ are deformation gradients as described above, $\mathbf{T}$ is the Cauchy stress, $\boldsymbol{\mu}$ can be biochemical or other fields and $\mathbf{x}$ and $\mathbf{X}$ describe positions in the undeformed and deformed body, respectively. The growth law $\dot{\mathbf{G}} = \mathcal{G}$ describes an active growth process which requires energy and adds material, whereas the elastic accommodation (geometrically described by $\mathbf{A}$) is passive, adds no mass and consumes no energy (unless plasticity is involved). The difficulty in stating growth laws is that the observed evolving form and mass of the body are a result of both growth and its elastic accommodation. There is a lack of data for constitutive relationships of biological tissues, let alone successfully retrieved growth deformations $\mathbf{G}$, which makes the growth law $\dot{\mathbf{G}}$ particularly difficult to access.

The study of growth laws is in its infancy. Even the question of whether stress or strain drives growth is not settled, every major review contains a debate on this topic((Taber, 1995; Ambrosi et al., 2011; Jones and Chapman, 2012). Here we take the view that growth laws should be functions of stress $\dot{\mathbf{G}} = \mathcal{G}\left(\mathbf{T}\right)$ as a stress-free configuration can be chosen as reference (a reference configuration for strain is less obvious).

The simplest stress-dependent growth law which incorporates the idea of homeostasis is a linear coupling with Cauchy stress

$$\dot{\mathbf{G}} = \mathbf{G}\mathbf{K}\left(\mathbf{T} - \mathbf{T}^*\right), \tag{50}$$

where $\mathbf{K}$ is a matrix of constant coefficients and $\mathbf{T}^*$ is the homeostatic stress. This was studied in (Ramasubramanian and Taber, 2008). The authors run finite element simulations on variations of two-dimensional beams glued in parallel, finding that (depending on choices of $\mathbf{K}$ and $\mathbf{T}^*$) the system may or may not equilibrate. They also study a delay differential equation version of (50) by replacing $\mathbf{T}\left(\mathbf{x}, t\right)$ with $\mathbf{T}\left(\mathbf{x}, t - \tau\right)$ where τ is a small delay, finding that depending on the delay and the stiffness ratio of beams glued in parallel, either no oscillations, stress oscillations or both stress and curvature oscillations of the beams occur. Finally, the authors study invagination of cylindrical and spherical, fluild-filled or empty shells. Again, they use a finite-element simulation in which the shell consists of passive material with a small striped active region in the shell driving the invagination. They reproduce shapes which somewhat resemble invagination. Overall, their analysis shows that (50) either leads to stationary states of the growing system, or to unrealistic ever-growing systems, depending on the choice of $\mathbf{K}$ and $\mathbf{T}^*$, but stability is not analysed systematically.

In (Taber, 2008, 2009), the previous growth law is modified by adding evolution equations for the target stress:

$$\dot{\mathbf{G}}\mathbf{G}^{-1} = \mathbf{A}\left(\mathbf{T} - \mathbf{T}^*\right) \qquad \dot{\mathbf{T}}^*\left(\mathbf{T}^*\right)^{-1} = \mathbf{B}\left(\mathbf{T} - \mathbf{T}^*\right),$$

where $\mathbf{A}, \mathbf{B}$ are fourth-order tensors with constant entries. The systems analysed in these papers are practically the same as in (Ramasubramanian and Taber, 2008) and the conclusions are similar: For some parameter choices of $\mathbf{A}, \mathbf{B}, \mathbf{T}^*$ all taken at time $(t = 0)$ $\mathbf{T}$ and $\mathbf{T}^*$ converge to the same equilibrium value. In other cases, they converge to different equilibrium values, and in yet other cases the stresses diverge, reaching no equilibrium at all. One fundamental problem with this model is that the equilibrium values depend on the initial condition $\mathbf{T}^*\ (t = 0)$ which should not be the case in robust biological systems.

Based on the thermodynamic argument briefly discussed above, and discussed in more detail in Section 3.5, a growth law driven by Eshelby stress is proposed in (Ambrosi and Guana, 2007)

$$\dot{\mathbf{G}} = \mathbf{K}\left(\mathbf{E} - \mathbf{E}^*\right)\mathbf{G}, \tag{51}$$

where $\mathbf{E}$ and $\mathbf{E}^*$ are Eshelby stress and homeostatic Eshelby stress, respectively, and $\mathbf{K}$ is a matrix of constant coefficients. The proposition that Eshelby stress drives growth is supported by thermodynamical arguments ((Epstein and Maugin, 2000; Ambrosi and Guana, 2007; Ambrosi and Guillou, 2007)).

In (Ambrosi and Guana, 2007), the authors take as their starting point a growth law inspired by cardiovascular mechanics, in which $\mathbf{G}$ is diagonal (*i.e.* no shear) and involves only the circumferential components of Cauchy stress. For diagonal $\mathbf{G}$ and small deformations they show that (51) coincides with (50) up to a sign. By showing results of particular numerical simulations (but without a general stability result), the authors achieve convergence to the homeostatic stress in the case of displacement boundary conditions but obtain a divergent (ever-growing) in the case of prescribed load. Finally, they show that by adding passive layers (analysing a three rather than one-layered system, with inner and outer layers passive), convergence towards homeostasis is also possible when the load is prescribed.

3.5 Thermodynamic restrictions of growth laws

The standard procedure for for deriving restrictions to constitutive laws is known as the Coleman-Noll procedure (Coleman and Noll, 1963). The framework which has been built up around is known as rational thermodynamics. While it is very appealing for its axiomatic construction and mathematical elegance, some of its assumptions are rather unorthodox and have been criticised as *e.g. 'adventurous'* (Maugin, 1999) and *'out of touch with physical reality'* (Lebon et al., 2008). However, for the treatment of thermoelastic solids, rational thermodynamics is quite elegant and produces equations which are consistent with elasticity and traditional (equilibrium) thermodynamics. For this reason, rational thermodynamics is widely used in the continuum mechanics community (Gurtin et al., 2010) and has found application to growing soft biological tissues.

Relevant ideas from rational thermodynamics The most important and useful objective of rational thermodynamics is to place restrictions on the constitutive relationships $\widehat{\Psi}$, $\widehat{\eta}$, $\widehat{\mathbf{T}}$ and $\widehat{\mathbf{q}}$ by means of the Clausius-Duhem inequality in the form (47). The standard procedure by which this is achieved is the Coleman-Noll procedure. Its objective is to restrict the constitutive relationships

$$\widehat{\Psi}\left(\mathbf{A}, \theta, \nabla\theta\right) \qquad \widehat{\eta}\left(\mathbf{A}, \theta, \nabla\theta\right) \qquad \widehat{\mathbf{T}}\left(\mathbf{A}, \theta, \nabla\theta\right) \qquad \widehat{\mathbf{q}}\left(\mathbf{A}, \theta, \nabla\theta\right). \tag{52}$$

It should be said that typically in solid mechanics literature unrelated to thermodynamics, the term 'constitutive relationship' relates to how strains are related to stresses, which is the terminology we use in Section 3.3. For hyperelastic materials, as described in Section 3.3, the constitutive relationship $\widehat{\mathbf{T}}\left(\mathbf{A}\right)$ can in fact be derived from a strain energy density $W(\mathbf{A})$. This is not assumed a priori in rational thermodynamics, but instead will be a result of the treatment of thermoelastic solids.[3] In the context of rational thermodynamics, constitutive relationships characterise a given material response. The Clausius-Duhem inequality, (47), with the constitutive assumptions incorporated and multiplied by J, is

$$J\rho\dot{\widehat{\Psi}} \leq J\widehat{\mathbf{T}} : \mathbf{F}\mathbf{F}^{-1} - J\rho\dot{\theta}\widehat{\eta} - \frac{J}{\theta}\widehat{\mathbf{q}} \cdot \nabla\theta + J\overline{\eta}. \tag{53}$$

It would be fair to ask at this point: If if in this context the Fourier heat flux $\widehat{\mathbf{q}}$ is considered a constitutive relationship, *i.e.* a material response, then why does this not apply to the heat source r in (43)? After all, r is a volumetric heat source (r stands for radiation) and $\widehat{\mathbf{q}}$ represents heat

[3]Both the fact that $\widehat{\Psi}$, $\widehat{\eta}$, $\widehat{\mathbf{T}}$ and $\widehat{\mathbf{q}}$ have the same arguments and the fact that $\nabla\theta$ appears as a constitutive argument (in addition to the obvious thermo-elastic parameters θ, $\mathbf{A}$) are the results of Truesdell's *principle of equipresence.* In the words of G. A. Maugin, "This is simply a precautionary medasure to avoid missing any significant dependence or coupling" ((Maugin, 1999), p. 65). See also (Gurtin et al., 2010) p. 230 for a comment on equipresence.

entering through a volume boundary, see (43). Similarly, if $\widehat{\mathbf{T}}$ is a constitutive relationship, why not the body force $\widehat{\mathbf{b}}$ in (40)? The answer is that rather than prescribing $\mathbf{b}$ and r as constitutive relationships about which restrictions are to be inferred, they are *defined* in terms of the linear momentum balance (40) and energy balance (43), respectively. In other words, $\mathbf{b}$ and r are free parameters making sure that the momentum and energy balance are always satisfied, ruling them out of the restriction process of constitutive laws. As r and $\mathbf{b}$ are usually supplied in applications ($\mathbf{b}$ might be gravity, for instance), rational thermodynamics takes a somewhat *'perplexing attitude'*, in the words of (Lebon et al., 2008), with regard to these balance laws.

To complete the chain of ideas of the Coleman-Noll procedure, we point out that (53) is expected to hold for all admissible thermodynamic processes, which means that it must hold for arbitrary deformation gradients and temperature fields. Since the momentum and energy balance laws are ruled out by the argument in the previous paragraph, the fact that (53) must hold for all $\mathbf{A}$ and θ puts restrictions on the constitutive relationships (52).

Coleman-Noll procedure for morphoelasticity We now use the constitutive assumptions (52) to evaluate the Clausius-Duhem inequality (53). We analyse separately two important terms, $\dot{\widehat{\Psi}}$ and $J\widehat{\mathbf{T}} : \dot{\mathbf{F}}\mathbf{F}^{-1}$. By the chain rule,

$$\dot{\widehat{\Psi}}\left(\mathbf{A},\theta,\nabla\theta\right) = \frac{\partial\widehat{\Psi}}{\partial\mathbf{A}} : \dot{\mathbf{A}} + \frac{\partial\widehat{\Psi}}{\partial\theta} : \dot{\theta} + \frac{\partial\widehat{\Psi}}{\partial\nabla\theta} \cdot \dot{\overline{\nabla\theta}}. \tag{54}$$

Evaluating $J\widehat{\mathbf{T}} : \dot{\mathbf{F}}\mathbf{F}^{-1}$ with $\mathbf{F} = \mathbf{AG}$, we obtain

$$J\widehat{\mathbf{T}} : \dot{\mathbf{F}}\mathbf{F}^{-1} = J\widehat{\mathbf{T}}\mathbf{A}^{-T} : \dot{\mathbf{A}} + J\mathbf{A}^{T}\mathbf{T}\mathbf{A}^{-T} : \dot{\mathbf{G}}\mathbf{G}^{-1}. \tag{55}$$

We now regroup the terms and divide by J_G, taking into account $J_A\rho = \rho_r$ and $J = J_A J_G$ to obtain

$$\begin{aligned}\left(\rho_r\frac{\partial\widehat{\boldsymbol{\Psi}}}{\partial\mathbf{A}} - J_A\widehat{\mathbf{T}}\mathbf{A}^{-T}\right) : \dot{\mathbf{A}} + \rho_r\left(\frac{\partial\widehat{\Psi}}{\partial\theta} + \eta\right)\dot{\theta} + \rho_r\frac{\partial\widehat{\Psi}}{\partial\nabla\theta}\cdot\dot{\overline{\nabla\theta}} \\ \leq \mathbf{A}^{T}\left(J_A\widehat{\mathbf{T}}\mathbf{A}^{-T}\right) : \dot{\mathbf{G}}\mathbf{G}^{-1} - \frac{J_A}{\theta}\widehat{\mathbf{q}}\cdot\nabla\theta + J_A\overline{\eta}.\end{aligned} \tag{56}$$

We have made the assumption of a constant reference density (see comment after (36)), which without loss of generality allows us to define $\rho_r := 1$ *e.g.* by rescaling mass or volume.

As pointed out in the previous section, (56) is expected to hold for all deformation gradients and all temperature fields. But it is easy to construct a deformation gradient $\mathbf{A}$ and a temperature field θ which at a particular point and time $\mathbf{X}_0, t_0$ take values $\mathbf{A}(\mathbf{X}_0, t_0)$, $\dot{\mathbf{A}}(\mathbf{X}_0, t_0)$, $\theta(\mathbf{X}_0, t_0)$, $\dot{\theta}(\mathbf{X}_0, t_0)$, $\nabla\theta(\mathbf{X}_0, t_0)$, $\dot{\overline{\nabla\theta}}(\mathbf{X}_0, t_0)$ that are independent.[4]. For this reason, the coefficients of $\dot{\mathbf{A}}$, $\dot{\theta}$ and $\dot{\overline{\nabla\theta}}$ must vanish. Taking into account $\rho_r = 1$, it follows that

$$\mathbf{P}_r := J_A \widehat{\mathbf{T}} \mathbf{A}^{-T} = \frac{\partial \widehat{\mathbf{\Psi}}}{\partial \mathbf{A}} \qquad \eta = -\frac{\partial \widehat{\Psi}}{\partial \theta} \qquad \frac{\partial \widehat{\Psi}}{\partial \nabla\theta} = 0. \tag{59}$$

This result demonstrates the elegance of rational thermodynamics. It shows that the first Piola-Kirchhoff $\mathbf{P}_r$ stress and the entropy are naturally derived from the thermodynamic potential $\widehat{\Psi}$ which is consistent with traditional nonlinear elasticity and thermodynamics. The Piola-Kirchhoff stress is given with respect to the stress-free configuration $\mathcal{B}_r$, as one would expect from traditional elasticity. It also shows that the free energy density is independent of the temperature gradient, suggesting (by equipresence, see footnote 3) that (52) should be updated to $\widehat{\Psi}(\mathbf{A}, \theta)$, $\widehat{\eta}(\mathbf{A}, \theta)$, $\widehat{\mathbf{T}}(\mathbf{A}, \theta)$, $\widehat{\mathbf{q}}(\mathbf{A}, \theta)$.

Taking into account (59), for (56) we obtain

$$0 \leq \mathbf{M} : \dot{\mathbf{G}}\mathbf{G}^{-1} + \frac{J_A}{\theta} \widehat{\mathbf{q}} \cdot \nabla\theta + J_A \overline{\eta}, \tag{60}$$

where we introduced the Mandel stress $\mathbf{M} := \mathbf{A}^T \mathbf{P}_r$.

Interpreting results of Coleman-Noll procedure For simplicity, we assume that temperature is spatially homogeneous, such that the $\widehat{\mathbf{q}} \cdot \nabla\theta$ term drops out, which leaves

$$\mathbf{M}_r : \dot{\mathbf{G}}\mathbf{G}^{-1} + J_A \overline{\eta} \geq 0. \tag{61}$$

[4]This is easy to show. We denote the values of $\mathbf{A}$, $\dot{\mathbf{A}}$, θ, $\dot{\theta}$, $\nabla\theta$, $\dot{\overline{\nabla\theta}}$ at the point $\mathbf{X}_0$, t_0 by underlined symbols $\underline{\mathbf{A}}$, $\underline{\dot{\mathbf{A}}}$, $\underline{\theta}$, $\underline{\dot{\theta}}$, $\underline{\nabla\theta}$, $\underline{\dot{\overline{\nabla\theta}}}$ We dropped the subscript m in θ for simplicity. The gradient is understood with respect to the reference variable $\mathbf{X}$, which we also dropped. Then

$$\mathbf{A}(\mathbf{X}, t) = \underline{\mathbf{A}} + (t - t_0)\underline{\dot{\mathbf{A}}} \quad \theta(\mathbf{X}, t) = \underline{\theta} + (t - t_0)\underline{\dot{\theta}} + \left[\underline{\nabla\theta} + (t - t_0)\underline{\dot{\overline{\nabla\theta}}}\right](\mathbf{X} - \mathbf{X}_0). \tag{57}$$

With this choice of $\mathbf{A}$ and θ, it simply follows

$$\mathbf{A}(\mathbf{X}_0, t_0) = \underline{\mathbf{A}} \quad \dot{\mathbf{A}}(\mathbf{X}_0, t_0) = \underline{\dot{\mathbf{A}}} \quad \theta(\mathbf{X}_0, t_0) = \underline{\theta} \quad \nabla\theta(\mathbf{X}_0, t_0) = \underline{\nabla\theta} \quad \dot{\overline{\nabla\theta}}(\mathbf{X}_0, t_0) = \underline{\dot{\overline{\nabla\theta}}}. \tag{58}$$

This shows that at $\mathbf{X}_0$, t_0, the quantities$\mathbf{A}$, $\dot{\mathbf{A}}$, θ, $\dot{\theta}$, $\nabla\theta$, $\dot{\overline{\nabla\theta}}$ can be chosen to have independent values $\underline{\mathbf{A}}$, $\underline{\dot{\mathbf{A}}}$, $\underline{\theta}$, $\underline{\dot{\theta}}$, $\underline{\nabla\theta}$, $\underline{\dot{\overline{\nabla\theta}}}$.

In the spirit of rational thermodynamics, restrictive statements about the growth law should only be the result of considering all possible deformations. We illustrate this idea with the simple growth law

$$\dot{\mathbf{G}} = \left(\widehat{\mathbf{T}} - \mathbf{T}^*\right)\mathbf{G}. \tag{62}$$

We assume that $\mathbf{G}$, $\dot{\mathbf{G}}$, $\widehat{\mathbf{T}}$ and $\mathbf{T}^*$ are diagonal and the material is incompressible ($J_A = 1$). In this case, the Mandel stress coincides with the Cauchy stress, $\mathbf{M}_r = \mathbf{T}$. Evaluating (61) yields

$$\widehat{\mathbf{T}} : \left(\widehat{\mathbf{T}} - \mathbf{T}^*\right) + \overline{\eta} = \left(\widehat{t}_1^2 + \widehat{t}_2^2\right) + \left(\widehat{t}_1 t_1^* + \widehat{t}_2 t_2^* + \overline{\eta}\right) \geq 0. \tag{63}$$

The first bracketed term is positive or zero for any deformation and therefore it cannot contribute to a violation of the Clausius-Duhem inequality. In the second bracketed term, however, $\widehat{t}_1 t_1^* + \widehat{t}_2 t_2^*$ can potentially become negative at some point in space and time $\mathbf{X}_0, t_0$, and therefore requires an external source or sink of entropy to satisfy the Clausius-Duhem inequality. This could be provided by some biochemically driven self-organisation process. This suggests that a growth law of the form $\dot{\mathbf{G}} = \widehat{\mathbf{T}}\mathbf{G}$ is in principle microscopically achievable at every time and position in the tissue without further input or supporting mechanisms. However, no microscopic physical process could achieve the contribution $-\mathbf{T}^*\mathbf{G}$ without further supply of entropy that is information.

4 Conclusions

Growth processes are abundant in nature, yet the rules governing them are inherently complex. Moreover, growth occurs in different forms, and the varying types of growth may be fundamentally different both on a biological and a mathematical level. Theoretical and experimental analyses to understand growth processes on a mechanical level have been developed over a number of years. Of particular interest is the residual stress that can be induced in a body through growth: how and why does it arises? What role does it play in material and structure properties?

We have presented in this notes a summary of some of the key ideas and challenges underlying morphoelasticity. We illustrated the basic concept by considering one-dimensional growth where interesting and challenging questions arise. We then developed the full three-dimensional theory, with the question of how to think about a residually stressed body culminating in the fundamental idea of morphoelasticity, the decomposition of the deformation tensor into growth and elastic deformations. We then explore the thr

Many problems remain unsolved in morphoelasticity, and the theory itself is far from complete. One of the big future challenges in the field is in morpho-elastodynamics. In particular, we discussed a continuously growing body as satisfying a relation

$$\dot{\mathbf{G}} = \mathcal{G}(\mathbf{G}, \mathbf{A}, \mathbf{T}, \ldots, ; \mathbf{X}, t). \tag{64}$$

Finding appropriate biologically derived growth laws $\mathcal{G}$ presents a fundamental challenge for mechanics, applied mathematics, and biophysics.

Acknowledgments: AG is a Wolfson/Royal Society Merit Award Holder and acknowledges support from a Reintegration Grant under EC Framework VII.

Bibliography

J.A. Adam and S.A. Maggelakis. Diffusion regulated growth characteristics of a spherical prevascular carcinoma. *B. Math. Biol.*, 52(4):549–582, 1990.

D. Ambrosi and F. Guana. Stress-modulated growth. *Math. Mach. Solids*, 12(3):319, 2007.

D. Ambrosi and A. Guillou. Growth and dissipation in biological tissues. *Continuum Mech. Therm.*, 19(5):245–251, 2007.

D. Ambrosi, GA Ateshian, EM Arruda, SC Cowin, J. Dumais, A. Goriely, GA Holzapfel, JD Humphrey, R. Kemkemer, E. Kuhl, et al. Perspectives on biological growth and remodeling. *J. Mech. Phys. Solids*, 59:863–883, 2011.

M. Ben Amar and A. Goriely. Growth and instability in elastic tissues. *J. Mech. Phys. Solids*, 53:2284–2319, 2005.

S. Budday, C. Raybaud, and E. Kuhl. A mechanical model predicts morphological abnormalities in the developing human brain. *Scientific reports*, 4, 2014.

H. M. Byrne and M. A. J. Chaplain. Growth of nonnecrotic tumors in the presence and absence of inhibitors. *Math. Biosci.*, 130(2):151–181, 1995.

Bernard D Coleman and Walter Noll. The thermodynamics of elastic materials with heat conduction and viscosity. *Archive for Rational Mechanics and Analysis*, 13(1):167–178, 1963.

S. C. Cowin. Tissue growth and remodeling. *Annu. Rev. Biomed. Eng.*, 6: 77–107, 2004.

P. Détienne and J. Thiel. Monograph on wapas in french guiana. *Bois et Forêts des Tropiques*, 216:43–68, 1988.

M. Epstein and G.A. Maugin. Thermomechanics of volumetric growth in uniform bodies. *Int. J. Plasticity*, 16(7-8):951–978, 2000.

Marcelo Epstein. *The elements of continuum biomechanics.* Wiley. com, 2012.

A.J. Franko and H.I. Freedman. Model of diffusion of oxygen to spheroids grown in stationary medium—I. Complete spherical symmetry. *B. Math. Biol.*, 46(2):205–217, 1984.

J.P. Freyer, N.H. Fink, P.L. Schor, J.R. Coulter, M. Neeman, and L.O. Sillerud. A system for viably maintaining a stirred suspension of multicellular spheroids during NMR spectroscopy. *NMR Biomed.*, 3:195–205, 1990.

A. Friedman. Cancer models and their mathematical analysis. *Lecture Notes in Math.*, 1872:223, 2006.

G. W. Gooday and A. P. J. Trinci. *Wall structure and biosynthesis in fungi. In: Gooday, G.W., Lloyd, D., Trinci, A.P.J. (Eds.), The Eukaryote Microbial Cell.* Cambridge University Press, London, 1980.

A. Goriely and M. Ben Amar. Differential growth and instability in elastic shells. *Phys. Rev. Lett.*, 94(19):198103, 2005. ISSN 0031-9007 (Print).

A. Goriely and D. E. Moulton. Morphoelasticity - a theory of elastic growth. In Oxford University Press, editor, *New Trends in the Physics and Mechanics of Biological Systems*, 2010.

A. Goriely and M. Tabor. Biomechanical models of hyphal growth in actinomycetes. *J. Theor Biol*, 222:211–218, 2003.

M.E. Gurtin, E. Fried, and L. Anand. *The mechanics and thermodynamics of continua.* Cambridge University Press, 2010.

G. Hamilton. Multicellular spheroids as an in vitro tumor model. *Cancer Lett.*, 131(1):29–34, 1998.

G. Helmlinger, P. A. Netti, H. C. Lichtenbeld, R. J. Melder, and R. K. Jain. Solid stress inhibits the growth of multicellular tumor spheroids. *Nature Biotech.*, 15:778–783, 1997.

R. J. Howard and B. Valent. Breaking and entering: host penetration by the fungal rice blast pathogen, *Magnaporthe grisea. Annu. Rev. Microbiol.*, 50:491–512, 1996.

J. D. Humphrey. Continuum biomechanics of soft biological tissues. *Proc. Roy. Soc. Lond. A*, 459:3–46, 2003.

G. W. Jones and S. J. Chapman. Modeling growth in biological materials. *SIAM Review*, 54(1):52–118, 2012.

E. Kröner. Alleganeine kontinuumstheorie der verzetsmugen and eigenspannuugen. *Arch. Ration. Mech. Anal.*, 4:273–334, 1960.

Georgy Lebon, David Jou, and José Casas-Vázquez. *Understanding non-equilibrium thermodynamics.* Springer, 2008.

E. H. Lee. Elastic-plastic deformation at finite strains. *Trans. ASME J. Appl. Mech.*, 54:1–6, 1969.

Th. Lessines, D. E. Moulton, and A. Goriely. Morphoelastic rods. Part II: A theory for two growing rods in interaction. *Preprint*, 2014.

Jerrold E Marsden and Thomas JR Hughes. *Mathematical foundations of elasticity*. Courier Dover Publications, 1994.

Gérard A Maugin. *The thermomechanics of nonlinear irreversible behaviors*. World Scientific, 1999.

D. E. Moulton and A. Goriely. Possible role of differential growth in airway wall remodeling in asthma. *J. Appl. Physiology*, 110(4):1003–1012, 2011.

D E Moulton, T Lessinnes, and A Goriely. Morphoelastic rods Part 1: A single growing elastic rod. *Journal of the Mechanics and Physics of Solids*, 61(2):398–427, 2012.

A. Ramasubramanian and L.A. Taber. Computational modeling of morphogenesis regulated by mechanical feedback. *Biomech. Model. Mechan.*, 7 (2):77–91, 2008.

E. K. Rodriguez, A. Hoger, and A. McCulloch. Stress-dependent finite growth in soft elastic tissues. *J. Biomech*, 27:455–467, 1994.

T. Roose, S.J. Chapman, and P.K. Maini. Mathematical models of avascular tumor growth. *SIAM Review*, 49(2):179, 2007.

M. T. Santini and G. Rainaldi. Three-dimensional spheroid model in tumor biology. *Pathobiology*, 67(3):148–157, 2000.

R. Skalak and A. Hoger. Kinematics of surface growth. *J. Math. Biol.*, 35: 869–907, 1997.

R. M. Sutherland. Cell and environment interactions in tumor microregions: the multicell spheroid model. *Science*, 240(4849):177, 1988.

L. A. Taber. Biomechanics of growth, remodeling and morphogenesis. *Appl. Mech. Rev.*, 48:487–545, 1995.

L. A. Taber and J. D. Humphrey. Stress-modulated growth, residual stress, and vascular heterogeneity. *J. Biomech. Eng.*, 123:528–535, 2001.

L.A. Taber. Theoretical study of beloussov's hyper-restoration hypothesis for mechanical regulation of morphogenesis. *Biomech. Model. Mechan.*, 7(6):427–441, 2008.

Larry A Taber. Towards a unified theory for morphomechanics. *Philosophical Transactions of the Royal Society A: Mathematical, Physical and Engineering Sciences*, 367(1902):3555–3583, 2009.

D. W. Thompson. *On Growth and Form: The Complete Revised Edition*. Dover, New York, 1992.

R. Vandiver and A. Goriely. Tissue tension and axial growth of cylindrical structures in plants and elastic tissues. *Europhys. Lett.*, 84:58004, 2008.

A. Yavari. A geometric theory of growth mechanics. *J. Nonlinear Sci.*, 20: 781–830, 2010.

MIX
Papier aus verantwortungsvollen Quellen
Paper from responsible sources
FSC® C105338

If you have any concerns about our products,
you can contact us on
ProductSafety@springernature.com

In case Publisher is established outside the EU,
the EU authorized representative is:
Springer Nature Customer Service Center GmbH
Europaplatz 3, 69115 Heidelberg, Germany

Printed by Libri Plureos GmbH
in Hamburg, Germany